LENK'S AUDIO HANDBOOK

Other McGraw-Hill Books of Interest

Other books by John Lenk

LENK'S VIDEO HANDBOOK (1990)
LENK'S LASER HANDBOOK (1991)
LENK'S RF HANDBOOK (1992)

Handbooks

Benson • AUDIO ENGINEERING HANDBOOK
Benson • TELEVISION ENGINEERING HANDBOOK
Benson and Whitaker • TELEVISION AND AUDIO HANDBOOK
Combs • PRINTED CIRCUITS HANDBOOK
Croft and Summers • AMERICAN ELECTRICIANS' HANDBOOK
Di Giacomo • DIGITAL BUS HANDBOOK
Fink and Beaty • STANDARD HANDBOOK FOR ELECTRICAL ENGINEERS
Fink and Christiansen • ELECTRONIC ENGINEERS' HANDBOOK
Hicks • STANDARD HANDBOOK OF ENGINEERING CALCULATIONS
Inglis • ELECTRONIC COMMUNICATIONS HANDBOOK
Kaufman and Seidman • HANDBOOK OF ELECTRONICS CALCULATIONS
Mee and Daniel • MAGNETIC RECORDING HANDBOOK
Stout and Kaufman • HANDBOOK OF OPERATIONAL AMPLIFIER CIRCUIT DESIGN
Tuma • ENGINEERING MATHEMATICS HANDBOOK
Williams and Taylor • ELECTRONIC FILTER DESIGN HANDBOOK

Other

Bartlett • CABLE TELEVISION TECHNOLOGY AND OPERATIONS
Luther • DIGITAL VIDEO IN THE PC ENVIRONMENT
Mee and Daniel • MAGNETIC RECORDING, VOLUMES I–III
Phillips • COMPACT DISC INTERACTIVE
Sherman • CD-ROM HANDBOOK
Whitaker • RF TRANSMISSION SYSTEMS

Greetings from the Villa Buttercup!
To my wonderful wife Irene,
Happy Anniversary. Thank you
for being by my side all these
years!
To my lovely family, Karen,
Tom, Brandon, and Justin.
And to our Lambie and Suzzie,
be happy wherever you are!
To my special readers, may
good fortune find your
doorways! Thank you for buying
my books and making me a best
seller!
This is book number 71.
Abundance!

CONTENTS

PREFACE

This is a "something for everyone" audio book. No matter where you are in electronics, this book provides basics, experimentation, simplified design, testing, and troubleshooting information that can be put to immediate use. *If you are an experimenter, student, or serious hobbyist*, the book provides sufficient information for you to design and build audio circuits "from scratch." The design approach here is the same as used in all of the author's best-selling books on simplified and practical design.

Design problems start with approximations or guidelines for the selection of all parts on a trial-value basis, assuming a specific design goal and a given set of conditions. Then, using these approximate values in experimental test circuits, the desired results (power output, rolloff, etc.) are produced by varying the test values.

Although operation of all circuits is described thoroughly where needed, mathematical theory is kept to a minimum. No previous design experience is required to use the design data and techniques described here. The reader need not memorize elaborate theories, or understand abstract math, to use the design data, which makes the book ideal for the *practical experimenter*.

With any audio circuit, it is possible to apply certain guidelines for the selection of component values. These rules are stated in basic equations, requiring only simple arithmetic for their solution. Specific design examples are given as required.

If you are a beginning technician or student, there is an entire chapter devoted to basic testing and troubleshooting audio equipment. A simple audio amplifier is used as an example. Not only does the chapter cover operation of the audio circuits, but it describes testing and troubleshooting approaches for the audio amplifier in step-by-step detail.

If you are an advanced technician or field-service engineer, there are seven chapters devoted to advanced testing and troubleshooting for a cross section of audio circuits and equipment. Each chapter includes a general description of the circuits or equipment, user controls, operating procedures and installation, circuit descriptions, typical testing and adjustment procedures, and step-by-step troubleshooting.

Chapter 1 is devoted to a review and summary of audio basics, from a practical standpoint.

Chapter 2 covers practical considerations for audio. The main concern is with audio power amplifiers (IC or discrete) where temperature-related problems can arise in design and service.

Chapter 3 describes theory and simplified, step-by-step design for audio circuits. The information in this chapter permits the reader to design audio circuits not readily available in IC form.

Chapter 4 describes basic testing and troubleshooting for audio equipment. The chapter starts with a review of typical audio test equipment and then goes on to describe test procedures that can be applied to audio circuits and devices. The chapter concludes with a summary of the basic troubleshooting approach for audio equipment, followed by a specific step-by-step troubleshooting example.

Chapter 5 describes the overall functions, user controls, operating procedures, installation, circuit theory, typical testing and adjustment procedures, and step-by-step troubleshooting for state-of-the-art solid-state and IC amplifiers and loudspeakers.

Chapters 6 through 10 provide coverage, similar to that of Chapter 5, for the audio sections of AM/FM tuners, tape cassettes, CD players, graphic equalizers and turntables, stereo TVs, and surround-sound systems.

Chapter 11 describes formats and circuits found in hifi audio-tape equipment such as VCRs and camcorders, including VHS hifi, S-VHS, Beta hifi, Super Beta and 8 mm. Also discussed are the formats and circuits used in digital audio tape (DAT) players. The chapter concludes with universal hifi audio-tape troubleshooting approaches.

John D. Lenk

ACKNOWLEDGMENTS

Many professionals have contributed their talents and knowledge to the preperation of this book. I gratefully acknowlege that the tremendous effort needed to make this book such a comprehensive work is impossible for one person and wish to thank all who have contributed, both directly and indirectly.

I wish to give special thanks to the following: Tom Roscoe, Dennis Yuoka, and Terrance Miller of Hitachi; Thomas Lauterback of Quasar; Donald Woolhouse of Sanyo; John Taylor and Matthew Mirapaul of Zenith; J. W. Phipps of Thomson Consumer Electronics (RCA); Pat Wilson and Ray Krenzer of Philips Consumer Electronics; and Joe Cagle and Rinaldo Swayne of Alpine/Luxman.

I also wish to thank Joseph A. Labok of Los Angeles Valley College for help and encouragement throughout the years.

And a very special thanks to Daniel Gonneau, Jim Fegen, Larry Jackal, Robert McGraw, Thomas Kowalczyk, Nancy Young, Suzanne Babeuf, Charles Decker, Charles Love, and Jeanne Glasser of the McGraw-Hill organization for having that much confidence in the author. I recognize that all books are a team effort and am thankful that I am working with the First Team!

And to my wife Irene, my research analyst and agent, I wish to extend my thanks. Without her help, this book could not have been written.

CHAPTER 1
AUDIO BASICS

This chapter provides a review and summary of audio basics from a practical standpoint. The information presented here provides the background necessary to understand the design, test, and troubleshooting data found in the remaining chapters. It is essential that you understand the basics to properly test and troubleshoot any audio equipment.

1.1 AUDIO-AMPLIFIER CLASSIFICATIONS

Audio amplifiers are classified in many ways. The two most common methods are by *operating-* or *bias-point* and *circuit connections*. Audio amplifiers are also classified by *function* or *purpose* (voltage amplifier, power amplifier, etc.). Before we get into classifications, let us review the basic audio-amplification principle.

1.2 AUDIO-AMPLIFICATION PRINCIPLE

Figure 1.1*a* shows a typical common-emitter (CE) audio circuit. Under *no-signal* (or *quiescent*) conditions, current flows in the input circuit (across R_1), causing a steady value of current to flow in the output circuit (across R_3).

A voltage is developed across R_1 during the first half (or alternation) of the audio signal applied to the input. This voltage, positive at the base end of R_1, adds to the bias voltage at the junction of R_1 *and* R_2, causing the base-to-emitter voltage (sometimes called V_{BE}) to increase.

Under these conditions, the voltage from collector to emitter (V_{CE}) increases but with the *phase inverted* (the collector goes negative when the base goes positive). Amplification occurs because the collector current (I_C) is *many times greater* than the base-emitter current (I_{BE}). When the second half of the audio signal is applied across R_1, the voltage across R_3 also alternates but in the opposite direction (a negative swing at the input produce a positive swing at the output and vice versa).

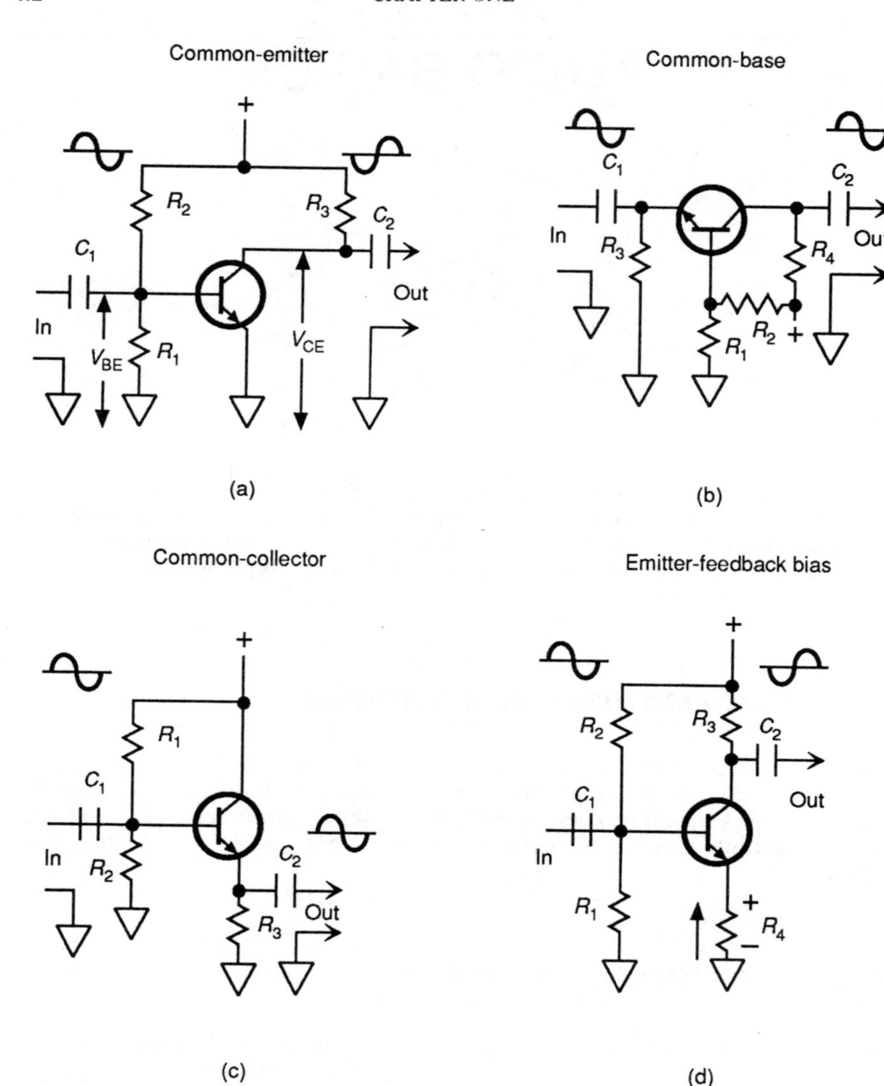

FIGURE 1.1 Audio-amplification and emitter-feedback basics. (*a*) Common emitter; (*b*) common base; (*c*) common collector; (*d*) emitter-feedback bias.

1.3 TRANSISTORS IN AUDIO AMPLIFIERS

The following general rules can be helpful in a practical analysis of how transistors operate in audio amplifiers. The rules apply primarily to a class A amplifier (Sec. 1.8.1) but also remain true for many other audio circuits. The rules are included here primarily for those totally unfamiliar with bipolar or two-junction transistors (and for those who have forgotten). FET audio circuits are discussed in Sec. 1.11.

In the NPN transistor, electrons flow from the emitter to the collector, so the *collector must be positive* in relation to the emitter. In PNP transistors, holes flow from emitter to collector, so the *collector must be negative* in relation to the emitter.

The *middle letter* in PNP or NPN applies to the base. The *first two letters* in PNP or NPN refer to the *relative bias polarities of the emitter* with respect to either the base or collector. For example, the letters PN (in PNP) show that the emitter is positive with respect to either the base or collector. The letters NP (in NPN) show that the emitter is negative with respect to both the base and collector.

The *dc electron-current flow* is always against the direction of the arrow in the emitter. If electron flow is into the emitter, electron flow is out from the collector. If electron flow is out from the emitter, electron flow is into the collector.

The collector-base junction is always reverse biased. The emitter-base junction is generally forward biased.

A *base-input voltage that opposes* or decreases the forward bias also decreases the emitter and collector currents. For example, a negative input to the base of an NPN, or a positive input to a PNP base, decreases *both* currents.

A *base-input voltage that aids* or increases the forward bias also increases the emitter and collector currents (positive to NPN, negative to PNP).

1.4 COMMON-EMITTER AUDIO AMPLIFIERS

The common-emitter (CE) circuit shown in Fig. 1.1*a* is the most widely used audio-amplifier configuration. The emitter is common to both the input and output circuits and is frequently called the grounded element (although the emitter is not always connected to ground).

To summarize the CE amplifier, the input signal is applied between base and emitter, and the output signal appears between emitter and collector. This provides a moderately low input impedance and a very high output impedance, with a 180° phase reversal between input and output. The CE amplifier produces the highest power gain of all three transistor amplifier circuits. The voltage and current gains are fairly high. CE amplifiers are most often used in audio work since there are current, voltage, resistance, and power gains (which we just happen to discuss next).

1.4.1 Audio Gain

Many terms are used to express gain of CE amplifiers, as well as other amplifier configurations. The terms are interrelated and are often interchanged (properly and improperly). To minimize confusion, the following is a summary of audio gain terms used in this book.

Alpha and Beta. The terms alpha and beta are applied to transistors connected in common-base (CB) and CE configurations, respectively. Both terms are a measure of current gain for the transistor (but not necessarily for the circuit). Alpha is always less than 1 (typically 0.9 to 0.99). Beta is more than 1 and can be as high as several hundred (or more). The relationships between alpha and beta are: alpha = beta/(beta + 1); beta = alpha/(alpha − 1).

The terms alpha and beta do not necessarily represent the current gain of the audio circuit in which the transistors are used. Instead, the current gain of the circuit *cannot be greater* than the alpha or beta of the transistor.

Current Gain. The term current gain can be applied to either the transistor or to the audio circuit and is a measure of *change* in current at the output for a given change in current at the input.

For example, when a 1-mA change in input current produces a 10-mA change in output current, the current gain is 10. Since alpha (CB) is always less than 1, there is no current gain in CB audio circuits (Sec. 1.5). Instead, there is a slight loss.

Resistance Gain. The ratio of output resistance (or impedance) divided by input resistance (or impedance) is the resistance gain. For example, if the input resistance is 1000 and the output resistance is 15,000, the resistance gain is 15. The input and output resistances (or impedances) depend on circuit values, as well as transistor characteristics.

Resistance gain, by itself, produces no usable gain for the audio circuit. However, resistance gain has a direct effect on the voltage and power gains, such as with the previous values (current gain of 10 and resistance gain of 15).

Assume that the input resistance is 1000 and a 1-mV signal is applied at the input. This results in an input current change of 1 μA. With a current gain of 10, the output current is 10 μA. This 10-μA current passes through a 15,000 output resistance to produce a voltage change of 150 mV. As a result, the 1-mV input signal produces a 150-mV output signal (a voltage gain of 150).

Voltage and Power Gain. Voltage gain is equal to the difference in output voltage divided by the difference in input voltage. Power gain is equal to the difference in output power divided by the difference in input power. Except in CB circuits, power gain is always higher than voltage gain, since power is based on the square of the voltage (power = E^2/R), as well as the square of current (I^2/R). Using the same values, the input power of the stage is 1×10^{-9}, the output power is 1.5×10^{-6}, and the power gain is 1500.

Voltage and Power Amplifiers. The function of a *voltage amplifier* is to receive an input signal consisting of a small voltage of definite waveform and to produce an output signal consisting of a voltage with the same waveform but much larger in amplitude. For example, the output produced by the pickup of an audio turntable is usually in the order of a few millivolts. The voltage amplifier of an audio system amplifies this voltage to produce a similar voltage that is large enough to operate a *power amplifier* which, in turn, operates the power-consuming loudspeakers.

Transistors designed for voltage amplification usually have high betas, with small current-carrying capability. Power transistors have large current-carrying capacity but relatively low betas. If the power involved exceeds about 1 W, the transistor (or IC) must be operated with *heat sinks* (as discussed in Chap. 2).

1.5 COMMON-BASE AUDIO AMPLIFIERS

The common-base circuit shown in Fig. 1.1*b* is not used extensively in audio work. One exception is where the audio source is at a low impedance.

To summarize the CB audio circuit, the input signal is applied between base and emitter (across R_3), and the output appears between base and collector (across R_4). This provides an *extremely low* input impedance and a very high output impedance. The output signal is *in phase* with the input. The CB amplifier produces high voltage gains and modest power gains even though there is no current gain. This is possible because of the resistance gain, as discussed in Sec. 1.4.1.

1.6 COMMON-COLLECTOR AUDIO AMPLIFIERS

The common-collector (CC) circuit shown in Fig. 1.1*c* is also known as an *emitter follower* since the output is taken from the emitter resistance, and the output follows the input (in phase relationship).

To summarize the emitter follower, the input signal is applied to the base (across R_2), and the output signal appears at the emitter (across R_3). This provides extremely high input impedance and a very low output impedance (usually set by the value of R_3). The output signal is *in phase* with the input. The emitter follower produces modest current gain (as well as power gain) even though there is no voltage gain. In general, the emitter-follower current gain (and power gain) is limited by the current gain (beta) of the transistor. The most common use of an emitter follower in audio work is to match the high impedance of a solid-state circuit to a low-impedance device (such as when an audio amplifier must be matched to a loudspeaker).

1.7 AUDIO BIAS NETWORKS

All audio circuits require some form of bias. As a minimum, the collector-base junction of any circuit must be reverse biased. That is, current should not flow between collector and base. Any collector-base current that does flow is a result of leakage or breakdown.

Under no-signal conditions, the emitter-base circuit of a solid-state amplifier can be forward, reverse, or zero biased (no bias). However, emitter-base current must flow under some condition of operation. The desired bias is produced by applying voltages to the corresponding transistor elements through bias networks, usually composed of resistors. The bias networks (or the resistors used to form the networks) serve more than one purpose. Typically, the bias network resistors (1) set the operating point, (2) stabilize the circuit at the operating points, and (3) set the approximate input-output impedances of the circuit as follows:

Operating point: Bias networks establish collector-base-emitter voltage and current relationships at the operating point of the audio circuit. (The operating point is also known as the quiescent point, Q-point, no-signal point, idle point, or static point.) Since transistors rarely operate at the Q-point, the basic bias networks are generally used as a reference or starting point for design. The actual circuit values are generally selected on the basis of dynamic circuit conditions (desired output voltage swing, expected input signal level, and so on).

Audio bias stabilization: In addition to establishing the operating point of an audio circuit, the bias networks must maintain the operating point in the presence of temperature and power-supply changes and possible transistor replacement. As discussed in the following paragraphs, a shift in operating point can produce distortion and a change of frequence response (two very undesirable effects in any audio circuit). The process of maintaining an audio circuit at a given operating point is generally referred to as *bias stabilization*. Improper bias stabilization can also produce another undesired effect known as *thermal runaway*.

Thermal runaway: Heat is generated when current passes through a transistor junction. If all heat is not dissipated by the case or heat sink (often an impossibility), the junction temperature rises. This, in turn, causes more current to flow through the junction even though the voltage, circuit values, and so on remain the same. With more current, the junction temperature increases even further, with a corresponding increase in current flow. The transistor burns out if the heat is not dissipated by some means.

In addition to the heat sinks described in Chap. 2, many audio circuits are provided with bias stabilization to prevent any drastic change in junction current, despite changes in temperature, voltage, and so on. This bias stabilization maintains the circuit at the desired operating point (within practical limits) and prevents thermal runaway.

Input-output impedances: The resistors used in bias networks also have the function of setting the input and output impedances of the circuit. In theory, the input-output impedances are set by many factors (transistor beta, transistor input-output capacitance, and so on). In practical audio circuits, the input-output impedances are set by the bias network resistors. For example, the output impedance of a CB or CE audio circuit is about equal to the collector resistor (or total resistance between the collector and power source).

1.7.1 Basic Bias-Stabilization Techniques

All bias-stabilization circuits use a form of *negative* or *inverse feedback*. That is, any change in transistor current produces a corresponding voltage or current change that tends to *offset* the initial change. This feedback not only offsets the undesired change but also tends to reduce and stabilize gain (when the feedback principle is used in an audio amplifier).

Typical Emitter-Feedback Bias Network. Figure 1.1*d* shows a typical emitter-feedback bias network (this is the most common audio bias circuit). Note that this circuit is essentially the same as the basic CE amplifier shown in Fig. 1.1*a* but with an emitter resistor to provide bias stabilization. The use of an emitter-feedback resistance in any audio circuit can be summed up as follows.

Base current (and, consequently, collector current) depends on the *differential* in voltage between base and emitter. If the differential voltage is lowered, less base current (and, consequently, less collector current) flows. The opposite is true when the differential is increased. All current flowing through the collector (ignoring collector-base leakage, I_{CBO}) also flows through the emitter resistor.

The voltage across the emitter resistor therefore depends (in part) on the collector current.

Should the collector current increase (for any reason), emitter current and the voltage drop across the emitter resistor also increase. This negative feedback tends to decrease the differential between base and emitter, thus lowering the base current. In turn, the lower base current tends to decrease the collector current and offset the initial collector-current increase.

1.7.2 Typical Audio Bias Networks

Figures 1.2 and 1.3 show typical bias networks found in a variety of audio circuits. Note that the equations in Figs. 1.2 and 1.3 hold true throughout the audio range (and typically up to about 100 kHz). Also note that emitter feedback is used in all of the bias circuits. Examples of simplified, practical design for audio circuits using these basic networks are given in Chap. 3.

As shown by the equations in Figs. 1.2 and 1.3, the approximate input and output impedance of the circuit are set by resistance ratios, as are voltage and current gain. This fact can be helpful in testing and troubleshooting audio equipment.

For example, in the circuit in Fig. 1.2*a*, if the value of R_2 is 10 times that of R_3, the output signal (at the collector) should be about 10 times the input signal (at the base). Of course, if the transistor does not have enough gain (beta) or the power supply does not allow sufficient collector-voltage swing, the output can be limited. *However, the resistance ratio does provide a guideline for troubleshooting* (if you find no gain or gain far less than the collector-emitter resistance ratio, the circuit is suspect).

Maximum Gain with Minimum Stability. The circuit in Fig. 1.2*a* provides the greatest possible gain but the least stability of all the bias circuits described here.

Maximum Gain with Improved Stability. The basic characteristics for the bias circuit in Fig. 1.2*b* are the same as for the circuit in Fig. 1.2*a* except that stability is increased. The increase in bias stabilization is brought about by connecting base resistance R_1 to the collector rather than to the supply.

In the circuit in Fig. 1.2*b*, if the collector current increases for any reason, the drop across R_2 increases, lowering the voltage at the collector. This lowers the base voltage and current, thus reducing the collector current. The feedback effect is combined with that produced by emitter resistor R_3 to offset any variation in collector current. However, gain for the circuit in Fig. 1.2*b* is only slightly less than for the circuit in Fig. 1.2*a*.

Maximum Stability. The bias circuit in Fig. 1.2*c* offers more stability than the circuits in Fig. 1.2*a* and *b* but with a trade-off of lower audio gain and lower input impedance.

As shown by the characteristics in Fig. 1.2*c*, the input impedance is about equal to R_2 (at frequencies up to about 100 kHz). Technically, the input impedance is equal to R_2 in parallel with $R_4 \times$ (beta + 1). In practice, unless the beta is very low, the $R_4 \times$ (beta + 1) factor is much greater than R_2. As a result, the value of R_2 (or slightly less) can be considered as the stage or circuit input impedance.

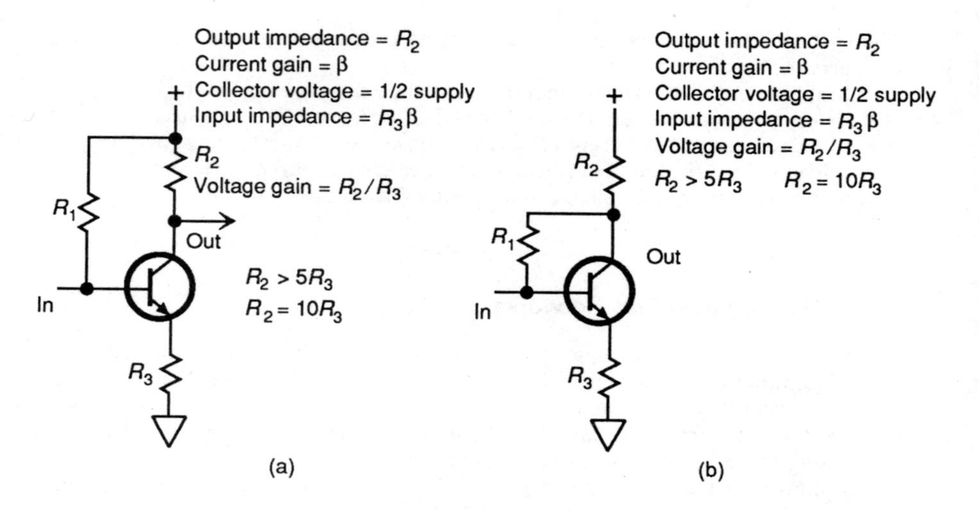

(S) stability = R_2/R_4
S = 20 for maximum gain
S = 10 for stability
S = 5 for power gain

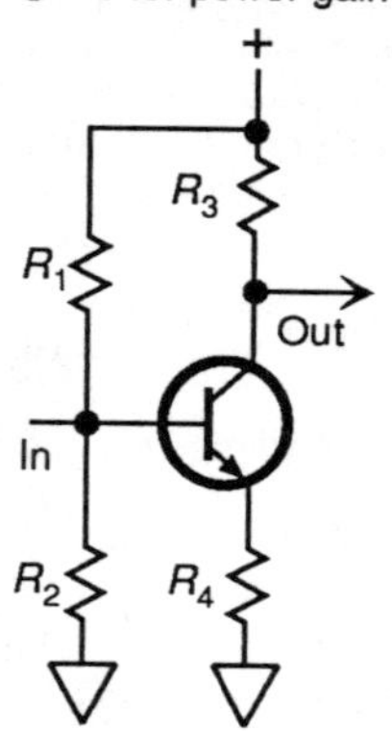

Output impedance = R_3
Current gain = R_2/R_4
Collector voltage = 1/2 supply
$R_3 = 10R_4$ $R_2 = 10R_4$ $R_3 > 5R_4$ $R_2 < 20R_4$
Input impedance = R_2
Voltage gain = R_3/R_4

(c)

Output impedance = R_3
Current gain = R_2/R_4
Collector voltage = 1/2 supply
$R_3 = 10R_4$ $R_2 = 10R_4$
Input impedance = R_2
Voltage gain = R_3/R_4
$R_3 > 5R_4$ $R_2 < 20R_4$

+
R_3
R_1
Out
In
D_1
R_2
R_4

(d)

FIGURE 1.2 Typical audio bias networks.

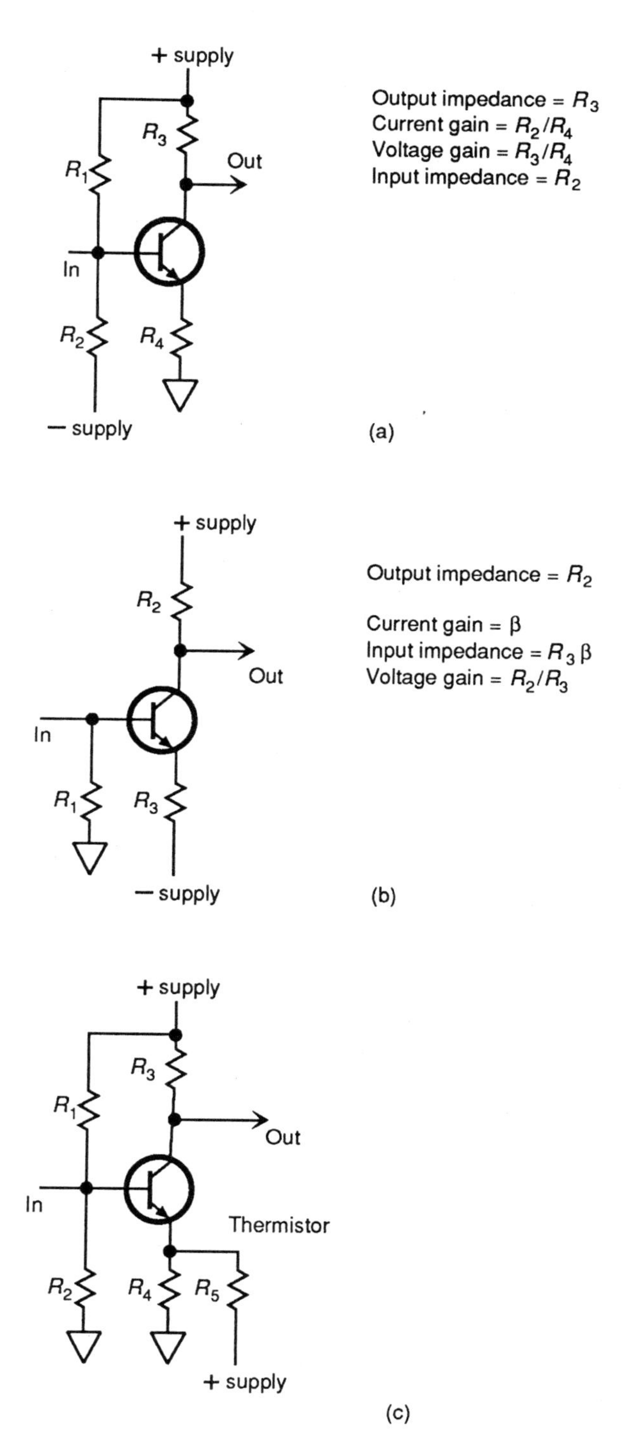

FIGURE 1.3 Audio bias networks with multiple power sources and thermistors.

Diode for Temperature Stability. The basic characteristics for the bias circuit in Fig. 1.2*d* are the same as for the circuit in Fig. 1.2*c* except that temperature stability is improved by diode D_1 connected between the base and R_2. D_1 is forward biased and is of the same material (silicon) as the base-emitter junction. Both D_1 and the junction are maintained at the same temperature. As a result, the voltage drops across D_1 and the junction are the same and remain the same with changes in temperature.

Positive and Negative Supply for Base Current. The basic characteristics for the bias circuit in Fig. 1.3*a* are the same as for the circuit in Fig. 1.2*c*. However, the circuit in Fig. 1.3*a* is used in those special applications that require a negative and positive voltage, each with respect to ground, to control base current. (Typically, the value of R_1 for the two circuits may be different because of the large current through R_2).

Positive and Negative Supply for Emitter-Collector Current. The circuit in Fig. 1.3*b* is used in those special applications where it is necessary to supply collector-emitter current from both a positive and negative source. Since the transistor is NPN, the collector is connected to the positive source through R_2, while the emitter is connected to the negative source through R_3.

If the sources are about equal, the circuit does not produce any voltage gain or audio amplification (unless emitter bypass is used, as described in Chap. 3). The basic characteristics of the circuit in Fig. 1.3*b* are essentially the same as for Fig. 1.2*a*, except that audio amplification (if any) is low because of the lower R_2 to R_3 ratio (R_3 must drop the entire negative source while R_2 typically drops the positive source to half, say from +20 to +10 V).

Thermistor for Temperature Stability. The basic characteristics for the bias circuit in Fig. 1.3*c* are the same as for the circuit in Fig. 1.2*c* except that the negative temperature coefficient (NTC) characteristics of a thermistor provide temperature compensation (the thermistor resistance decreases with temperature increases and vice versa). Usually, the thermistor is mounted near the transistor so that both devices are at the same temperature.

Transistor emitter and collector currents tend to rise with temperature increases. The same rise reduces the thermistor resistance, increasing current flow and increasing voltage drop across R_4 and R_5. This increases the reverse bias to the emitter-base junction, decreasing the emitter and collector currents (to offset the current increase produced by the temperature rise). The action is reversed if the temperature decreases.

1.8 OPERATING POINT OF AUDIO CIRCUITS

The following is a brief summary of the three basic operating-point classifications (A, B, AB) normally associated with audio circuits, as well as the class C operating point (usually found in RF circuits).

From a practical troubleshooting standpoint, note that the base-collector junction is always reverse biased at the operating point (in all four classes) as well as under all signal conditions. No base-collector current flows (with the possible exception of reverse leakage current I_{CBO}). On the other hand, the base-emitter junction is biased so that base-emitter current flows under certain conditions and possibly under all conditions. When base-emitter current flows, emitter-collector current also flows. If you find any audio circuit with measurable base-collector current flow, and/or no base-emitter flow, under any condition, the circuit is suspect. This is also true where there is no emitter-collector flow with some base-emitter flow.

1.8.1 Class A Audio

As shown in Fig. 1.4*a*, a class A amplifier operates only over the *linear portion* of the transistor curve. (The curve represents the relationship between base voltage, or input, and collector current, or output.) At no point of the input signal cycle does the base become so positive or negative that the transistor operates on the nonlinear portion of the curve. The transistor collector current is never cut off nor does the transistor ever reach saturation.

The main advantage of the class A amplifier is the relative lack of distortion. The output waveform follows that of the input, except in amplified form. However, with any class of audio circuit, there is some distortion (Sec. 1.9).

The main disadvantages of class A circuits are relative inefficiency (lower power output for a high-power input that must be dissipated by the transistor) and the inability to handle large signals. Rarely is a class A amplifier over about 35 percent efficient. If the power input to a class A amplifier is 1 W (generally, the maximum power dissipation capability of a single transistor), the output is less than 0.3 W.

The peak-to-peak (p-p) output voltage swing of a class A amplifier is limited to less than the total supply voltage. Since the output voltage must swing both positive and negative, the peak output is less than one-half the supply voltage. For example, assume that the supply is 20 V, and the amplifier is biased so that the Q-point collector voltage is one-half the supply, or 10 V. Under these conditions, the output voltage cannot exceed ±10 V.

If distortion is to be at a minimum (the usual reason for class A amplifiers), the output is usually about ±5 V (with a 20-V supply). This keeps the transistor on the linear portion of the curve. (Typically, the curve becomes nonlinear near the cutoff and saturation points.) However, this can be determined only from an actual test of the circuit.

The input voltage swing of a class A audio amplifier is limited by the output voltage swing and the voltage-amplification factor. For example, if the output is limited to ±10 V and the voltage amplification factor is 100, the input is limited to ±0.1 V (100 mV).

Because of these limitations, class A audio circuits are generally used as voltage amplifiers rather than power amplifiers. Typically, a class A amplifier stage is used ahead of a power amplifier stage.

1.8.2 Class B Audio

As shown in Fig. 1.4*b*, a class B amplifier operates only on one-half of the input signal. Class B operation is produced when the base-emitter bias is set so that the operating point coincides with the transistor cutoff point. When the input signal voltage is zero, there is no flow of collector current. During one half-cycle of the signal voltage, the collector current rises to the peak and then falls back to zero in step with the input-signal half-cycle. During the opposite half-cycle, there is no

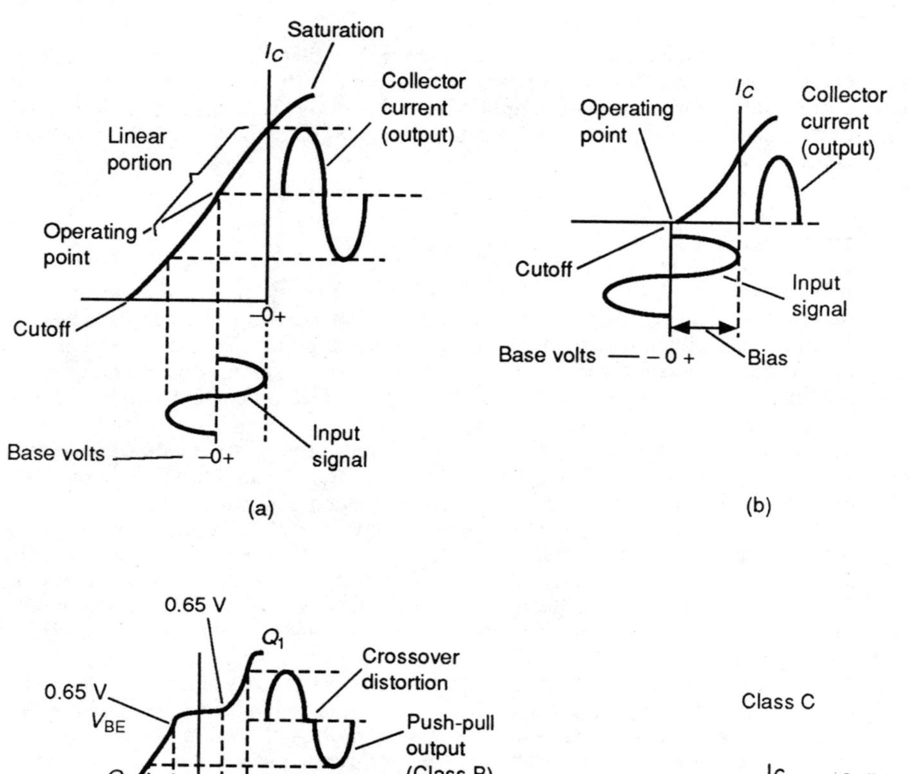

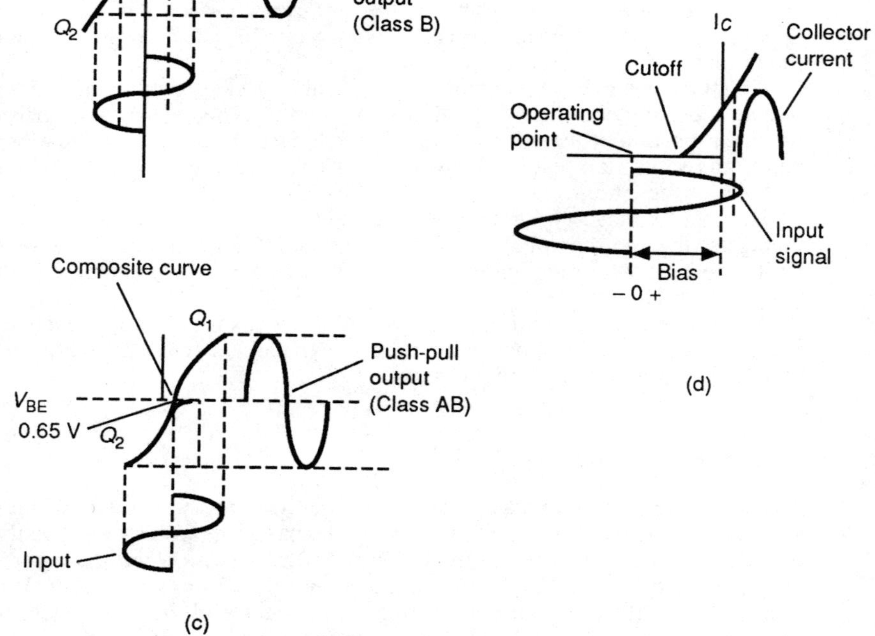

FIGURE 1.4 Audio operating-point classifications.

collector current since the reverse bias is always greater than the transistor cutoff voltage. Collector current flows only during half of the input-signal cycle.

There is considerable distortion if a single transistor is operated as class B. This is because the waveform of the resulting collector current resembles that of one input half-cycle and does not resemble the complete waveform at the input. Class B is generally used when *two transistors are connected in push-pull.* This makes it possible to reconstruct an output waveform that resembles the full waveform of the input.

The peak output voltage swing of a class B amplifier is slightly less than the supply voltage. Since the output appears only on half-cycles, it is possible to operate class B amplifiers at a higher current (or power) rating than class A, all other factors being equal.

As an example, if a transistor is capable of 0.3-W dissipation (without damage) as class A, the same transistor can be operated at 0.6 W as class-B since the transistor is conducting collector current only half the time. Of course, this is a theoretical example. In practice, there are factors that limit class B power dissipation to something less than twice that of class A.

Also note that the peak output of a class B amplifier is equivalent to the p-p output of a class A amplifier. So, if two transistors are connected in push-pull and operated as class B, the output voltage can be *twice* that of class A.

Because of these voltage and power factors, class B amplifiers are generally used as power amplifiers rather than voltage amplifiers. In a typical audio amplifier using discrete components (yes, they still exist), two push-pull transistors are operated in class B, preceded by a single class A stage. The class A stage provides voltage amplification, whereas the class B stage produces the necessary power amplification.

1.8.3 Class AB Audio

Class AB operation (a trade-off between A and B) is used in audio amplifiers to minimize the effects of *crossover distortion* as shown in Fig. 1.4*c*. In true class B, the transistor remains cut off at very low signal inputs and turns on abruptly with a large input signal. In class B push-pull operation, during the instantaneous pause when one transistor stops conducting and the other transistor starts conducting, the output waveform is distorted. This instantaneous cutoff of collector current can also set up large *voltage transients* equal to several times the supply voltage, possibly resulting in transistor breakdown.

To minimize these undesired effects using class AB, the transistors are forward biased just enough for a small amount of collector current to flow at the Q-point, and there is no cutoff or abrupt turn on. The combined collector currents produce a composite curve essentially linear at the crossover point, resulting in a faithful reproduction of the input (at least as far as the crossover point is concerned). Of course, class AB is less efficient than class B, since more current must be used.

Some hifi audio circuits use diodes in series with the collector or emitter leads of the push-pull transistors. Because the voltage must reach a certain value (typically 0.65 V for silicon diodes) before the diode conducts, the collector-current curve is rounded (not sharp) at the crossover point.

1.8.4 Class C Circuits

Figure 1.4*d* shows the characteristic curve of a typical class C amplifier. Note that the transistor is reverse biased *well below* the cutoff point and that there is collector-current flow for *only a portion of half* the input signal. The resulting collector current is a pulse, the duration of which is considerably less than a half-cycle of the input signal.

Obviously, the waveform of the output signal cannot resemble that of the input signal. Nor can this resemblance be restored by push-pull operation (as with B or AB). Class C is limited to those applications where distortion is of no concern (which usually means RF circuits rather than audio circuits).

1.9 AUDIO DISTORTION

A small amount of distortion (where the output is not identical to the input) is generally present in all amplifiers. However, audio amplifiers are usually designed to keep distortion within acceptable limits. In some special cases, audio amplifiers contain circuits to introduce a form of distortion (such as compression or expansion). This is generally to offset or compensate for distortion already present in the signal.

The following paragraphs describe three basic types of distortion common to all audio circuits. Any of these, either separately or in combination, may be present in an amplifier in addition to crossover distortion (Sec. 1.8) and *intermodulation distortion* and/or *harmonic distortion* (Chap. 4).

1.9.1 Amplitude-Distortion Basics

Amplitude distortion occurs in the transistor and is the result of operating the transistor over the nonlinear portion of the characteristic curve. The usual remedy is to use a base-emitter bias that places the operating point well within the linear portion of the curve, preferably at the center of the linear portion.

From a practical troubleshooting standpoint, always check the operating point of an amplifier stage first when distortion is present. The operating point may have shifted because of component aging (or failure). This problem is discussed further in Chap. 4.

In addition to a correct operating point, the amplitude of the input signal must be small enough so that the positive and negative half-cycles do not drive the transistor beyond the linear portion. An overdriven amplifier (used to get maximum gain) almost always results in some amplitude distortion.

Generally, a low input signal and proper operating point (at the center of the linear portion) mean that gain must be sacrificed. Consequently, a low-distortion audio amplifier may require two (or more) stages to get the same gain as an overdriven amplifier.

1.9.2 Frequency-Distortion Basics

Frequency distortion rarely if ever occurs because the input signal is at a single frequency. Instead, the input signal usually contains components of several frequencies, making the signal waveform somewhat complex.

In addition to transistors and ICs, an amplifier circuit is composed of resistors, capacitors, and possibly inductances (coils and transformers). Capacitors and inductances have reactance. Since reactance is a function of frequency, the different frequencies of the signal encounter different reactances. The high and low frequencies can be impeded in *different* degrees or amounts, producing distortion of the signal waveform from the original.

For example, assume that an input signal is a complex waveform composed of three frequencies, 10, 100, and 1000 Hz, all of the same amplitude. Since capacitive reactance increases with a decrease in frequency, the 10-Hz signal is attenuated more than the 100-Hz signal and much more than the 1000-Hz signal. Even though the transistor amplifies all three frequencies equally, the signals are no longer of equal amplitude, and the output waveform is different from the input.

Audio amplifiers are usually designed to eliminate unwanted capacitance and inductance, thus minimizing the effects of frequency distortion. Likewise, compensating components may be introduced into the circuit.

1.9.3 Phase-Distortion Basics

Phase distortion in an amplifier occurs because the input signal contains components of different frequencies. When a signal flows through a capacitor or an inductor, the signal encounters a phase shift. The degree of phase shift is a function of frequency. The high- and low-frequency components of the signal are phase shifted by different amounts. These different phase shifts cause a distortion of the signal waveform.

As is the case of frequency distortion, the phase distortion in an audio amplifier may be minimized by proper design to eliminate unwanted capacitance and inductance. Such procedures are discussed in Chap. 3.

1.10 DECIBEL BASICS

The decibel, or dB, is widely used in audio work to logarithmically express the *ratio* between two power or voltage levels. A decibel is one-tenth of a bel (the bel is too large for most practical applications).

Although there are many ways to express a ratio, the decibel is used in audio work for two reasons: (1) the decibel is a convenient unit to use for all types of amplifiers, and (2) the decibel is related to the reaction of the human ear and is thus well suited for use with audio amplifiers. The human ear does not hear sounds in direct power ratios. Humans can listen to ordinary conversation quite comfortably and yet be able to hear thunder (which is taken to be 100,000 times louder than conversation) without damage to the ear. This is because the response of the human ear to sound waves is approximately proportional to the logarithm of the sound-wave energy and is not proportional to the energy.

The increase in power of an audio amplifier can be expressed as

$$\text{Power gain in dB} = 10 \log_{10} \frac{\text{power output}}{\text{power input}} \quad \text{or} \quad 10 \log_{10} \frac{P_2}{P_1}$$

Usually, P_2 represents power output and P_1 represents power input. If P_2 is greater than P_1, there is a power gain, expressed in positive decibels (+dB). With P_1 greater than P_2, there is a power loss, expressed in negative decibels

(−dB). Whichever is the case, the ratio of two powers (P_1 and P_2) is taken and the logarithm of this ratio is multiplied by 10. As a result, a power ratio of 10 = 10-dB gain, a power ratio of 100 = 20-dB gain, a power gain of 1000 = 30-dB gain.

1.10.1 Doubling Power Ratios

Each time the power of an audio amplifier is doubled, there is a power gain of +3 dB. For example, if the volume control of an audio amplifier is turned up so that the power rises from 4 to 8 W, the gain is up +3 dB. Increasing the power output further to 16 W produces another gain of +3 dB, with a total power gain of +6 dB. If the power output is reduced from 4 to 2 W, the gain is down −3 dB.

1.10.2 Adding and Subtracting Decibels

When several audio stages are connected so that one stage works into another (stages connected in *cascade*), the gains for each stage are *multiplied*. For example, if three stages, each with a gain of 10, are connected, there is a total power gain of 10 × 10 × 10, or 1000. When using the decibel system to calculate this gain, the decibel gains are *added* (or subtracted). Using our example, the decibel power gain is 10 + 10 + 10, or +30 dB. Similarly, if two audio stages are connected, one of which has a gain of +30 dB and the other a loss of −10 dB, the net result is +30 −10, or +20 dB gain (a power ratio of 100).

1.10.3 Voltage Ratios

The decibel system is also used to compare the voltage input and output of an audio amplifier, using the function

$$\text{Voltage gain in dB} = 20 \log_{10} \frac{\text{voltage output}}{\text{voltage input}} \quad \text{or} \quad 20 \log_{10} \frac{V_2}{V_1}$$

The ratio of the two voltages is taken, and the logarithm of this ratio is multiplied by 20.

Unlike power ratios (which are independent of input and output impedances) voltage ratios hold true only when input and output impedances are equal.

In audio circuits where the input and output impedances differ, voltage ratios are calculated as follows:

$$20 \log_{10} \frac{V_2 \sqrt{R_2}}{V_1 \sqrt{R_1}}$$

where R_1 is the input impedance and R_2 is the output impedance.

Like power ratios, if the voltage output is greater than the input, there is a decibel gain (+dB). If the output is less than the input, there is a voltage loss (−dB).

Each time the voltage of an audio amplifier is doubled, there is a voltage gain of +6 dB. Conversely, if the voltage output is half of the input voltage, there is a voltage loss of −6 dB. To get the net effect of several voltage amplifiers working together, add the decibel gains (or losses) of each amplifier or stage.

1.10.4 Decibel Meter and Volume Unit

Decibels are often used with specific reference levels. The most common reference levels in audio work are the *volume unit* (VU) and the *decibel meter* (dBm).

When the VU is used, it is assumed that the zero level is equal to 0.001 W (1 mW) across a 600-W impedance. Therefore,

$$\mathrm{VU} = 10 \log \frac{P_2}{0.001} = 10 \log \frac{P_2}{10^{-3}} = 10 \log 10^3 P_2$$

where P_2 is the output power.

Both dBm and VU have the same zero level. However, a dBm scale is (generally) used when the signal is a sine wave (normally 1 kHz) whereas the VU is used for complex audio waveforms.

1.11 FIELD-EFFECT TRANSISTORS IN AUDIO

Most audio circuits do not make extensive use of field-effect transistors (FETs). One exception is where the circuit requires a very high input impedance. Another exception is where FETs are used in AM/FM tuners (a standard part of any present-day audio system, as discussed in Chap. 6). The following paragraphs summarize those FET characteristics that apply to audio circuits.

1.11.1 Basic FET Amplifier Characteristics

Figure 1.5 shows the basic FET circuit and operating characteristics. Both junction FET (JFET) and metal-oxide semiconductor FET (MOSFET) devices are found in audio circuits (to a limited extent). The MOSFET is also called an insulated-gate FET (IGFET) in some literature.

Both JFETs and MOSFETs operate on the principle of drain-source current (I_D) controlled by gate-source voltage (V_{GS}). However, there is considerable difference in gate characteristics. The input to a JFET audio amplifier acts like a reverse-biased diode, whereas the input to a MOSFET amplifier is similar to a small capacitor. Often, MOSFET amplifiers are connected in cascade without coupling capacitors between stages.

The FET has three distinct characteristic regions, only two of which are operational for amplifiers. As shown in Fig. 1.5*b* the FET operates in the *ohmic* or *resistance* region when V_{GS} is below the pinchoff-voltage (V_P) level. (The ohmic region is not generally used for amplifiers, except in special cases, such as some rare audio volume-control schemes.) When V_{GS} is above V_P, at any point up to the avalanche breakdown region, or $V_{(BR)DSS}$, the FET is in the *constant-current* region, which is the region most used for amplifier circuits. The third region,

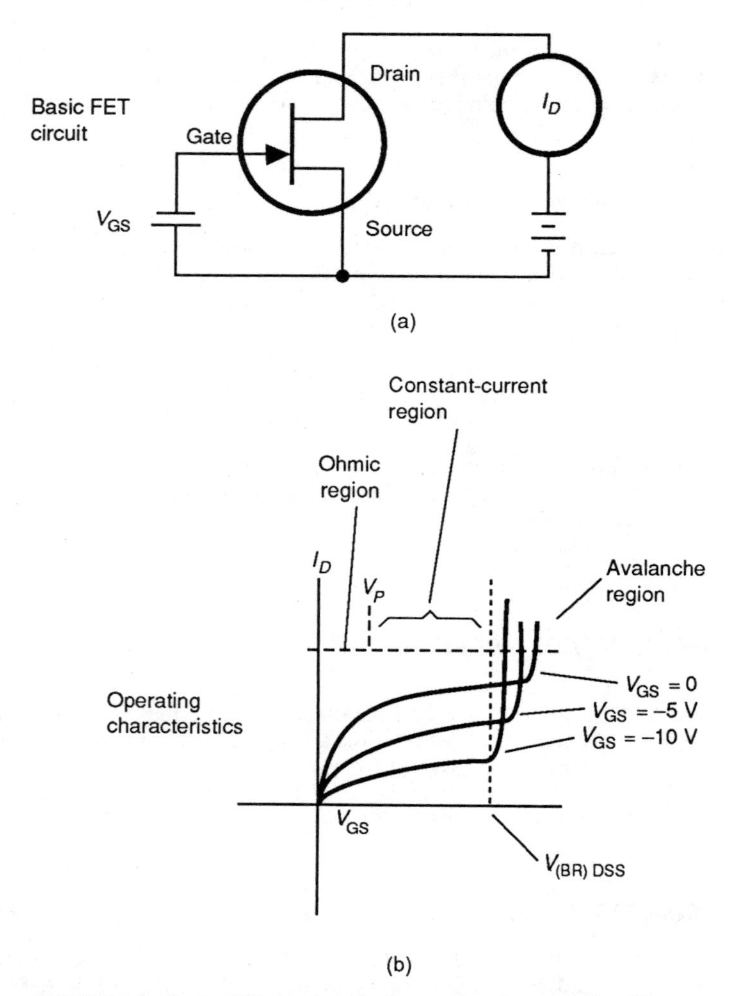

FIGURE 1.5 Basic FET circuit (*a*) and operating characteristics (*b*).

above $V_{(BR)DSS}$, is the *avalanche* region where the FET is not operated in practical audio circuit.

1.11.2 Zero Temperature Coefficient

An important characteristic of all FETs is the ability to operate at a zero-temperature-coefficient (0TC) point. This means that if the gate source is biased at a specific voltage and is held constant, the drain current, or I_D, does not vary with changes in temperature. This characteristic makes for very stable amplifier circuits, particularly when the FET is used in the input stage of the circuit.

Although the 0TC point varies from one FET to another and depends on a number of factors, the *approximate* 0TC point can be found when 0.63 is subtracted from V_P. For example, if V_{GS} must be at 1 V to produce pinchoff, or V_P, the FET will operate at the 0TC point when V_{GS} is 0.37 V.

Sometimes FET datasheets will show the V_{GS} value to get 0TC. A more practical method is to heat and cool the FET with soldering tool and coolant (can of Freon) while adjusting V_{GS} to get a constant I_D (in the presence of temperature changes).

Of course, it is not always practical to operate a FET at 0TC. For example, assume that the required V_{GS} to produce 0TC is 0.37 V, and the FET is to operate as an audio amplifier with 0.5-V input signals.

CHAPTER 2
PRACTICAL CONSIDERATIONS FOR AUDIO

This chapter is devoted to mounting techniques and thermal considerations for audio components, particularly transistors and ICs. The main concern is with *audio power amplifiers* where thermal problems can arise in both design and service. This chapter summarizes such problems and their solutions. The practical information in this chapter is particularly helpful if it becomes necessary to replace or reinstall *heat sinks* found on many audio power amplifiers (both IC and discrete component).

2.1 TEMPERATURE-RELATED PROBLEMS

The three most critical parameters for transistors used in audio circuits are current gain, collector leakage, and power dissipation. These parameters change with temperature. To compound the problem, a change in parameters can also affect temperature (for example, an increase in current gain or power dissipation results in a temperature increase).

When power dissipation is over about 1 W for a single transistor or circuit, *heat sinks* (or special component mounting provisions) are used to offset the effects of temperature (such as the thermal-runaway effect described in Sec. 1.7). For example, if a transistor (or diode, rectifier, or IC) is used with a heat sink (or is mounted on a metal chassis that acts as a heat sink), an increase in temperature (from any cause) can be dissipated into the surrounding air.

2.1.1 Effects of Temperature

Collector leakage (I_{CBO}) increases with temperature. As a guideline, collector leakage doubles with every 10°C increase in temperature for germanium transistors and doubles with every 15°C/V increase for silicon transistors. Also, always consider the possible effects of a different collector voltage when approximating collector leakage at temperatures other than those on the datasheet (as discussed in Chap. 3).

Current gain (h_{fe}) increases with temperature. As a guideline, current gain doubles when the temperature is raised from 25 to 100°C for germanium transistors and doubles when the temperature is raised from 25 to 175°C for silicon. If the datasheet does not specify a maximum operating temperature (or there is no datasheet), do not exceed 100°C for germanium or 200°C for silicon.

The power-dissipation capabilities of a transistor (or diode-rectifier and IC) are directly related to temperature and must be carefully considered when designing or servicing any audio circuit. For example, do not apply power during service with the heat sinks removed or you will quickly learn the effects of temperature on transistor power dissipation.

Audio components (particularly transistors) often have some form of *thermal resistance* specified to show the power-dissipation capabilities. Thermal resistance can be defined as the *increase in temperature of the component junction* (with respect to some reference) divided by the power dissipated, or °C/W.

In audio power transistors, thermal resistance is normally measured from the component junction to the case, resulting in the term θJC. When the case is bolted directly to the mounting surface with a built-in threaded bolt or stud, the term θMB (thermal resistance to mounting base) or θMF (thermal resistance to mounting flange) is used. For audio power components where the junction is mounted directly on a header or pedestal, the total internal thermal resistance from junction to case (or mount) varies from about 50°C/W to less than 1°C/W.

Note that some audio transistor (and IC) datasheets specify a *maximum case temperature* rather than θJA. As discussed in Sec. 2.1.3, maximum case temperature can be combined with heat-sink thermal resistance to find maximum power dissipation.

2.1.2 Practical Heat-Sink Considerations

Commercial heat sinks are available for various transistor case sizes and shapes (Sec. 2.2). Such heat sinks are especially useful when the transistors (or ICs) are mounted in sockets which provide no thermal conduction to the chassis or PC board. Commercial heat sinks are rated in terms of thermal resistance, usually in terms of °C/W.

When heat sinks involve the use of washers, the °C/W factor usually includes the thermal resistance between the case and sink, or θCA. With a washer, only the sink-to-ambient, θSA, thermal-resistance factor is given. Either way, the thermal-resistance factor represents a temperature increase (in °C) divided by wattage dissipated.

For example, if the heat-sink temperature rises from 25 to 100°C (a 75°C increase) when 25 W is dissipated, the thermal resistance is 75/25 = 3. This can be listed on a datasheet as θSA, or simply as 3°C/W.

All other factors (such as transistor size, mounting provisions, etc.) being equal, the heat sink with the lowest thermal resistance is best (a heat sink with 1°C/W is better than a 3°C/W heat sink).

To operate an audio circuit at full power, there should be no temperature difference between the case and ambient air. This occurs only when the thermal resistance of the heat sink is zero and the only thermal resistance is that between the junction and case. It is not practical to manufacture a heat sink with zero resistance. However, the greater the ratio θJC/θCA, the nearer the maximum power limit (set by θJC) can be approached.

When transistors are mounted on heat sinks, some form of *electrical insulation* is provided between the case and heat sink (unless a grounded-collector circuit is used). Because good electrical insulators are (usually) good thermal insulators, it is difficult to provide electrical insulation without introducing some thermal resistance between case and heat sink.

The most common materials for electrical insulation of heat sinks are mica, beryllium oxide (Beryllia), and anodized aluminum. The properties of these three materials for the case-to-heat-sink insulation of a TO-3 case are compared as follows:

Material	Thickness (in)	°C/W	Capacitance (pF)
Beryllia	0.063	0.25	15
Anodized aluminum	0.016	0.35	110
Mica	0.002	0.40	90

As shown, any insulation between collector and the chassis (such as produced by a washer between the case and heat sink) also results in capacitance between the two metals. This capacitance can be a problem in RF design but is usually of no concern in audio work.

2.1.3 Calculating Power Dissipation

For practical design, the no-signal dc collector voltage and current can be used to calculate power dissipation when a transistor is operated under steady-state conditions (such as in an audio amplifier). Other calculations must be used for pulse operating conditions but generally do not apply to audio. In theory, there are other currents that produce power dissipation (collector-base leakage current, emitter-base current, etc.). However, these can be ignored, and power dissipation (in watts) can be considered as the dc collector voltage times the collector current.

Once the power dissipation is calculated, the *maximum power dissipation* must be found. Under steady-state conditions, the maximum dissipation capability depends on (1) the sum of the series thermal resistances from the transistor junction to ambient air, (2) the maximum junction temperature, and (3) the ambient temperature.

Assume that it is desired to find the *maximum power dissipation* of a transistor under the following conditions: a maximum junction temperature of 200°C (typical for a silicon power transistor used in audio), a junction-to-case thermal resistance of 2°C/W, a heat sink with a thermal resistance of 3°C/W, and an ambient temperature of 25°C.

First, find the total junction-to-ambient thermal resistance:

$$2°\mathrm{C/W} + 3°\mathrm{C/W} = 5°\mathrm{C/W}.$$

Next, find the maximum permitted power dissipation:

$$\frac{200°\mathrm{C} - 25°\mathrm{C}}{5°\mathrm{C/W}} = 35\ \mathrm{W} \qquad \text{(maximum)}$$

If the same transistor is used without a heat sink, but under the same conditions and with a TO-3 case (which is rated at 30°C/W), the maximum power can be calculated as follows: First find the total junction-to-ambient thermal resistance

$$2°C/W + 30°C/W = 32°C/W$$

Next, find the maximum permitted power dissipation

$$\frac{200°C - 25°C}{32°C/W} = 5\ W \qquad \text{(approximate)}$$

Some audio power-transistor datasheets specify a *maximum case temperature* rather than a maximum junction temperature. Assume that a maximum case temperature of 130°C is specified instead of a maximum junction temperature of 200°C. In that event, subtract the ambient temperature from the maximum permitted case temperature: 130°C − 25°C = 105°C.

Then divide the case temperature by the heat-sink thermal resistance: 105°C/3°C = 35 W maximum power.

2.2 HEAT-SINK AND COMPONENT MOUNTING TECHNIQUES

This section summarizes mounting techniques for metal-packaged power transistors found in audio circuits.

2.2.1 Interface Thermal Resistance

The interface thermal resistance is often referred to as the case-to-sink resistance and is sometimes listed as RθCS. Table 2.1 shows the approximate interface thermal resistance (in °C/W) for a few transistor and diode cases. The table also shows recommended hole and drill sizes, as well as torque for the mounting nuts or screws. These values can be used when replacing audio component during service (in the absence of specific service literature instructions).

As shown in Table 2.1, interface thermal resistance changes quite drastically for different mounting conditions. For example, assume that a TO-3 case is involved. If the case is mounted (on a heat sink or chassis) with an insulator and no thermal compound or lubrication is used, the interface thermal resistance is 1.45°C/W. If a thermal compound is used, the resistance drops to 0.8°C/W. If circuit conditions make it possible to eliminate the insulator, the thermal resistance drops to 0.1°C (with thermal compound) or 0.2°C (without compound).

2.2.2 Fastening Techniques

The various types of transistor and diode packages shown in Table 2.1 require different fastening techniques. Mounting details for stud, flat-base, press-fit, and disk-type transistors are shown in Fig. 2.1. Of course, there are many other types of fastening techniques for transistors used in audio amplifiers. The following notes supplement the few examples shown here.

With any of the mounting schemes, the *screw heads should be free of grease* to prevent inconsistent torque readings when tightening nuts. Maximum allow-

TABLE 2.1 Interface Thermal Resistance Characteristics

Package type			Interface thermal resistance (°C/W)					
				Metal to metal		With insulator		
JEDEC outline number	Description	Recommended hole and drill size	Torque in-lb	Dry	Lubed	Dry	Lubed	Type
DO-4	10-32 stud 7⁄16 hex	0.118, no. 12	15	0.41	0.22	1.24	1.06	3-mil mica
DO-5	1⁄4-28 stud 11⁄16 hex	0.25, no. 1	30	0.38	0.20	0.89	0.70	5-mil mica
DO-21	Pressfit, 1⁄2	See Fig. 2.1*c*	—	0.15	0.10	—	—	—
TO-3	Diamond	0.14, no. 28	—	0.20	0.10	1.45	0.80	3-mil mica
						0.80	0.40	2-mil mica
						0.40	0.35	Anodized aluminum
TO-66	Diamond	0.14, no. 28	—	—	0.50	—	—	—
TO-83	1⁄2-20 stud	0.50	130	—	0.10	—	—	—

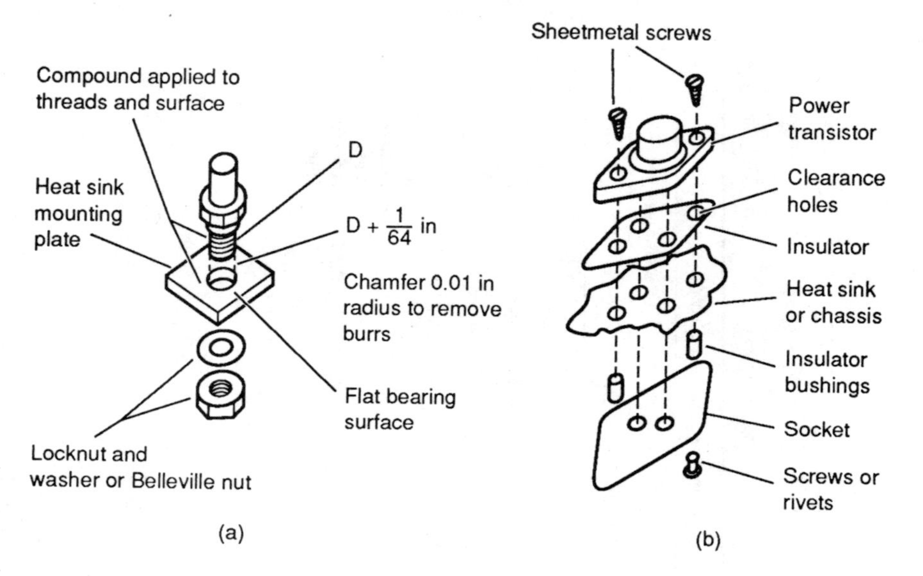

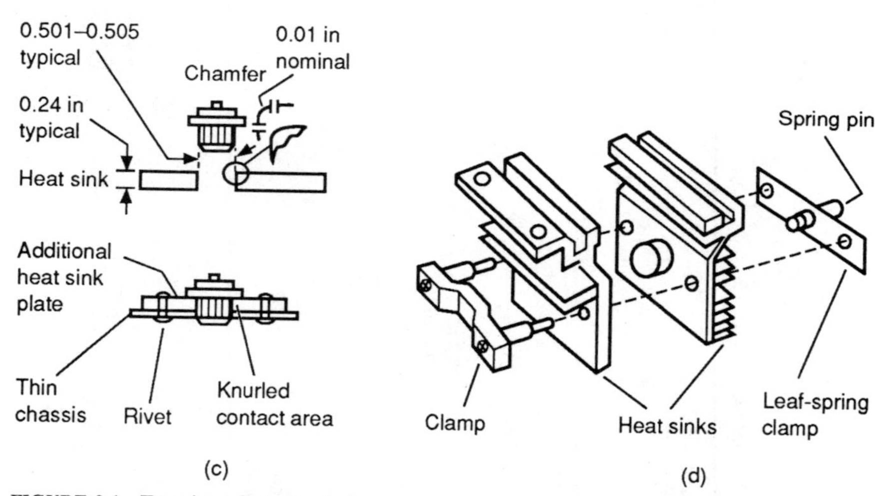

FIGURE 2.1 Transistor fastening techniques.

able torque should always be used to reduce thermal resistance. However, take care not to exceed the torque rating of parts. Excessive torque applied to disk- or stud-mounted parts can cause damage to the semiconductor die.

To prevent galvanic action from occurring when components are used with aluminum heat sinks in a corrosive atmosphere, many devices are nickel or gold plated. Take precautions not to mar the surface.

With press-fit components (Fig. 2.1*c*), the hole edge must be chamfered as shown to prevent shearing off the knurled edge of the component during press-in.

The pressing force should be applied evenly on the shoulder ring to avoid tilting or canting the device case in the hole during the pressing operation. Also, thermal compound (Sec. 2.2.4) should be used to ease the component into the hole.

With the disk-type mounting (Fig. 2.1*d*), a self-leveling type of mounting clamp is often used to keep the contacts parallel, with even distribution of pressure on each contact area. A swivel-type clamp or a narrow leaf spring in contact with the heat sink is often used.

When reinstalling the component, apply the clamping force smoothly, evenly, and perpendicular to the disk-type package (to prevent deformation of the device or the sink-mounting surfaces). Use the correct clamping force as specified in the service literature.

2.2.3 Preparing the Heat-Sink Mounting Surface

In general, the heat sink should have a flatness and finish comparable to that of the component. For the typical experimenter or hobbyist, the heat-sink surface is satisfactory if the surface *appears flat* against a straightedge and is free of any deep scratches. During the manufacture of commercial audio equipment, it may be necessary to measure the actual flatness with special tools and indicators.

Many commercial or off-the-shelf heat sinks require spot-facing. In general, milled or machined surfaces are satisfactory if prepared with tools in good working order.

The surface must be free from all dirt, film, and oxide (freshly bared aluminum forms an oxide layer in a few seconds). Unless used immediately after machining, it is a good practice to polish the mounting area with no. 000 steel wool, followed by an acetone or alcohol rinse. Thermal grease should then be applied *immediately*. The same is true when reinstalling heat sinks.

Many aluminum heat sinks are *black anodized* for appearance, durability, performance, and economy. Anodizing is an electrical and thermal insulator that offers resistance to heat flow. As a result, anodizing should be removed from the mounting area.

Another aluminum finish is *iridite* (chromate acid dip), which offers low resistance because of the thin surface. For best results, the iridite finish must be cleaned of oils and films that collect in the manufacture and storage of the sinks.

Some heat sinks are *painted* after manufacture. Paint of any kind has a high thermal resistance (compared to metal). For that reason, it is essential that paint be removed from the heat-sink surface where the component is attached.

2.2.4 Thermal Compounds

Thermal compounds (also called *joint compounds* or *silicon greases*) are a formulation of fine zinc particles in a silicon oil that maintain a grease-like consistency with time and temperature. These compounds are used to fill air voids between mating surfaces and thus improve contact between the component and heat sink.

Compounds can be applied in a very thin layer with a spatula or lintless brush (wiping lightly to remove excess compound) or by applying a small amount of pressure to spread the compound (removing any excess compound after the mounting is complete). It may be necessary to use a cloth moistened with acetone or alcohol to remove all excess compound.

A typical compound has a resistivity of about 60°C-in/W, compared to about 1200°C-in/W for air. The following are some often recommended thermal compounds.

Dow Corning, Silicon Heat Sink Compound 340

General Electric, Insulgrease

Wakefield, Thermal Compound Type 1201

Astrodyne, Conductive Compound 829

Emerson & Cuming, Inc., Eccotherm TC-4

2.3 POWER DISSIPATION PROBLEMS IN IC COMPONENTS

The basic rules for audio-circuit ICs regarding power dissipation and thermal considerations are essentially the same as those for discrete transistors and diodes, as discussed in Sec. 2.1.

The maximum allowable power dissipation (usually specified as PD or *maximum device dissipation* on IC datasheets) is a function of the maximum storage temperature T_S, the maximum ambient temperature T_A, and the thermal resistance from the semiconductor chip to case. The basic relationship is PD = $(T_S - T_A)$/thermal resistance.

Some IC datasheets list only the maximum power dissipation for a given ambient temperature and then show a *derating* factor in terms of maximum power decrease for a given increase in temperature. For example, a typical IC might show a maximum power dissipation of 110 mW at 25°C, with a derating factor of 1 mW°C/W. If such an IC is operated at 100°C, the maximum power dissipation is 100 − 25, or a 75°C increase, 110 − 75 = 35 mW.

In the absence of specific datasheet information, the following typical temperature characteristics can be applied to the basic IC package types. *No IC should have a temperature in excess of 200°C.*

Ceramic flat pack

Thermal resistance = 140°C/W

Maximum storage temperature = 175°C

Maximum ambient temperature = 125°C

Dual-in-line package (ceramic)

Thermal resistance = 70°C/W

Maximum storage temperature = 175°C

Maximum ambient temperature = 125°C

Dual-in-line package (plastic)

Thermal resistance = 150°C/W

Maximum storage temperature = 85°C

Maximum ambient temperature = 75°C

TO-5 style (metal can) package

Thermal resistance = 140°C/W

Maximum storage temperature = 200°C

Maximum ambient temperature = 125°C

2.3.1 Audio Power ICs

Power ICs (such as found in audio power amplifiers) generally use some form of metal package. The datasheets for these power ICs usually list sufficient information to select the proper heat sink. Also, the datasheets or other literature often make recommendations about mounting for the power IC. Always follow the IC manufacturer's recommendations. In the absence of such data and to make the reader familiar with the terms used, the following sections summarize considerations for power ICs.

2.3.2 Maximum Power for ICs

From a design standpoint, an IC is a complete, predesigned, functioning circuit that cannot be altered in regard to power dissipation. If the supply voltages, input signals, and ambient temperature are at the recommended levels, power dissipation is within the capabilities of the IC. With the possible exception of data required to select or design heat sinks, the amplifier designer or experimenter need only follow the datasheet recommendations. Of course, derating may be needed if the IC is operated at higher temperatures.

2.3.3 Thermal Resistance for ICs

In audio power ICs, thermal resistance is normally measured from the semiconductor chip to the case. On those ICs where the cases are bolted directly to the mounting surfaces with a threaded bolt or stud (similar to that shown in Fig. 2.1*a*), the thermal resistance is measured from the chip to the mounting stud or flange.

2.3.4 Thermal Runaway in Power ICs

Audio power ICs are subject to thermal runaway as described for transistors. However, unlike discrete transistors, most ICs have built-in circuits to prevent thermal runaway. A typical arrangement is to connect a diode (internal) in the reverse-bias circuit for one or more of the transistors in the IC. When temperature increases, diode resistance changes to offset the initial change in current caused by temperature increase. Unless specifically noted on the datasheet, the IC user need not be concerned so long as the datasheet limits are observed.

CHAPTER 3
SIMPLIFIED AUDIO DESIGN

This chapter provides a review and summary of audio design from a very practical standpoint. The information is presented primarily for the serious experimenter who must design audio circuits not always available in IC form (and those audio circuits found between ICs in present-day audio equipment). While this book makes no pretense at being an engineering design text, the information presented here provides the background necessary to understand the test and troubleshooting described in the remaining chapters. If you understand what is required in design of audio circuits, you have a starting point for troubleshooting such circuits (when the circuits fail to perform as designed).

3.1 FREQUENCY LIMITATIONS IN AUDIO CIRCUITS

Every audio component has some impedance (and reactance) and is thus *frequency sensitive*. Four major components are used in discrete audio-circuit design: transistors, resistors, capacitors, and inductances. Let us examine how the impedances and reactances of these components affect audio circuits.

3.1.1 Transistor Frequency Limitations

Capacitive reactance decreases with an increase in frequency and vice versa. Consider the effect of this on a common-emitter audio amplifier such as shown in Fig. 3.1*a*. As frequency increases, the capacitive reactance drops, producing a short across the input and output, and increases attenuation of the signal. At some frequency, the attenuation equals the transistor amplification, so there is no gain. At higher frequencies, the attenuation exceeds amplification, and there is a loss, even though the transistor may still operate. However, the problem is of no concern at frequencies up to about 20 kHz and of little concern up to about 100 kHz.

3.1.2 Resistor Frequency Limitations

In the audio range, resistors offer relatively few problems since resistors attenuate signals equally. Only at very high frequencies, where resistor leads and body

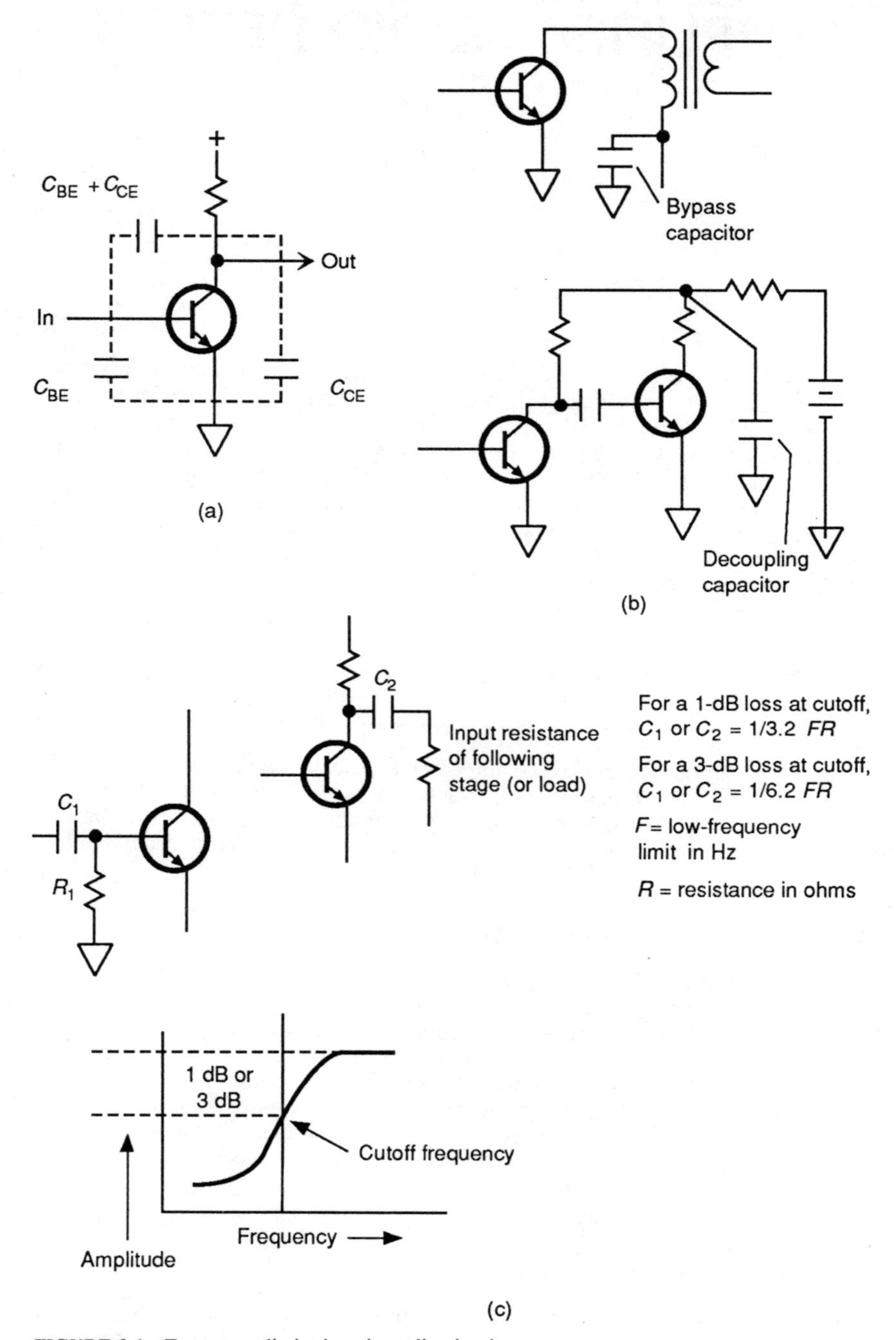

FIGURE 3.1 Frequency limitations in audio circuits.

could produce some kind of reactance, is there any particular concern about frequency limits imposed by resistors.

3.1.3 Capacitor Frequency Limitations

Capacitors have three major functions in audio circuits: bypass, decoupling, and coupling. Figure 3.1*b* shows both bypass and decoupling functions. From a practical standpoint, the functions of bypass and decoupling capacitors are the same, and the terms are interchanged. In either case, the main concern is that the *reactance be low at the lowest frequency involved.*

As an example, assume that the lowest frequency involved is 100 Hz, and the minimum required reactance is 100 Ω. This requires a capacitor value of about 10 μF. If the frequency is decreased to 10 Hz, the capacitor value must be raised to 160 μF to keep the reactance below 100 Ω.

Figure 3.1*c* shows the effect of coupling capacitors on audio circuits. The values of *coupling capacitors depend on the low-frequency limit* at which the circuit operates and on the resistances with which the capacitors operate. (The high-frequency limit need not be considered since capacitive reactance decreases with increases in frequency to pass the signals.)

In Fig. 3.1*c*, C_1 forms a high-pass *RC* filter with R_1, while C_2 forms another high-pass filter with the input resistance of the following stage (or the load). When the capacitive reactance of C_1 or C_2 drops to about one-half the resistance, the output from the corresponding filter drops to about 90 percent of the input (or about a 1-dB loss).

As shown by the equations in Fig. 3.1*c*, the approximate value of C_1 and C_2 can be found when the resistance and desired low-frequency cutoff are known (both the desired drop and frequency).

3.1.4 Inductance Frequency Limits

As discussed in Sec. 3.2, coils are sometimes used in place of the collector resistor as a load (to permit the collector to be operated at a higher voltage). Likewise, transformers are used for coupling between stages (mostly in older audio equipment) to provide impedance matching.

The inductive reactance of coils and transformers increases with frequency. At the high end of the audio range (about 15 to 20 kHz), the reactance of a typical transformer drops to a few ohms. This low impedance acts as a short across the line and attenuates the signal.

3.2 AUDIO COUPLING METHODS

All audio amplifiers require some form of coupling. Even a single-stage audio amplifier must be coupled to the input and output devices. If more than one stage is involved, there must be *interstage* coupling. Figure 3.2 shows the four basic coupling methods for audio circuits. The following is a summary of these methods.

The outstanding characteristic of a *direct-coupled amplifier* (Fig. 3.2*a*) is the ability to amplify direct current and low-frequency signals. One of the disadvan-

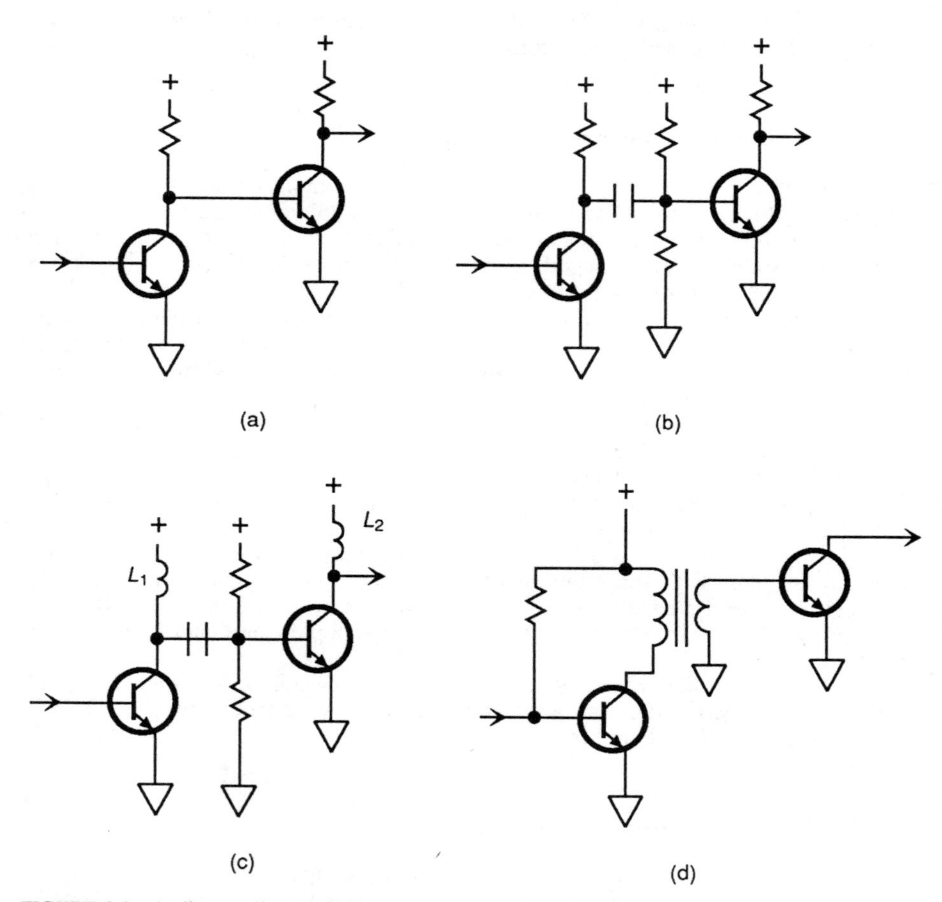

FIGURE 3.2 Audio coupling methods.

tages of direct coupling is the inability to tell the difference between a signal change and a power-supply change.

The main advantage of *capacitor* (or *RC*) *coupling* (Fig. 3.2*b*) is uniform amplification over nearly the entire audio range since resistor values are independent of frequency changes. However, as discussed in Sec. 3.1.3, *RC*-coupled amplifiers have a low-frequency limit imposed by reactance of the coupling capacitor.

The advantage of *impedance coupling* (Fig. 3.2*c*) over *RC* coupling is that the ohmic resistance of the load inductor (L_1 and L_2) is less than that of a load resistor. (For a power supply of a given voltage, there is a higher collector voltage.) The main disadvantage of impedance coupling is frequency discrimination.

The *transformer-coupled amplifier* (Fig. 3.2*d*) has essentially the same advantages and disadvantages as the impedance-coupled circuit. (The transistor collectors can be operated at higher voltages, but there are frequency limitations.) However, the impedances of a transformer-coupled circuit are set by the transformer primary and secondary windings. Transformer coupling is most effective when the final amplifier output must be fed to a low-impedance load. For example, the impedance of a typical loudspeaker is on the order of 4 to 16 Ω, whereas the output impedance of a transistor stage is several hundred (or thousand) ohms. A transformer at the output of an audio amplifier can offset the obviously undesired effects of such a mismatch.

3.2.1 Coupling versus Audio Frequency Response

Figure 3.3 is a simplified frequency-response graph or curve. (A more comprehensive graph, as well as the procedures for producing such graphs, is discussed in Chap. 4.) The graph in Fig. 3.3 is provided to illustrate the effects of coupling methods on audio amplifier frequency response. Such response is measured by amplifier gain at various frequencies in the audio range.

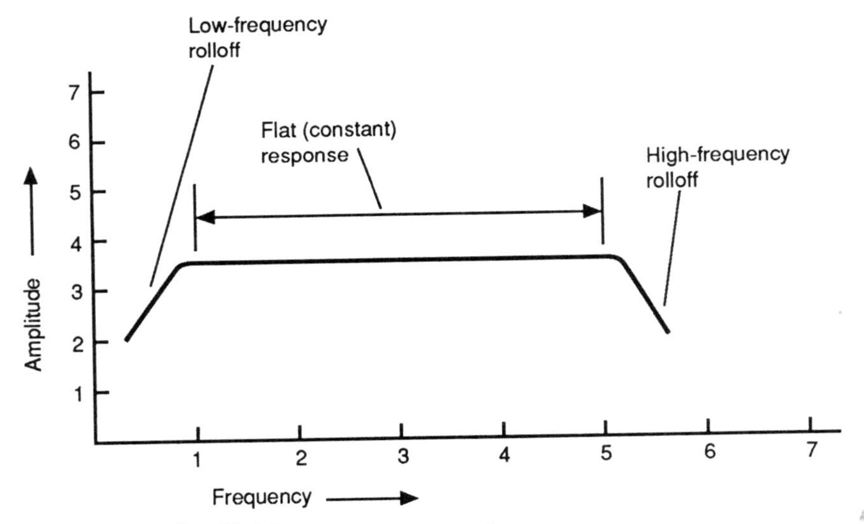

FIGURE 3.3 Simplified frequency-response graph.

Note that the gain falls off at very low frequencies. In an *RC*-coupled amplifier, this drop in gain (generally called *rolloff*) is caused by the reactance of the coupling capacitors. With impedance- or transformer-coupled amplifiers, the low-frequency rolloff is caused by the very low inductive reactance which acts as a short across the signal path.

The gain also falls off at the higher frequencies. In *RC*-coupled amplifiers, this high-frequency rolloff is caused by the output capacitance of the first stage, the input capacitance of the second stage (Fig. 3.1*a*), and any stray capacitance in the coupling circuit. These capacitances bypass some of the signal to ground.

The higher the frequency is the smaller the capacitive reactance becomes and the greater the amount of signal so bypassed. As a result, the overall gain drops as frequency increases. With impedance- or transformer-coupled amplifiers, the high-frequency rolloff is produced by the large inductive reactance that attenuates the signal.

To sum up, *RC* coupling produces the lowest gain and transformer coupling, the highest. As a guideline, three stages of *RC*-coupled amplification produce about the same gain as two stages of comparable transformer-coupled amplification. *RC* coupling produces the least frequency distortion. Transformer coupling has the added advantage of impedance matching. Note that transformer coupling is used primarily where power is required. In present-day audio equipment, transformer coupling is replaced with power ICs, as discussed throughout this book.

3.3 HOW TRANSISTOR RATINGS AFFECT AUDIO PERFORMANCE

The characteristics of discrete-component audio circuits are directly related to the ratings of capabilities of the transistors involved. For example, the current gain of an individual amplifier stage can be no greater than the maximum possible current gain of the transistor in that stage. Whether the problem is one of discrete-component amplifier design or amplifier testing and troubleshooting, it is always helpful (and often necessary) to know the characteristics of the circuit transistors. These characteristics are given in transistor datasheets. This section provides a summary of the most important characteristics found on transistor datasheets (for the practical experimenter and troubleshooter).

Maximum voltage is often listed as V_{CBO}. Actually, this is a test voltage rather than an operating or design voltage. However, for design purposes, the figure can be considered as the *absolute maximum* collector voltage. In a typical class A audio circuit, the collector voltage is half the source voltage at the normal operating point. Since the collector voltage rises to or near the source voltage when the transistor is at or near cutoff, *never* connect the collector to a source higher than the maximum, even through a resistor.

Always allow for some variation in source voltage when an electronic power supply is used. Although a battery does not deliver more than the rated voltage, any electronic power supply is subject to some voltage variation.

Another factor that affects maximum voltage is temperature. (Breakdown occurs at a lower voltage when temperature is increased.) Refer to the temperature-related design characteristics discussed in Chap. 2.

Always consider any *input signal* that may be applied to the base-emitter junction, in addition to the normal operating bias. In practical design, it is often necessary to select a bias (base-emitter voltage) on the basis of input signal rather than on some arbitrary point on the transistor curve.

Collector current is often listed as I_C and always increases with temperature. As discussed in Chap. 2, temperature also increases with current (possibly causing thermal runaway). Therefore, do not operate any transistor at or near the maximum current rating.

In practical design, it is the power dissipated in the collector circuit (rather than a given current) that is of major concern. The power dissipated is found by multiplying the collector voltage by the collector current. As discussed in Chap. 2, the maximum power dissipated is usually less than 1 W for transistors operated without heat sinks.

The power-dissipation capabilities of a transistor in any circuit are closely associated with the temperature range. It may be necessary to use the datasheet *temperature-derating* factors discussed in Chap. 2.

Datasheet *small-signal characteristics* can be defined as those where the ac signal is small compared to the dc bias. For example, h_{fe}, or forward-current transfer ratio (also known as *ac beta* or *dynamic beta*), is properly measured by noting the change in collector alternating current for a given change in base alternating current, without regard to static base and collector currents. Use small-signal characteristics as a starting point for simplified or practical design, not as hard-and-fast design rules.

Datasheet *high-frequency characteristics* are especially important in the design of RF, IF, and video amplifiers but not too important for most audio work. Most present-day transistors will operate at full capability well beyond the audio range.

Datasheet *switching characteristics* are vital in the design of pulse amplifiers but again are not so important for audio. A possible exception is where an audio amplifier must pass pulse or square-wave signals. (As discussed in Chap. 4, pulse and square-wave signals are often used to test audio amplifiers.)

Typical switching characteristics shown on transistor datasheets are defined in Fig. 3.4*a*. These factors (delay, storage, rise, and fall times) determine the operating limits for any circuit that must pass pulses or switched signals. For example, if an audio amplifier uses a transistor with a 20-ns rise time to pass a 15-ns pulse, the pulse is hopelessly distorted. Likewise, if there is a 1.5-μs delay added to a 1-μs pulse, an absolute minimum of 2.5 μs is required before the next pulse can occur. This means that the amplifier can pass pulses with a maximum repetition rate of 400 kHz ($1/2.5^{-6}$ = 400,000 Hz). Actually, the maximum repetition rate is lower since there is some "off" time between pulses.

Datasheet *direct-current characteristics*, while important in the design of basic bias circuits (Chap. 1), do not have too critical an effect on operation of an audio amplifier. The dc characteristics are primarily test values rather than design parameters.

3.3.1 Transistor Characteristics at Different Frequencies

Datasheets specify most transistor characteristics at some given frequency. Unless the circuit is to operate at that exact frequency, it is essential to know the characteristic at other frequencies (in the frequency range where the circuit is used). The following paragraphs describe methods for converting from one transistor characteristic to another at different frequencies.

Datasheet Terms. Before we get into the characteristics, let us define some typical terms found on transistor datasheets. Figures 3.4*b* and 3.4*c* define the most often used frequency terms.

Common-Base Terms. The quantity h_{fbo} (the value of h_{fb} at 1 kHz) remains constant as frequency is increased until a top limit is reached. After the top limit, h_{fb} begins to drop rapidly as shown in Fig. 3.4*b*. The frequency at which a significant drop in h_{fb} occurs provides a basis for comparison of the expected frequency performance of different transistors. This frequency is known as f_{ab} and is defined as that frequency at which h_{fb} is 3 dB below h_{fbo}.

Note that the typical curve of Fig. 3.4*b* has the following significant characteristics:

1. At frequencies below f_{ab}, h_{fb} is nearly constant and about equal to h_{fbo}.
2. h_{fb} begins to decrease significantly in the region of f_{ab}.

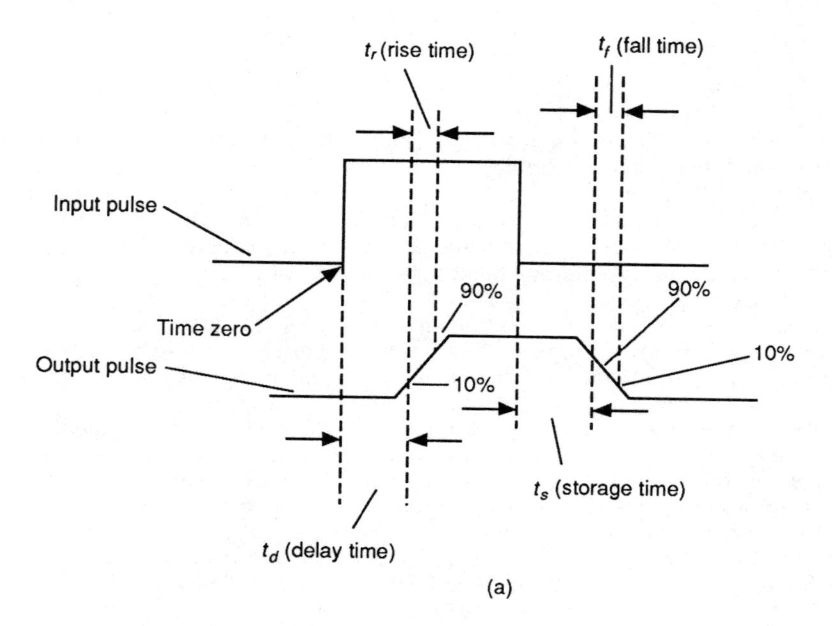

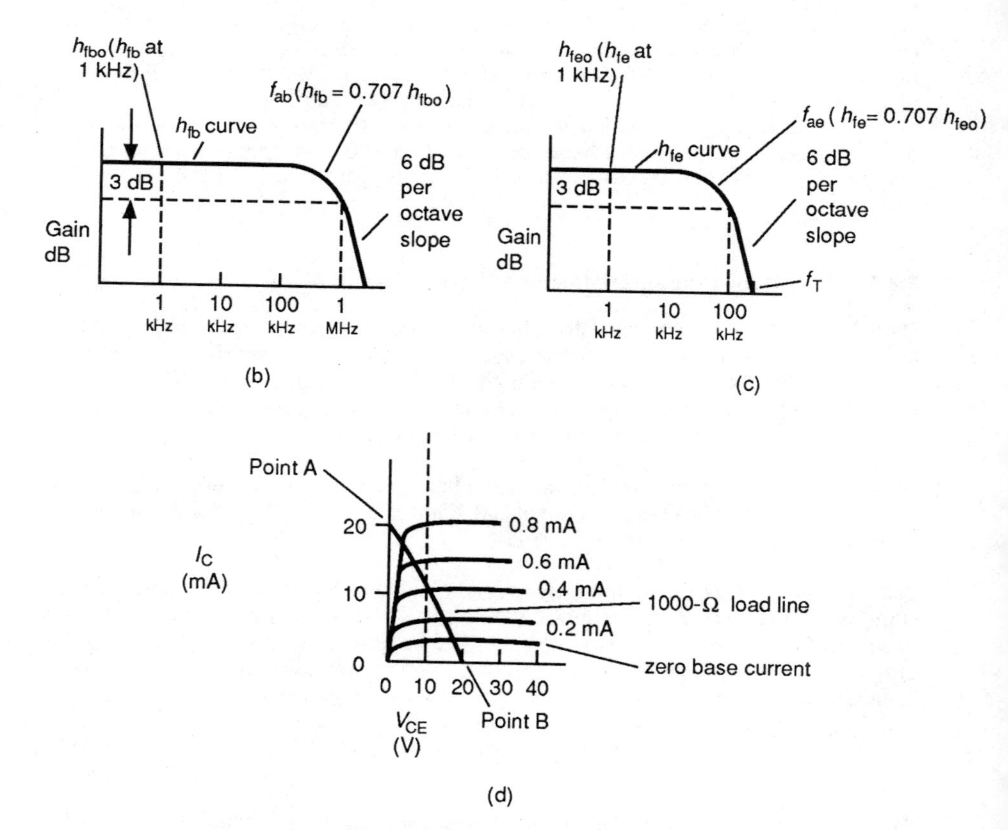

FIGURE 3.4 Transistor frequency and switching characteristics.

3. Above f_{ab}, the rate of decrease for h_{fb} (with increasing frequency) approaches 6 dB/octave.

Common-Emitter Terms. The common-emitter characteristic that corresponds to f_{ab} is f_{ae}, the common-emitter current-gain cutoff frequency. This f_{ae} is the frequency at which h_{fe} (beta) has decreased 3 dB below h_{feo}. Figure 3.4*c* shows a curve of h_{fe} versus frequency for a transistor with an f_{ae} of 100 kHz.

The curve in Fig. 3.4*c* has the same significant characteristics as those described for Fig. 3.4*b*. That is, h_{fe} is considered to be decreasing at a rate of 6 dB/octave at f_{ae}. These characteristics allow such a curve to be constructed for a particular transistor by knowing only h_{feo} and f_{ae}. With the curve constructed, h_{fe} can be determined at any frequency. Also, if f_{ae} is not known, a curve can also be constructed if h_{feo} and h_{fe} at any frequency above f_{ae} are known.

Gain-bandwidth product, or f_T (Fig. 3.4*c*) is a common-emitter term sometimes specified on datasheets instead of f_{ae}. This f_T is the frequency at which gain drops to 1 (0 dB). f_T can be approximated when f_{ae} is multiplied by h_{feo} ($f_T = f_{ae} \times h_{feo}$).

The term f_{max} is the frequency at which common-emitter power gain is equal to 1. (Although current gain is equal to 1 at f_T, there may still be some power gain at f_T because of different input and output impedances, as discussed in Chap. 1.)

Conversion of Terms. It is often necessary to convert from one set of terms to another (say from common base to common emitter) since all datasheets are not alike. The following rules summarize the conversion process:

To find beta when alpha is given, beta = alpha/(1 − alpha)

To find alpha when beta is given, alpha = beta/(1 + beta)

To find h_{feo} when h_{fbo} is given, $h_{feo} = h_{fbo}/(1 - h_{fbo})$

To find h_{fbo} when h_{feo} is given, $h_{fbo} = h_{feo}/(1 + h_{fbo})$

To find f_{ae} when f_{ab} is given, $f_{ae} = K(1 - h_{fbo})f_{ab}$

To find f_{ab} when f_{ae} is given, $f_{ab} = f_{ae}/K(1 - h_{fbo})$

Note that the constant K refers to a phase-shift factor. The K constant usually ranges between 0.8 and 0.9, but can be as low as 0.6 for some transistors.

Different Frequencies. Three frequency-related common-emitter characteristics are of particular importance to design: (1) h_{fe} at some frequency other than specified on the datasheet, (2) f_T, and (3) f_{max}. The following three rules summarize the procedures for finding these three characteristics when they are not available on the datasheet:

1. *To find* h_{fe} *at a particular frequency*:

h_{fe} is approximately equal to h_{feo} when the frequency of interest is less than f_{ae}.

h_{fe} is approximately equal to 0.7 h_{feo} when the frequency of interest is near f_{ae}.

h_{fe} decreases at a rate of 6 dB/octave and is approximately equal to f_T/frequency when the frequency of interest is above f_{ae}.

h_{fe} is equal to 1 (unity gain) at f_T.

2. *To find* f_T:

When h_{feo} and f_{ae} are given, $f_T = h_{feo} \times f_{ae}$.

When h_{fbo} and f_{ab} are given, $f_T = h_{fbo} \times f_{ab} \times K$.

3. *To find* f_{max}: As discussed, f_{max} is a common-emitter characteristic and requires that the common-emitter power gain be established at some frequency on the sloping portion of the gain versus frequency curve (Fig. 3.4*c*). With such a gain established,

$$f_{max} = \text{frequency of measurement} \times \sqrt{\text{power gain in magnitude}}$$

3.3.2 Load Lines

Many datasheets show transistor characteristics by means of curves that are reproductions of displays obtained with a curve tracer. The collector voltage-current curves shown in Fig. 3.4*d* are typical. Load lines can be superimposed on the datasheet curves to show how the transistor will perform in an audio circuit.

As an example, assume that a collector load of 1 k is used with the transistor shown in Fig. 3.4*d*. When the collector current reaches 20 mA (base current about 0.8 mA), the collector voltage drops to zero. Likewise, when the collector current drops to zero, collector voltage rises to 20 V (base current zero). If a load line is connected between these two extreme points (marked A and B on Fig. 3.4*d*), the instantaneous collector voltage and collector current can be obtained for any base current along the line.

Load lines provide only approximate transistor gain. For example, with a base-current change from 0.2 to 0.4 mA (a 0.2-mA change), the collector current changes from about 4 to 9 mA (a 5-mA change). This represents an approximate current gain of 25 (magnitude).

Using datasheet curves and load lines to find transistor characteristics has some drawbacks. The datasheet curves are "typical" for a transistor of a given type and represent an average gain (or in some cases, the minimum gain). Also, as discussed, transistor gain depends on temperature and frequency. The selection of load values, bias values, and operating point on the basis of static gain curves is subject to error. If the transistor beta shifts, the operating point must shift, requiring different bias and load values.

As discussed in Sec. 3.4, the problem of variable-gain transistor audio circuits can be overcome by means of *feedback to stabilize the gain*. With feedback, audio-circuit characteristics depend primarily on the relationship of circuit values rather than on transistor gain characteristics.

3.4 BASIC AUDIO-AMPLIFIER DESIGN

Figure 3.5 shows a basic single-stage audio amplifier. Note that the circuit is similar to the bias circuits discussed in Chap. 1 (Fig. 1.1*d*). Note also that a bypass capacitor C_3 is shown connected across the emitter resistor R_4. Capacitor C_3 is required only under certain conditions, as discussed in Sec. 3.5.

The input audio signal adds to, or subtracts from, the bias voltage across R_1. Variations in bias voltage cause corresponding variations in base current, collector current, and the drop across collector resistor R_3. The collector voltage (or

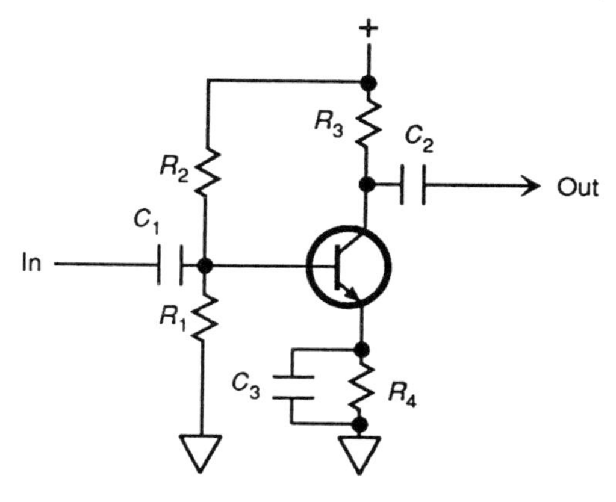

Input impedance = R_1 $\quad$ $R_3 > 5R_4$ $\quad$ $R_1 = 10R_4$
Output impedance = R_3 $\quad$ $R_3 = 10R_4$ $\quad$ $R_1 < 20R_4$
Current gain = R_1/R_4 $\quad$ $R_4 = 100$ to $1000\ \Omega$
Voltage gain = R_3/R_4
Collector voltage = 0.5 × supply voltage (as adjusted by R_2)

Stability (S) = current gain = R_3/R_4
$S = 20$ for high gain $\quad$ $C_1/C_2 = 1/3.2FR$ for 1 dB
$S = 10$ for stability $\quad$ $C_1/C_2 = 1/6.2FR$ for 3 dB
$S = 5$ for power gain

Stability ( ) = current gain

FIGURE 3.5 Basic audio-amplifier circuit design parameters.

circuit output) follows the input signal waveform, except that the output is inverted in phase.

Variations in collector current also cause variations in emitter current. This produces a change of voltage drop across R_4 and a change in the base-emitter bias relationship. The change in bias that results from the voltage drop across R_4 tends to cancel the initial bias change caused by the input signal and serves as a form of negative feedback to increase stability (and limit gain), as discussed in Chap. 1.

Note that this form of emitter feedback (or current feedback) is known as *stage feedback*, or *local feedback*, since only one stage is involved. *Overall feedback* or *loop feedback* is sometimes used where several stages are involved, as discussed in Sec. 3.6.

The outstanding feature of the circuit in Fig. 3.5 is that characteristics (gain, stability, impedance) are determined (primarily) by circuit values rather than by transistor characteristics (beta).

3.4.1 Audio-Amplifier Design Considerations

The circuit in Fig. 3.5 is shown with an NPN transistor. Reverse the power-supply polarity if a PNP transistor is used.

If a maximum source voltage is specified in the design problem, the maximum peak-to-peak output is set. For class A operation, the collector is operated at about one-half the source voltage. This permits the maximum positive and negative swing of output voltage. (The peak-to-peak output voltage cannot exceed the source voltage.)

Generally, the absolute maximum peak-to-peak output can be between 90 and 95 percent of the source. For example, if the source is 20 V, the collector operates at 10 V (Q-point) and swings from about 1 to 9 V. However, there is *less distortion* if the output is one-half to one-third of the source.

If a source voltage is not specified, two major factors should determine the value: (1) the maximum collector-voltage rating of the transistor and (2) the desired output voltage (or the desired collector voltage at the operating point).

The maximum collector-voltage rating must not be exceeded. Choose a source voltage that does not exceed 90 percent of the maximum rating. This allows a 10 percent safety factor. Any desired output voltage (or collector Q-point voltage) can be selected within these limits.

If the circuit is to be battery operated, choose a source voltage that is a multiple of 1.5 V.

If a peak-to-peak output voltage is specified, double the collector Q-point voltage.

For minimum distortion, use a source that is 2 to 3 times the desired output voltage.

If the input and/or output impedances are specified, the values of R_1 and R_3 are set, as shown by Fig. 3.5. However, certain limitations for R_1 and R_3 are imposed by the trade-off between gain and stability. For example, for a stage current gain of 10 and nominal stability, R_1 should be 10 times R_4 but should never be greater than 20 times R_4 (for maximum current gain and minimum stability) or less than 5 times R_4 (for minimum current gain and maximum stability). Since R_4 is between 100 and 1000 Ω in a typical class A circuit, R_1 (and the input impedance) should be between 500 (100 × 5) and 20,000 (1000 × 20).

If the input and/or output impedances are not specified, try to match the impedances of the previous stage and the following stage, where practical. This provides *maximum power transfer*.

The values of C_1 and C_2 depend on *the low-frequency limit* at which the circuit operates and on the resistances with which the capacitors operate. As shown in Fig. 3.1*c*, either the 1- or 3-dB low-frequency cutoff point can be selected.

The *minimum ac beta* of the transistor should be higher than the desired gain even though gain for the circuit is set by circuit values. Since the circuit in Fig. 3.5 is designed for maximum gains of 20, any transistor with a minimum beta of 20 (in the audio-frequency range) should be satisfactory.

3.4.2 Audio-Amplifier Design Example

Assume that the circuit in Fig. 3.5 is to be used as a single-stage audio amplifier. The desired output is 3 $V_{p\text{-}p}$ with a 2000-Ω impedance. The input impedance is to be 1000 Ω. The input signal is 0.3 $V_{p\text{-}p}$. This requires a voltage gain of 10. The low-frequency limit is 30 Hz, with a high-frequency limit of 100 kHz. Minimum distortion is desired. (The circuit should not be overdriven.) The source voltage and transistor type are not specified, but the circuit is to be battery operated.

Supply Voltage and Operating Point. The 3-V output can be obtained with a 4.5-V battery. However, for minimum distortion, the supply should be 2 or 3 times the desired output, or between 6 and 9 V. A 9-V battery provides the maximum insurance against distortion. The collector operating point should be 9/2, or 4.5 V.

Load Resistance and Collector Current. The value of R_3 should provide the output impedance of 2000 Ω. With a 4.5-V drop across R_3, the collector current is 2.25 mA (4.5/2000).

Emitter Resistance, Current, and Voltage. To provide a voltage gain of 10, the value of R_4 should be one-tenth of R_3, or 200 Ω (2000/10). If the test produces a gain slightly below 10, try reducing the value of R_4 to the next lower standard value of 180 Ω. The current through R_4 is the collector current of 2.25 mA, plus the base current. Assuming a stage current gain of 10, the base current is 0.225 mA (2.25/10). The combined currents through R_4 are 2.475 mA (2.25 + 0.225). This produces a drop of 0.495 V across R_4 (2.475/200). For practical design, this can be rounded off to 0.5 V.

Input Resistance and Current. The value of R_1 should provide the input impedance of 1000 Ω and should be at least 5 times R_4, which makes the 1000/200-Ω relationship correct. This relationship provides maximum circuit stability. The base voltage is 0.5 V higher than the emitter voltage, or 1 V (0.5 + 0.5). With a 1-V drop across R_1, the current through R_1 is 1 mA (1/1000).

Base Resistance and Current. The value of R_2 should be sufficient to drop the 9-V source to 8 V so that the base is 1 V above ground. The current through R_2 is the current through R_1 of 1 mA plus the base current of 0.225 mA, or 1.225 mA (1.0 + 0.225). The resistance required to produce an 8-V drop with 1.225 mA is 6530 Ω (8/1.225). Use a 6500-Ω standard resistance as the trial value.

Coupling Capacitors. The value of C_1 forms a high-pass filter with R_1. The high limit of 100 kHz can be ignored. The low-frequency limit of 30 Hz requires a capacitance value of about 10 μF (if a 1-dB drop at 30 Hz is required). This is found by 1/(3.2 × 30 × 1000).

The value of C_2 is found in the same way except that the resistance value is the load resistance R_3 (2000 Ω). For a 1-dB drop at 30 Hz, the value of C_2 should be about 5.2 μF (or 6 μF for practical design). This is found by 1/(3.2 × 30 × 2000).

The voltage values of C_1 and C_2 should be 1.5 times the maximum voltage involved, or 13.5 V (9 × 1.5). Use 15-V capacitors for practical design.

Transistor Selection. Some circuits must be designed around a given transistor. In such cases, the source voltage, collector current, power dissipation, and so on must be adjusted accordingly. In this example, a 2N337 transistor is available and can be used provided that the transistor meets the circuit requirements.

The following is a comparison of 2N337 characteristics and circuit requirements. (Note that any other transistor can be used if the transistor meets the same requirements.)

2N337 characteristics	Circuit requirements
Collector voltage (max): 45 V	9-V source
Collector current (max): 20 mA	2.25-mA nominal (at saturation, collector current is 4.5 mA; 9/2000)
Power dissipation (max): 125 mW (The 2N337 must be derated 1 mW/°C above 25°C ambient temperature)	Approximately 10 mW (4.5 V × 2.25 mA)
Ac beta (as h_{fe}) (min): 19 (typical): 55	10, minimum gain at low frequency
Beta (as h_{fe} at 100 kHz): 25	10, minimum gain at 100 kHz

3.5 EMITTER BYPASS

Figure 3.5 shows a bypass capacitor C_3 across emitter resistor R_4. This permits R_4 to be removed from the circuit as far as the signal is concerned but leaves R_4 in the circuit (with regard to direct current). With R_4 removed from the signal path, the voltage gain is about equal to R_3/dynamic resistance (of the transistor), and the current gain is about equal to the ac beta of the transistor. (Emitter bypass permits the highly temperature-stable dc circuit to remain intact, while providing a high signal gain.)

3.5.1 Design Considerations for Emitter Bypass

An emitter-bypass capacitor creates some problems. Transistor input impedance changes with frequency and from transistor to transistor, as does beta. This means that current and voltage gains can only be approximated. When the emitter resistance is bypassed, the circuit input impedance is about equal to beta times the transistor input impedance, so circuit input impedance is even more subject to variation (and is unpredictable). Generally, emitter bypass is used only in those cases where high voltage gain must be obtained from a single stage.

The value of C_3 is found by capacitance = 1/(6.2*FR*) where capacitance is in farads, *F* is the low-frequency limits in Hz, and *R* is the input impedance of the transistor in ohms.

3.5.2 Emitter Bypass Design Example

Assume that C_3 is to be used as an emitter bypass for the circuit described in Sec. 3.4 to increase voltage gain. All of the circuit values remain the same, as does the low-frequency limit of 30 Hz.

Further assume that a 2N337 transistor is to be used and that the dynamic input resistance is 50 Ω (obtained from the datasheet). This provides a voltage gain of about 40 (2000/50). The desired 3-V output can then be obtained with a 0.075-V input rather than the 0.3-V input required for the previous example.

The low-frequency limit of 30 Hz requires a C_3 capacitance value of 107 μF (110 μF for practical design). This is found by 1/(6.2 × 30 × 50). The voltage value of C_3 should be 1.5 times the maximum voltage involved, or 1.5 V (1 × 1.5). Use a 3-V capacitor for practical design.

3.6 MULTISTAGE AUDIO CIRCUITS

When stable voltage gains greater than about 20 are required and it is not practical to bypass the emitter resistor of a single stage, two or more transistor amplifier stages can be used in *cascade* (where the output of one transistor is fed to the input of a second).

In theory, any number of audio circuits can be connected in cascade to increase voltage gain. In practice, the number of stages is usually limited to three. The overall gain of the amplifier is equal (approximately) to the cumulative gain of each stage, multiplied by the gain of the adjacent stage.

As an example, if each stage of a three-stage amplifier has a gain of 10, the overall gain is (approximately) 1000 (10 × 10 × 10). Since it is possible to design a very stable single stage with a gain of 10 and adequately stable stages with gains of 15 to 20, a three-stage amplifier can provide gains in the 1000 to 8000 range. Generally, this is more than enough voltage gain for most practical applications. Using the 8000 figure, a 1-μV input signal (say from a low-voltage transducer or delicate electronic device) can be raised to the 8-mV range, while maintaining stability in the presence of temperature and power-supply variations.

IC Amplifiers. In present design, IC amplifiers are generally used when it is not practical to get sufficient gain with one or two transistors. For this reason, we do not concentrate on multistage amplifiers in this book. However, certain factors must be considered when using more than one transistor as an amplifier. The following paragraphs summarize these factors.

3.6.1 Basic Considerations for Multistage Amplifiers

Any of the single-stage amplifiers described in this chapter can be connected to form a two- or three-stage *voltage amplifier*. For example, the basic stage can be connected to two like stages in cascade. The result is a highly temperature-stable amplifier. Since each stage has its own feedback, the gain is precisely controlled and very stable.

It is also possible to mix stages to get some given design goal. For example, a three-stage amplifier can be designed using a highly stable, unbypassed amplifier for the first stage and two bypassed amplifiers for the remaining stages. Assuming a gain of 10 for the unbypassed stage and gains of 30 for the bypassed stages, this results in an overall gain of 9000. Of course, with bypassed stages, the gain depends on the transistor characteristics and is therefore unpredictable.

Distortion and Clipping. As with any high-gain amplifier, the possibility of *overdriving* a multistage amplifier is always present. If the maximum input signal is known, check this value against the overall gain and the maximum allowable output signal swing.

As an example, assume an overall gain of 1000 and a supply voltage of 20 V.

Typically, this implies a 10-V operating point (for the output collector) and a 20-V_{p-p} output swing from 0 to 20 V. In practical work, a swing from about 1 to 19 V is more realistic. Either way, a 20-mV input (p-p) multiplied by a gain of 1000 can drive the final output to the limits and possibly into distortion or clipping.

Feedback. The most precise control of gain is obtained when each stage of a multistage amplifier has feedback (local or stage feedback). However, such feedback is often unnecessary. Instead, *overall feedback* (or *loop feedback*) can be used, where part of the output from one stage is fed back to the input of the previous stage.

Usually, feedback is through a resistor (to set the amount of feedback), and the feedback is from the final stage to the first stage. However, it is possible to use feedback from one stage to the next (second stage to first, third stage to second, and so on).

Feedback Phase Inversion. There is a problem of phase inversion when using loop or overall feedback. In a common-emitter amplifier, the phase is inverted from input to output. If feedback is between two stages, the phase is inverted twice, resulting in *positive feedback*. This usually produces oscillation. In any event, positive feedback does not stabilize gain.

The phase-inversion problem can be overcome, when multiple stages are involved, by connecting the output collector of the second stage back to the emitter of the first stage. This produces the desired *negative feedback*.

As an example, if the base of the first stage is swinging positive, the collector of that stage swings negative, as does the base of the second stage. The collector of the second stage swings positive, and this positive swing can be fed back to the emitter of the first stage. A positive input at the emitter has the same effect as a negative input at the base. This produces the desired negative feedback to stabilize gain.

Low-Frequency Cutoff. Unless direct coupling is used, coupling capacitors must be used between stages, as well as at the input and output. Such capacitors form a low-pass *RC* filter with base-to-ground resistance. Each stage has its own low-pass filter. In multistage amplifiers, the effects of these filters are cumulative. (If each filter causes a 1-dB drop and there are three filters, the result is a 3-dB drop at that frequency in the final output.)

3.6.2 Direct-Coupled and Differential Multistage Amplifiers

One method of eliminating the *RC*-filter problem created by coupling capacitors is to use direct coupling. This eliminates the interstage coupling capacitor as well as some of the interstage resistances. In addition to the direct-coupled amplifier, a number of multistage circuits can be used to provide voltage, current, and power amplification at audio frequencies. The following paragraphs summarize such circuits.

Direct-Coupled with Like Transistors. Figure 3.6*a* shows a two-stage direct-coupled audio amplifier using two NPN transistors. This circuit has a tendency to become unstable. For example, if the collector current of Q_1 varies because of power-supply or temperature changes, Q_2 amplifies these changes and adds changes (because of possible variations in Q_2 characteristics). For this reason, it

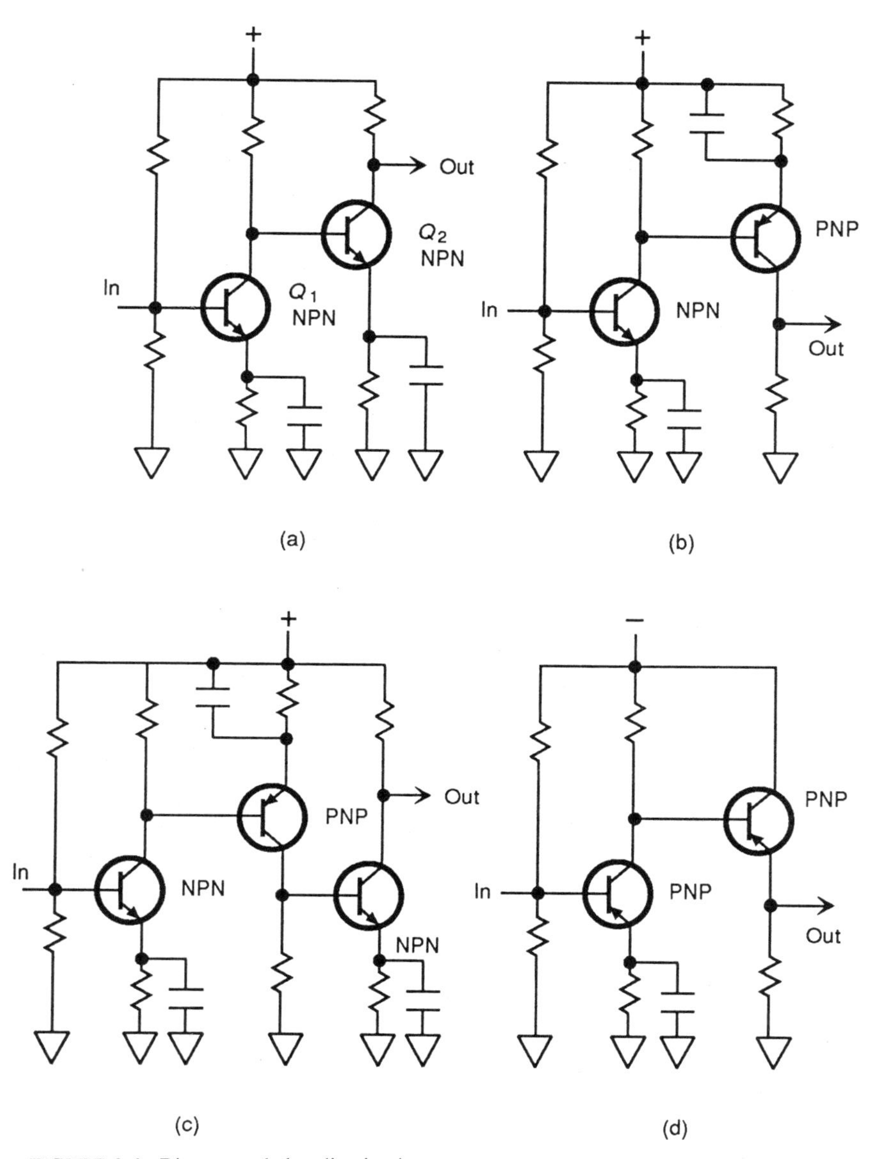

FIGURE 3.6 Direct-coupled audio circuits.

is difficult to cascade more than two stages of direct coupling where both transistors are of the same type.

Direct-Coupled with Complementary Transistors. Figure 3.6*b* shows a more stable direct-coupled audio amplifier (known as a *complementary amplifier*). This circuit is much more stable because any change in collector current (caused by temperature and power-supply variations and so on) is opposed by an equal change in collector current in the following stage. More than two complementary stages can be used. Of course, NPN and PNP transistors must be used alternately as shown in Fig. 3.6*c*.

Direct-Coupled Unlike Stages. Figure 3.6*d* shows how to use direct coupling between unlike stages (between a common-emitter amplifier and an emitter follower in this case).

Darlington Compounds. Figure 3.7 shows the basic Darlington circuit (known as the Darlington compound) together with some practical versions of the circuit. The Darlington compound is an emitter-follower driving a second emitter-follower. As discussed in Chap. 1, an emitter-follower provides no voltage gain but can provide considerable power gain.

The main reason for using a Darlington compound in audio work is to produce high current (and power) gain. For example, Darlington compounds (often in IC form) are used as audio drivers to raise the power of a signal from a voltage amplifier to a level suitable to drive a final power amplifier. Darlingtons are also used as a substitute for a driver section (or to eliminate the need for a separate driver).

Emitter-Coupled Amplifier. Figure 3.8*a* shows the basic emitter-coupled amplifier. This circuit, or one of the many variations, is similar to that of the phase inverter or phase splitter described in Sec. 3.9, in that two 180° out-of-phase signals or outputs can be taken from one input. Unlike the single-stage inverter in Sec. 3.9, the emitter-coupled amplifier can be used when a low-voltage output must be amplified to drive a push-pull output stage.

Power-Complementary Amplifier. The basic complementary amplifier can be used in audio power applications when a third power transistor is added. Such an arrangement is shown in Fig. 3.8*b* where a PNP/NPN pair is followed by a power PNP transistor.

Using the values shown, and assuming that Q_3 is capable of handling about 5 W, the circuit in Fig. 3.8*b* can deliver about 5 W of stable power at audio frequencies. Although the voltage gain is low (about 10), the power gain is high (about 100,000) since there is a large difference between input and output impedances.

The input impedance is about equal to the value of R_2. The output impedance is equal to the transformer impedance. If the loudspeaker has an 8-Ω impedance and T_1 has a 1:1 ratio, the output impedance is 8 Ω.

The circuit is adjusted for class A operation by R_1. In practice, the value shown (30 k) is used as a starting point and is adjusted for the desired class A operation and/or a specific operating point (collector voltage of Q_3) under no-signal conditions. Note that Q_3 must be operated with a heat sink since the power dissipation is in excess of 5 W.

Series and Complementary Output. Three classic direct-coupled circuits are often found in discrete-component audio amplifiers: the series, quasi-complementary, and full-complementary output. In present design, these circuits are usually found in IC form (or have been replaced by similar circuits in IC form).

Figure 3.9*a* shows two typical *series-output amplifiers* used in audio systems. One configuration requires two power supplies but omits the coupling capacitor

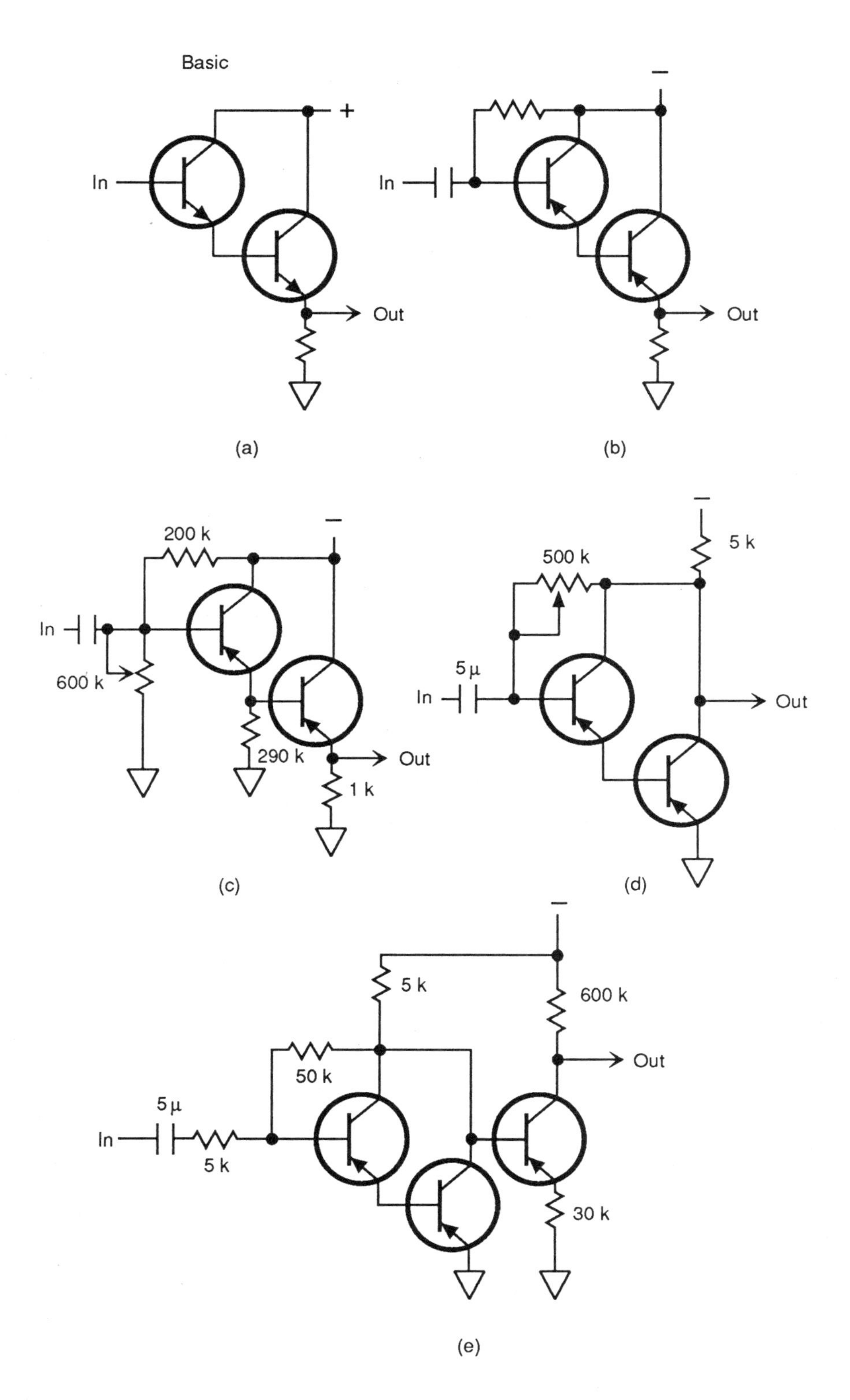

FIGURE 3.7 Darlington audio circuits.

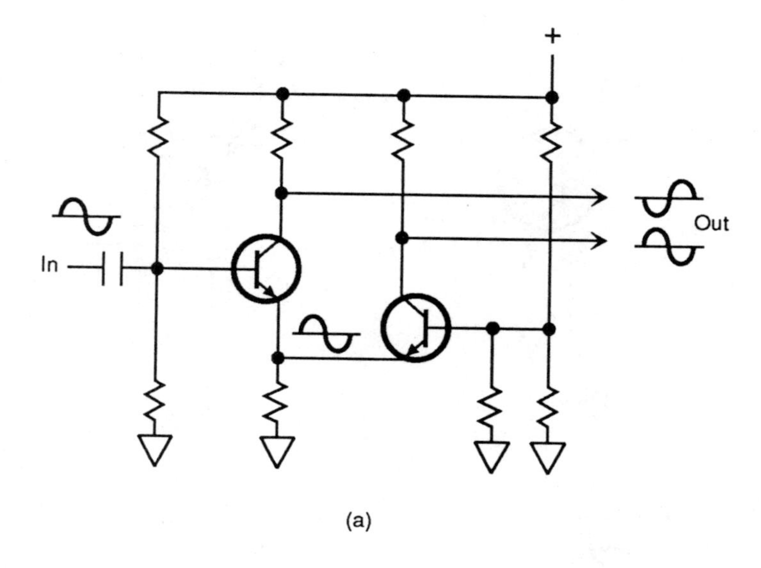

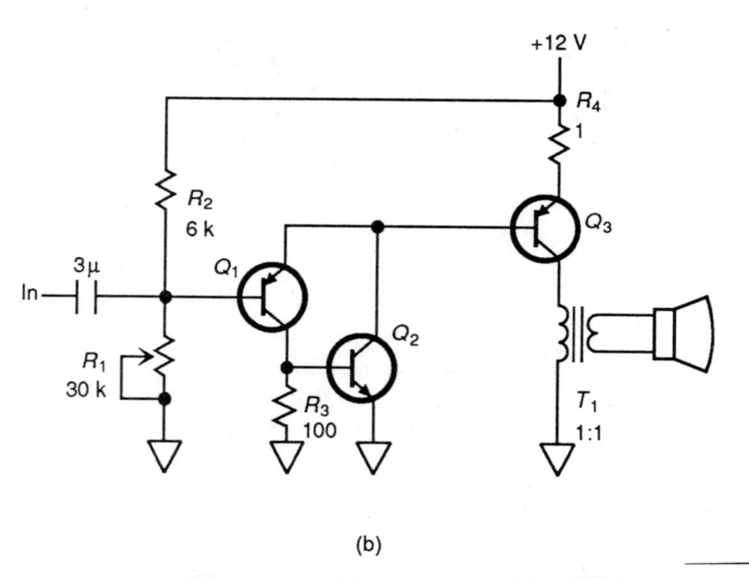

FIGURE 3.8 (*a*) Emitter-coupled (*b*) and power-complementary audio circuits.

to the load (often a pair of loudspeakers). This configuration provides better low-frequency response (since there is no capacitor) but can be inconvenient because of the two power supplies. The configuration with a single power supply has reduced low-frequency response since the coupling capacitor forms a high-pass filter with the load resistance.

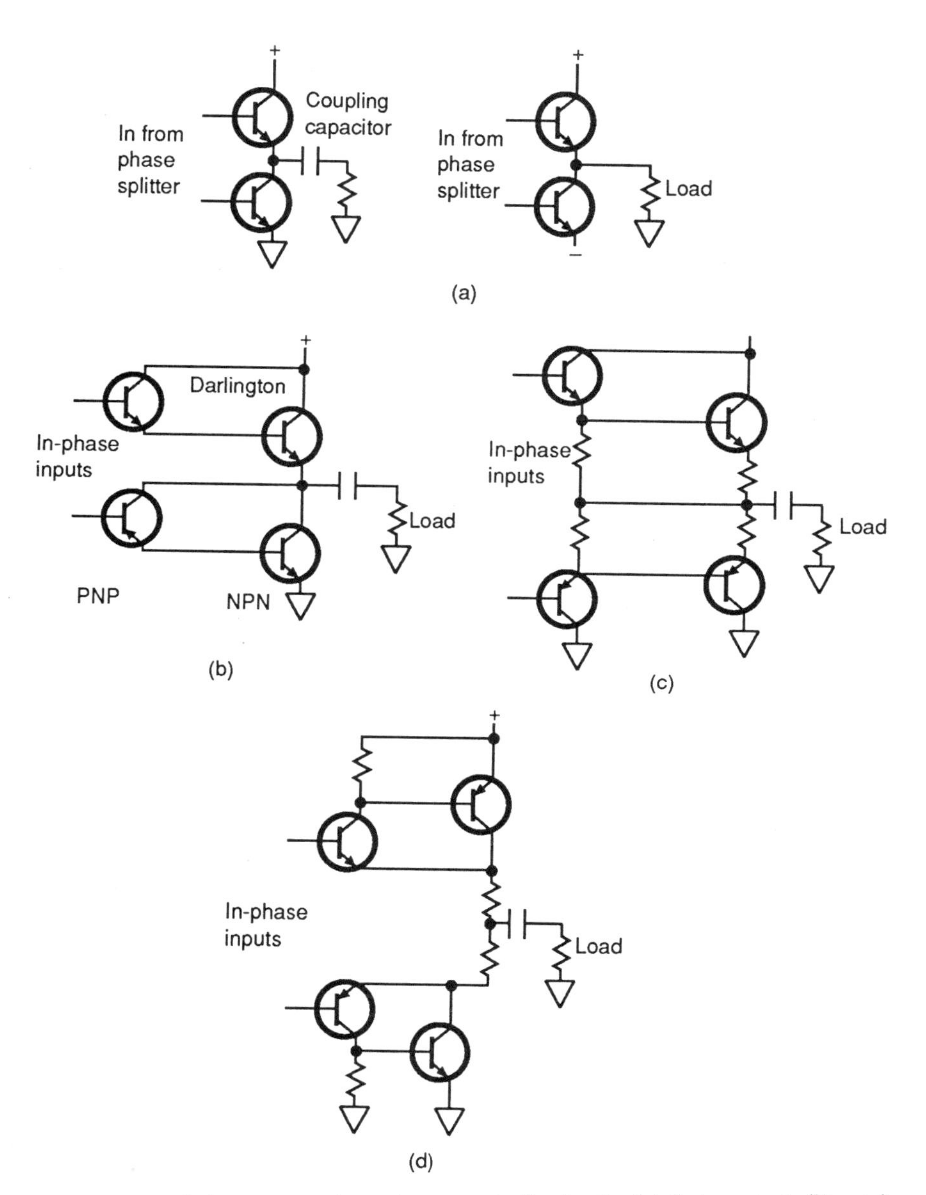

FIGURE 3.9 Series-output and complementary audio circuits. (*a*) Series output; (*b*) quasi-complementary output; (*c*) dual-Darlington output; (*d*) direct-coupled output.

Figure 3.9*b* shows a *quasi-complementary* output. Note that both base signals can be in phase so that a phase inverter or emitter-coupled amplifier is not needed (as is the case with the circuits in Fig. 3.9*a*).

Figure 3.9*c* and *d* shows two versions of the *full-complementary amplifier* output. Either version has an advantage over the quasi-complementary circuit in that both halves of the circuit are identical. This makes it easier to match both halves (for positive and negative signals) to minimize distortion that could be caused by uneven amplification of the signal.

As in the case of series-output circuits, the load-coupling capacitor can be omitted if two power supplies are used (one positive and one negative). As a guideline, a 2000-μF capacitor working into a 4-Ω load (such as a 4-Ω loudspeaker) produces an approximate 3-dB drop at 20 Hz (a 10-V output drops to 7 V at frequencies below 20 Hz).

Note that you can tell if an audio amplifier uses the load-coupling capacitor by unplugging and replugging the amplifier power cord (with the amplifier power switch off). There is a characteristic "pop" in the loudspeakers when the power cord is plugged in (when a load-coupling capacitor is used).

Differential-Amplifier Basics. Figure 3.10*a* shows a basic differential amplifier. The circuit responds differently to a common-mode audio signal than to single-ended signals. Signals common to both inputs (such as radiated signals) are known as *common-mode signals*. The ability of a differential amplifier to prevent conversion of a common-mode signal into a difference signal (which produces an output) is expressed as the common-mode rejection ratio (CMR or CMRR).

A common-mode signal (such as power-line pickup) drives both bases in phase with equal-amplitude voltages, and the circuit acts as though the transistors are in parallel to cancel the output. In effect, one transistor cancels the other.

Non-common-mode signals are applied to either of the cases (Q_1 or Q_2). The *inverting input* is applied to the base of Q_2, and the *noninverting input* is applied to the base of Q_1. With a signal applied only to the inverting input, and the noninverting input grounded, the output is an amplified and inverted version of the input. For example, if the input is a positive pulse, the output is a negative pulse. If the noninverting input is used, with the inverting input grounded, the output is an amplified version of the input (but without inversion).

The emitter resistor introduces emitter feedback to *both* transistors simultaneously. This reduces the common-mode signal gain without reducing the differential signal gain in the same proportion.

Figure 3.10*b* shows a more practical differential amplifier (packaged in IC form). The circuit is basically a single-stage differential amplifier (Q_2 and Q_4) with input emitter-followers (Q_1 and Q_5) and constant-current source Q_3 in the emitter-coupled circuit. Note that the single emitter resistor in Fig. 3.10*a* is replaced by Q_3 in the circuit in Fig. 3.10*b* and that direct coupling is used throughout the IC (which is typical for ICs).

Common-Mode Definitions. One manufacturer defines common-mode rejection (CMR or sometimes CM_{rej}), or the common-mode rejection ratio (CMRR), as the ratio of differential gain (usually large) to common-mode gain (usually a fraction). That is, the amplifier may have a large gain of differential signals (different signals at each input or one input grounded and the opposite input with a signal) but with little gain (or possibly a loss) of common-mode signals (same signal at both inputs).

Another manufacturer defines CMR as the relationship of change in the output

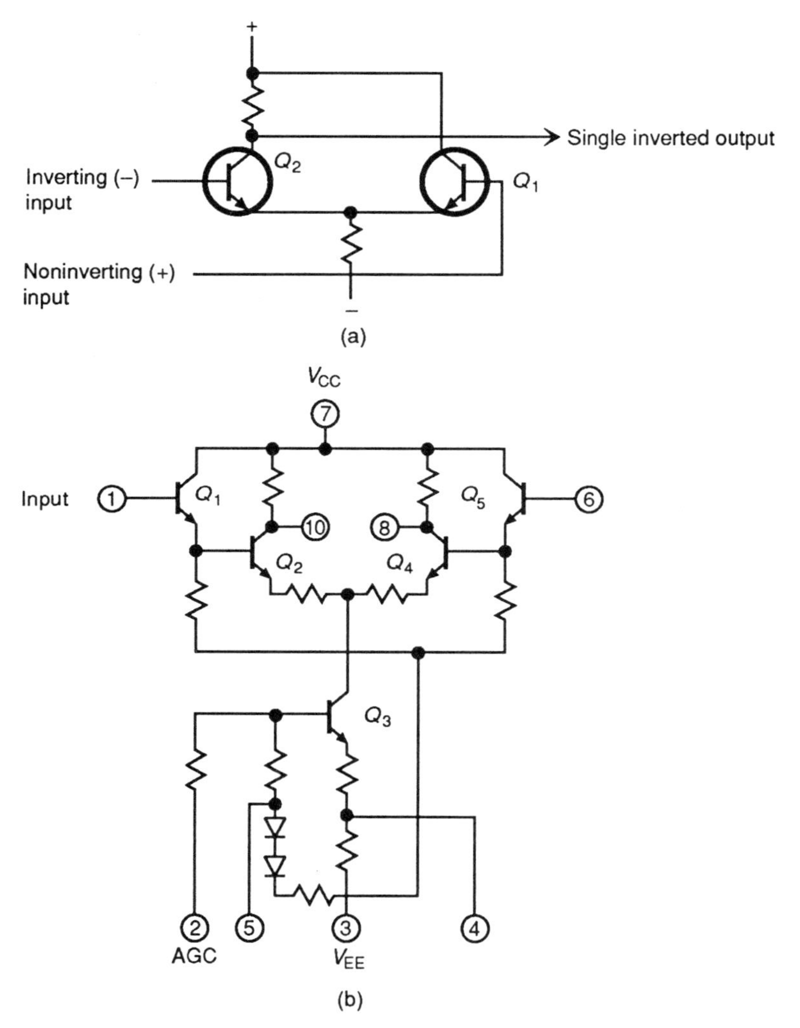

FIGURE 3.10 Differential audio circuits.

voltage to the change in the input common-mode voltage producing the change, divided by the gain (open-loop gain, or the gain without feedback).

As an example, using the latter definition, assume that the common-mode input (applied to both terminals simultaneously) is 1 V, the resultant measured output is 1 mV, and the open-loop gain is 100. The CMR is then

$$\frac{\text{Output/input}}{\text{Open-loop gain}} = \text{CMR} \qquad \frac{0.001/1}{100} = 100{,}000 = 100 \text{ dB}$$

Another way to calculate CMR is to divide the output signal by the open-loop gain to find an *equivalent differential input signal*. Then the common-mode input signal is divided by this equivalent different input signal. Using the same figures as in the previous CMR calculation,

$$\frac{0.001}{100} = 0.00001 \qquad \text{equivalent differential input signal}$$

$$\frac{1}{0.00001} = 100{,}000 \qquad \text{or 100 dB}$$

CMR is an indication of the *degree of circuit balance* of the differential stages, since a common-mode input signal should be amplified identically in both halves of the circuit. A large output for a given common-mode input is an indication of large imbalance or poor CMR. If there is an imbalance, a common-mode signal becomes a differential signal after passing the first stage (or at the output of a single-stage differential amplifier).

As with amplifier gain, CMR usually decreases as frequency increases. However, as a guideline, the CMR of any differential amplifier should be *at least 20 dB greater* than the gain at any frequency (within the limits of the amplifier).

3.7 TRANSFORMER COUPLING

Figure 3.11 shows the schematic of a classic transformer-coupled audio-amplifier circuit. This basic circuit (or one of the many variations) was used extensively in audio equipment of all types (home entertainment, transmitter modulator, TV audio amplifier, and the like). However, in present design, the IC power amplifier described in Chap. 5 has largely replaced the transformer-coupled amplifier. For that reason, we do not concentrate on transformer coupling here. However, the following paragraphs summarize the characteristics of transformer coupling (in case you happen to find such a circuit when troubleshooting older equipment).

The circuit in Fig. 3.11 has a class A input or *driver stage* and a class B push-pull *output stage*. The class A stage provides both voltage and power amplification as needed to raise the low input signal to a level suitable for the class B power-output stage.

The class A stage is transformer coupled or *RC* coupled, as needed. Transformer coupling is used at the input where a specific impedance-match problem must be considered in design. The class A input stage can be driven directly by the signal source or can be used with a preamplifier for very low-level signals. When required, a high-gain voltage amplifier is used as a *preamplifier*.

The push-pull output stage may be operated as a class B amplifier, where the transistors are cut off at the operating point and draw collector current only in the presence of an input signal. Class B is the most efficient operating mode for audio amplifiers since the least amount of current is drawn (and no current where there is no signal). However, class B operation can result in crossover distortion, as discussed in Chap. 1.

The efficiency of an amplifier is determined by the ratio of collector input-to-output power. An amplifier with 70 percent efficiency produces 7-W output for a 10-W input (with power input being considered as collector source voltage multiplied by total collector current).

Typically, class B amplifiers can be considered as 70 to 80 percent efficient.

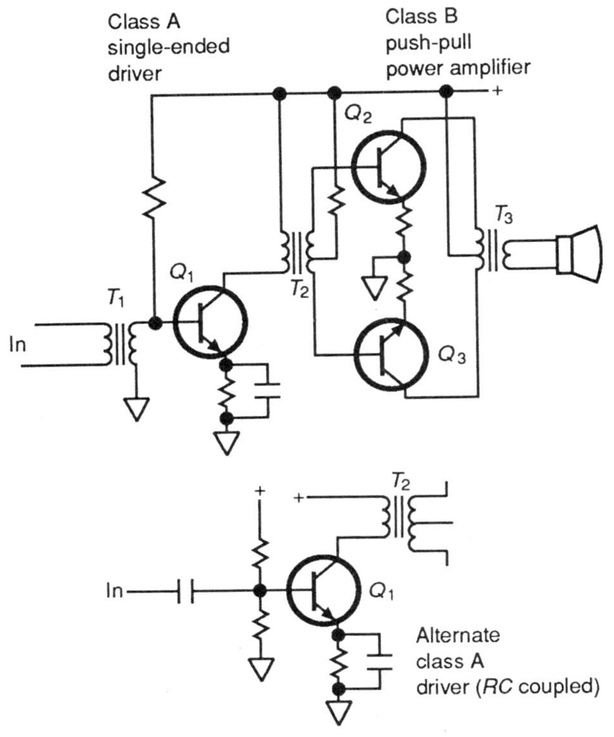

FIGURE 3.11 Transformer-coupled audio circuits.

Class A amplifiers are typically in the 35 to 40 percent efficiency range, with class AB amplifiers showing 50 to 60 percent efficiency.

In all cases, any amplifier circuit that produces an increase in collector current (at the operating point) produces corresponding lower efficiency. This results in a trade-off between efficiency and distortion.

The efficiency produced by a class of operation also affects the heat-sink requirements. Any design that produces more collector current at the operating point requires a greater heat-sink capability (Chap. 2). As a guideline, a class A amplifier requires double the heat-sink capability of a class B amplifier, all other factors being equal.

3.8 AUDIO PHASE INVERTERS

One advantage of a transformer-coupled amplifier (Sec. 3.7) is that two signal voltages (180° out of phase) can be taken from a center-tapped transformer winding (such as T_2 and T_3 in Fig. 3.11). The same result can be produced with a transformerless phase inverter such as shown in Fig. 3.12.

In a common-emitter amplifier with a resistive load, the collector and emitter

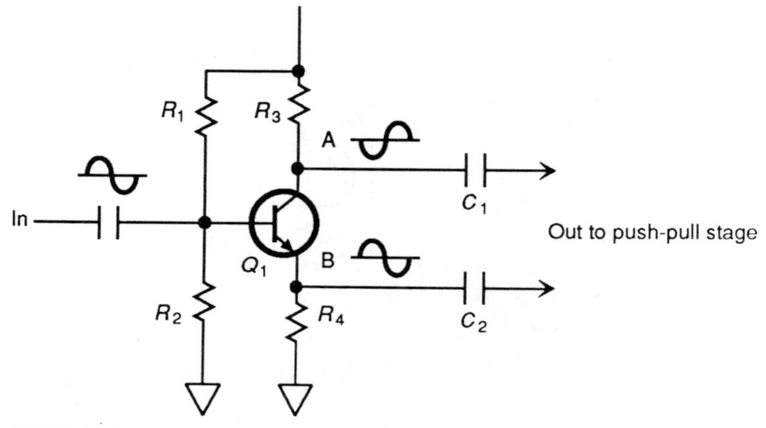

FIGURE 3.12 Audio phase inverters.

are 180° out of phase with each other. If the input signal voltages for a push-pull stage are obtained from these two points (collector and emitter), the necessary 180° out-of-phase relationship is produced. Also, since approximately the same collector current flows through R_3 and R_4 (if the resistances of R_3 and R_4 are equal), the voltage drops are equal. Point A becomes as negative as point B becomes positive.

In the circuit in Fig. 3.12, resistors R_1 and R_2 form the voltage divider that forward biases the emitter-base junction of Q_1. The collector resistor R_3 and emitter resistor R_4 are equal in value, as are the coupling capacitors C_1 and C_2. Note that R_4 is unbypassed. This provides inverse-current feedback that reduces distortion and stabilizes gain.

The main advantage of the phase-inverter stage is to eliminate the need for a transformer. This results in a smaller, lighter, and less expensive amplifier and eliminates the transformer's magnetic field (which can produce distortion of the signal unless the transformer is properly shielded).

3.9 IC AUDIO

Many of the circuits discussed throughout this chapter are available in IC form. Virtually all of the audio circuits and equipment covered in the remaining chapters include one or more ICs in addition to discrete components between the ICs. Each of the ICs is discussed in the related chapter.

CHAPTER 4

BASIC TESTING AND TROUBLESHOOTING FOR AUDIO EQUIPMENT AND CIRCUITS

This chapter describes testing and troubleshooting audio equipment. It starts with a review of typical audio test equipment and then goes on to describe test procedures that can be applied to audio circuits and devices. The chapter concludes with a summary of the basic troubleshooting approach for audio equipment, followed by a specific step-by-step troubleshooting example.

4.1 *TYPICAL AUDIO TEST EQUIPMENT*

The test equipment used in audio service is basically the same as that used in other fields of consumer electronics (TV, VCR, camcorder, etc.). That is, most service procedures are performed using meters, generators, scopes, counters, power supplies, and assorted clips, patch cords, and so on. So, if you have a good set of test equipment suitable for other electronic service, you can probably service most audio equipment. However, there are certain items that can provide considerable help when troubleshooting audio devices. We discuss each of these in the following paragraphs.

4.1.1 Matching Test Equipment to the Circuits

No matter what test instrument is involved, try to match the capabilities of the test equipment to the circuit being serviced. For example, if pulses, square waves, or complex waves are to be measured, a peak-to-peak meter can possibly provide meaningful indications, but a scope is the logical instrument. Or you can try a really novel approach and use the test instrument recommended in the service literature.

4.1.2 Wow and Flutter Meters

Wow and flutter are turntable or tape-transport *speed fluctuations* that can cause a quivering or wavering effect in the sound during play. In the case of a tape system, wow and flutter can occur both during record and playback. Wow and flutter are virtually nonexistent in CD players.

The longer fluctuations in sound (below about 3 to 6 Hz) are called wow. Shorter fluctuations (typically from 3 to 6 Hz up to 20 Hz) are called flutter.

Although wow and flutter problems are common to all turntables and cassette decks, it is only when the problems go beyond a certain tolerance that they are objectionable. So, when troubleshooting a complaint of "excessive wow and flutter," first check the actual amount.

The basic method for measuring wow and flutter is to measure the reproduced frequency of a precision tone or signal that is recorded on a test record or tape. Any deviation in frequency of the reproduced or playback tone, from the prerecorded tone, is an indication of wow and flutter.

In the case of a tape system, you can record your own precision signal instead of using a test tape. However, remember that *both* the recorded and playback signals are subject to wow and flutter. This may lead you to think that there is a serious wow and flutter problem when the actual fluctuation is within tolerance.

There are four primary standards for wow and flutter measurement: JIS (Japan), NAB (USA), CCIR (France), and DIN (Germany). With these standards, wow and flutter is measured by playing a test record or tape which has been prerecorded. JIS, NAB, and CCI use a prerecorded test tone of 3 kHz, whereas DIN uses 3.15 kHz.

A frequency counter (with very good resolution) can be used to monitor the test tone during play. However, a commercial wow and flutter meter is usually more convenient and accurate. Even though the counter may be quite accurate, the problem is one of resolution.

As an example, the wow and flutter tolerance for a typical consumer-electronics turntable is 0.025 percent. This means that a playback frequency of 3000.75 Hz is acceptable, but 3000.76 Hz is not acceptable.

In addition to the accuracy and resolution problems, there is a matter of *weighting*, or test conditions under which wow and flutter are measured. Each standard has a separate system of weighting. Likewise, each standard measures wow and flutter in different values. Peak values are used for CCIR and DIN. NAB uses mean values, while JIS uses rms values. Fortunately, all of the weightings and values are accounted for in a commercial wow and flutter meter.

Typical Wow and Flutter Meter Characteristics. The typical wow and flutter meter will have several ranges (at least six) to provide good resolution from about 0.003 to 10 percent. Usually, the variations in playback (or reproduced frequency) are separated from the test tone and then measured.

The frequency of the test tone is displayed on a frequency counter (at least four digit). The percentage of wow and flutter is indicated on an analog meter. Filters are used to separate both wow and flutter components (separately, if desired) from the test signal. A mode selector permits the user to select measurement of flutter only, wow only, or combined wow and flutter (both weighted and unweighted).

Most wow and flutter meters can accommodate a wide range of input signal levels, permitting measurement at nearly any desired point in the audio equipment. For example, one meter has two selectable sensitivities of 0.5 to 100 mV and 5 mV to 30 V. These sensitivity levels permit measurement of small signals directly from tape heads and phono cartridges if desired.

Some meters have a level monitor that turns on if the input signal level is adequate for wow and flutter measurements. No other measurements or adjustments are required. The level monitor is a time-saving feature for setting up tests.

Other typical conveniences include built-in crystal-controlled oscillators that provide very stable sources of 3 and/or 3.15 kHz. These signals are used for recording wow and flutter test tapes.

Most wow and flutter meters have auxiliary output jacks that provide outputs to other instruments, such as scopes or chart recorders, if desired. These jacks provide a dc voltage proportional to rotational speed error or deviation and both ac and dc voltages proportional to the wow and flutter meter reading.

The ac wow and flutter voltage is used with an auxiliary scope. The dc wow and flutter voltage is used with a chart recorder (or perhaps an digital memory scope) to make a copy of wow and flutter measurements. The dc rotational-speed error-deviation voltage can be used with either a chart recorder or a scope.

Note that rotational-speed error is not to be confused with wow and flutter. Speed error (if any) means that the record or tape is not being driven at the correct speed. For example, speed tolerance for a typical consumer turntable is 0.003 percent. This means that a rotational speed of 45.00135 rpm (for a 45-rpm record) is acceptable. Generally, speed error implies a constant value or one that changes slowly, whereas wow and flutter implies constant fluctuation.

For added versatility, many wow and flutter meters have a built-in frequency counter that may be used (independently from wow and flutter measurements) for general measurement of audio frequencies. Typically, either a crystal-controlled or line-frequency (50 or 60 Hz) time base can be selected for the counter.

Basic Wow and Flutter Measurement. We need not go into detailed operation of any particular meter here. Such information is available in the instruction manual. However, for those totally unfamiliar with wow and flutter measurement, we will run through a typical procedure. Figure 4.1*a* shows the basic test connections between the meter and the turntable or cassette.

1. Select the type of measurement standard, JIS, NAB, CCIR, or DIN.
2. Use a prerecorded wow and flutter test tape or record on the equipment to be tested.
3. If you want to record a test tape, connect the oscillator output terminals to the input of the tape recorder as shown in Fig. 4.1*a*. Then set the 3- or 3.15-kHz switch to on (3 kHz for JIS, NAB, or CCIR; 3.15 kHz for DIN).
4. Remember that wow and flutter present during record are also present during playback and are added to the playback wow and flutter. Under these conditions, measured wow and flutter represents the *total* wow and flutter introduced for record and playback. *One-half* the measured value represents wow and flutter for playback only.
5. Connect the output of the equipment under test (cassette deck or turntable) to the input terminal as shown in Fig. 4.1*a*.
6. The level-monitor indicator (if any) should turn on if the signal is of adequate level and it contains the 3- or 3.15-kHz test signal. If the level-monitor indicator does not turn on, change the sensitivity switch setting. If the level monitor still does not turn on, check to make sure that there is some signal present and that the signal is at the correct test frequency. (It is possible that you are not on the correct portion of the test record or tape or that there is no signal at the point being measured.)

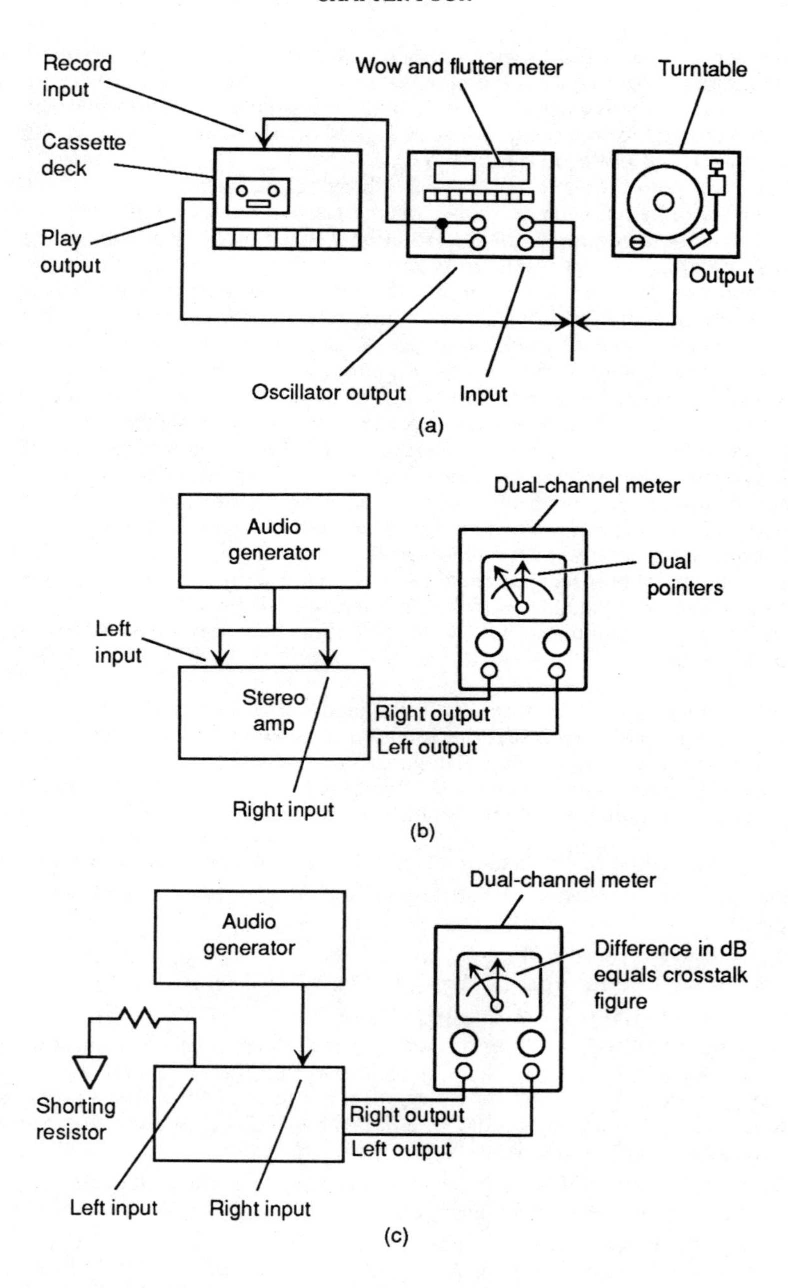

FIGURE 4.1 Basic test connections for wow and flutter and stereo measurements.

7. Check that the frequency display shows the wow and flutter test frequency (3 or 3.15 kHz). Note that if the signal being monitored is at a high level, the level monitor may turn on at frequencies other than 3 or 3.15 kHz (on some meters).
8. Set the weight-function switch for the desired frequency component:

   Use *weighted* for a combination of wow and flutter (with a typical frequency response of about 0.2 to 200 Hz).

   Use *wow* for wow components only (0.5 to 6 Hz for NAB or JIS; 0.3 to 6 Hz for CCIR or DIN).

   Use *flutter* for flutter components only (with a typical frequency response of about 6 to 200 Hz).

   Use *unweighted* for an unweighted combination of wow and flutter (with 0.5 to 200 Hz for NAB or JIS and 0.3 to 200 Hz for CCIR or DIN).
9. Set the range for the highest obtainable meter reading, *without going off scale*. Read the percentage of wow and flutter from the meter, using a scale that corresponds to the range.

4.1.3 Audio Voltmeters

In addition to making routine voltage and resistance checks, the main functions for voltmeters in audio troubleshooting are to measure frequency response and trace audio signals from input to output. Many technicians prefer scopes for these procedures. The reasoning is that scopes also show distortion of the waveforms during measurement or signal tracing. Other technicians prefer the simplicity of a meter, particularly in such procedures as voltage-gain and power-gain measurements. So the choice of meter versus scope is really up to you.

It is possible that you can get by with any ac meter (even a multimeter, analog or digital) for all audio work. However, for accurate measurements, the author recommends a wideband ac meter, preferably a *dual-channel* model. The dual-channel feature makes it possible to monitor both channels of a stereo system simultaneously. This is particularly important for stereo frequency-response and crosstalk measurements.

Typical Audio Voltmeter Characteristics. The typical audio voltmeter will have several ranges (at least 10, possibly 12) covering audio voltages from about 0.2 mV to 100 V. The usual wideband frequency permits measurements from 5 Hz to 1 MHz. High sensitivity permits measurements down to 30 μV (and lower in some meters).

Audio meters have two voltage scales: a dB scale (0 dB = 1 V) and a dBm scale (0 dBm = 1 mW across 600 Ω). These scales are used for making relative measurements (gain, attenuation, etc.) Typical ranges are −90 to +40 for dB and −90 to +42 for dBm.

Most audio meters have built-in low-distortion amplifiers used to drive the meter movements. This allows the meter to be used as a calibrated, high-gain, preamp. Typically, the output is calibrated at 1 V rms for a full-scale reading.

Generally, audio meters use absolute mean-value (average) sensing but are calibrated to read the rms value of a sine-wave voltage. The input voltage is capacitively coupled, permitting measurement of an ac signal superimposed on a dc voltage. A typical 10-mΩ input impedance on all ranges, with low shunting capacitance, assures minimum circuit loading to the audio circuit under test.

Audio meters are dual channel, capable of measuring two voltages simultaneously, and have dual pointers (often red and black) for convenient direct comparison of two levels (such as is required during stereo balance measurements.

The input attenuators of present-day meters are electronically switched, providing higher reliability, improved S/N ratio, and less crosstalk than with direct rotary-switch methods. Often, the attenuators can be *interlocked* to permit a single control to select the range for both channels (or independent ranges may be selected for each channel).

Basic Audio Measurements and Signal Tracing. Present-day audio meters are suitable for all forms of audio measurement and signal tracing. It is assumed that you are already familiar with such tests as frequency response, voltage gain, power output and gain, power bandwidth, load sensitivity, feedback, dynamic output impedance, and dynamic input impedance. However, for those totally unfamiliar with these tests, we run through some basic procedures in Secs. 4.5 and 4.6.

It is also assumed that you can trace audio signals from input to output through audio circuits, using a meter or scope. If not, you may have a problem troubleshooting modern audio equipment. Before you panic, we have a step-by-step procedure for basic audio-circuit troubleshooting in Sec. 4.10.

Stereo Measurements. A meter with two input channels applied to a dual-pointer scale is more useful for direct-comparison measurements than two separate voltmeters. An excellent example is measurement of left- and right-channel characteristics of stereo equipment. To show you the advantages of a dual-channel, wide-band meter, we describe typical stereo-crosstalk and stereo-response measurement (with a dual-channel meter) in the following paragraphs.

Figure 4.1*b* shows the basic test connections for *stereo-response measurements*. With equal audio signals applied to the left- and right-channel inputs of the stereo amplifier, set the controls to the interlock position, and select the desired range.

It may be necessary to set a ground-mode switch for minimum hum. For a stereo amplifier that has no connection common to the left and right channels, this usually means setting the ground-mode switches to *open*. In most other cases (where there is some connection common to both channels), use the *ground* position of the ground-mode switch.

With the connections made and controls set, vary the frequency of the audio generator. Note any difference in frequency response between left and right channels, as indicated by *unequal deflection* of the two pointers. Even those not thoroughly familiar with frequency-response measurements will see how much simpler a dual-pointer meter is to use (than two separate meters).

Figure 4.1*c* shows the basic test connections for *stereo-crosstalk measurements*. With an audio signal applied to one channel and the other channel shorted, set the controls to the independent position, and select the desired range. Again, it may be necessary to set the ground-mode switch for minimum hum.

Note that the value of the shorting resistor should be equal to the input impedance of the amplifier. However, this should be checked against the value recommended in the service manual.

With the connections made and controls set, note the difference in dB between the signal level measured on the channel with the shorted input and the signal level on the channel with the signal input. This is the crosstalk figure. Crosstalk is generally specified in dB (such as −40 dB).

For a complete crosstalk measurement, reverse the input connections to the stereo amplifier, and repeat the measurement. There should be no substantial difference in the meter readings with the input connections reversed. Make certain that the signal level, frequency, and all other input conditions are identical when the connections are reversed.

4.1.4 Distortion Meters

If you are already in audio and stereo service work, you probably have distortion meters (and know how to use them effectively). There are two schools of thought on distortion meters. One insists that you must have at least one type of distortion meter. You simply cannot run an audio shop without such an instrument. The other school says you can probably get by without a distortion meter, as a practical matter. If distortion is severe, you will hear it. If the distortion is below a level where it can be heard, you can generally forget the problem.

The author has no recommendations on either side of the distortion-meter argument. However, remember that if you have a customer with a "golden ear," an accurate distortion meter is an excellent tool for settling "discussions" concerning his or her audio system's performance, especially after service. If you are not already aware of this, all audio-system owners have a "golden ear."

There are two types of distortion measurement: harmonic and intermodulation. We do not describe any particular meter here. Instead, we include descriptions in Sec. 4.6 of how harmonic and intermodulation distortion measurements are made.

4.1.5 Scopes

The scopes used for audio work should have the same characteristics as for other electronic service. If you have a good scope for TV and VCR work, use that scope for all audio troubleshooting and measurements. If you are considering a new scope, remember that a dual-channel instrument permits you to monitor both channels of a stereo system (as is the case with a dual-channel voltmeter).

A scope has the advantage over a meter in that the scope can display such common audio circuit conditions as distortion, hum, ripple, and oscillation (we describe how in Secs. 4.6 through 4.10). However, the meter is easier to read.

4.1.6 FM-Stereo Generator

An FM-stereo generator is essential for troubleshooting the FM-stereo tuners and receivers described in Chap. 6. This is because an FM generator simulates the very complex modulation system used by FM-stereo broadcast stations. Without an FM-stereo generator, you are totally dependent on the constantly changing signals from such stations, making it impossible to adjust FM tuners and receivers or to measure response after adjustment. So you can put an FM-stereo generator high on your list of test equipment recommended for consumer audio troubleshooting.

4.1.7 TV-Stereo Generator

As in the case of FM, a TV-stereo generator is essential for troubleshooting and adjusting the audio section of stereo-TV (MTS and MCS) sets described in Chap.

10. Again, this is because a TV-stereo generator simulates the complex modulation system used by stereo-TV broadcast stations. While it is possible to troubleshoot a stereo-TV set without a TV-stereo generator, it is difficult (if not impossible) to adjust the audio-section circuits after troubleshooting.

4.1.8 Shop-Standard Stereo Amplifier with Speakers

You should have at least one good shop-standard stereo amplifier and speaker system. This permits you to check all audio components passing through the shop against a known standard.

Remember that the amplifier must be capable of handling a wide range of inputs. For example, the line output of a typical cassette deck is about 500 mV, while a turntable produces considerably less (the output from a pickup or cartridge is generally less than 5 mV). The output from a CD player (which may or may not be adjustable) is typically 2 V.

4.1.9 Digital Test Equipment

Many modern audio components use some form of microprocessor (sometimes called a microcomputer) in the system-control circuits. More than one microprocessor is used in some cases. While some manufacturers recommend a *logic probe*, it is generally possible (and more convenient) to monitor all microprocessor signals with a scope. Although a probe is easy to use, the probe indication proves only the presence or absence of pulses. A scope indication also shows pulse amplitude and waveshape.

4.1.10 Test Records, Cassettes, and Disks

Many audio component manufacturers provide test records, cassettes, and disks as part of the recommended test equipment and/or tools. Some manufacturers even recommend the test device of another manufacturer. There are also some standard test devices available.

These test devices are essentially standard records, cassettes, or CDs with several very useful signals recorded at the factory using very precise test equipment and signals. You play the test device on the turntable, cassette deck, or CD player being serviced and note the response and/or use the signals to perform alignment and adjustment. With the proper test device, you can often eliminate the need for your own signal sources (signal generators, audio generators, etc.).

One major problem with test devices is the lack of standardization. You will probably need several test devices if you are going to service many different models of audio components. The test and adjustment procedures found in some service literature call for signals not available on all test devices. The only way around this problem is to use the recommended test device in all cases. Of course, you can use any known-good test device for a final, after-service check of the component, but this does not give you the necessary signals to perform all test and adjustment procedures.

4.2 TOOLS AND FIXTURES FOR AUDIO TROUBLESHOOTING

Most turntables, cassette decks, and CD players have special tools and fixtures recommended for field service. We describe such tools in the related chapters. Generally, tools are available from the audio component manufacturer. In some cases, complete tool kits are made available (at least to authorized factory service centers).

There are certain tools and fixtures used at the factory for both disassembly and service of the equipment. These factory tools are not available for field service (not even to factory service centers, in some cases). This is the manufacturer's subtle way of telling service technicians that they should not attempt any adjustments (electrical or mechanical) not recommended in the service instructions. The author strongly recommends that you take this subtle hint.

In addition to special tools, the mechanical sections of turntables, cassette decks, and CD players are disassembled, adjusted, and reassembled with common hand tools, such as wrenches and screwdrivers. Keep in mind that much audio equipment is manufactured to Japanese metric standards, and your tools must match. For example, you will need metric-sized Allen wrenches, and Phillips screwdrivers with the Japanese metric point.

4.3 SAFETY PRECAUTIONS

It is assumed that you are familiar with all of the standard safety precautions for electronic troubleshooting and the use of test equipment. Such precautions include checking for leakage that can cause chassis and metal covers to be "hot," handling electrostatically sensitive (ES) devices, replacement of leadless components, repair of PC boards, avoiding mechanical shock and vibration, not altering equipment, not tampering with interlocks, checking all warnings and safety notices, using care when connecting leads, and avoiding high-voltage problems.

Most service literature includes all necessary safety precautions for a particular type of equipment. For example, CD players have a laser diode that produces an invisible but potentially dangerous light beam. Because of this possible service hazard, the subject of laser safety is discussed fully in Chap. 8. Always compare any safety precautions found in this book with those in the service literature.

4.4 CLEANING, LUBRICATION, AND GENERAL MAINTENANCE

There is considerable disagreement among audio equipment manufacturers concerning the need for periodic maintenance or routine checks. At one extreme, a certain manufacturer recommends replacement of a few parts after a given number of playing hours or playing times. For example, they recommend that the optical pickup of a CD player be replaced at 5000 h of playing time and that all motors in turntables and cassette decks be replaced after 10,000 plays (who counts?). At the other extreme, another manufacturer recommends no periodic replacement, cleaning, lubrication, or adjustment of any kind. "Fix it if it breaks down" is the rule. (It is fair to say that this rule will probably be observed religiously.)

Somewhere between these two extremes, other manufacturers recommend adjustment (electrical and/or mechanical) only as needed to put the equipment back in service or when certain parts or assemblies have been replaced. However, most audio equipment manufacturers recommend a complete checkout, using the recommended test record, cassette tape, or CD disk, after any service.

The author has no recommendations in the areas of periodic maintenance except that you follow the manufacturer's recommendations. The author also realizes that the general public regards audio equipment in the same way it does TV and VCR equipment (that is, "take it in for service when it breaks down").

We describe some typical cleaning and lubrication of components in the related chapters. Again, compare this information with cleaning and lubrication information found in the service literature.

4.5 BASIC AUDIO TROUBLESHOOTING APPROACHES

The remainder of this chapter is devoted to troubleshooting of basic audio circuits, primarily audio amplifiers. Before presenting a step-by-step example of amplifier troubleshooting, we discuss basic troubleshooting approaches for audio circuits, the test procedures normally associated with audio amplifier troubleshooting, and some practical notes on analysis of basic amplifier circuits.

Much of the information presented here is basic since most solid-state audio circuits are relatively simple. The techniques discussed here are of the most benefit to those readers totally unfamiliar with audio troubleshooting. These techniques serve as a basis for understanding the more complex IC (or combination solid-state and IC) audio equipment covered in the remaining chapters of this book.

4.5.1 Signal Tracing

The basic troubleshooting approach for an amplifier involves signal tracing. The input and output waveforms of each stage are monitored on a scope or meter. Any stage showing an abnormal waveform (in amplitude, waveshape, and so on) or the absence of an output (with a known-good input signal) points to a defect in that stage. Voltage and/or resistance measurements on all elements of the transistor (or IC) are then used to pinpoint the problem.

A scope is the most logical instrument for checking amplifier circuits (both complete amplifier systems or a single amplifier stage). The scope can duplicate every function of a meter in troubleshooting, and the scope offers the advantage of a visual display. Such a display can reveal common audio-amplifier conditions (distortion, hum, noise, ripple, and oscillation).

When troubleshooting amplifier circuits with signal tracing, a scope is used in much the same manner as a meter. A signal is introduced into the input by a gen-

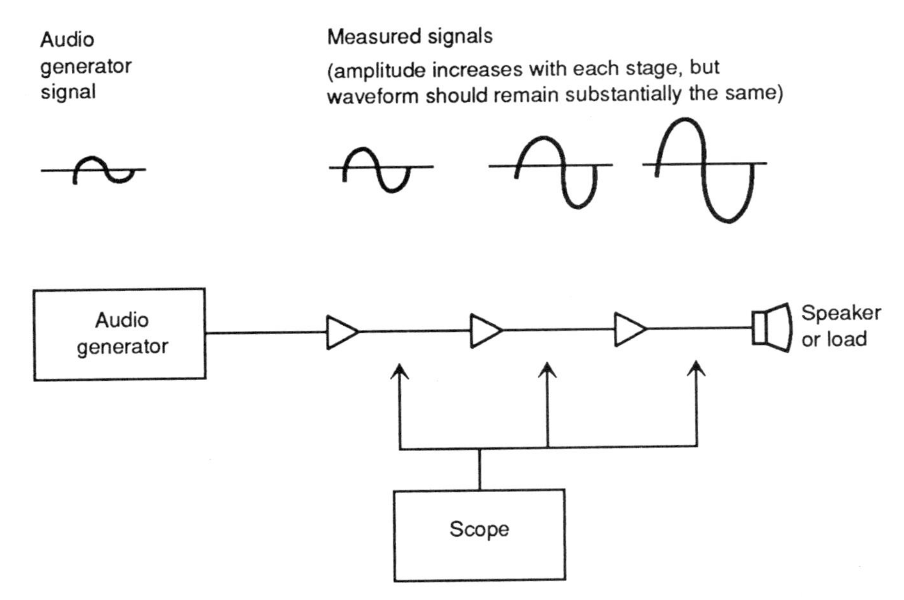

FIGURE 4.2 Basic audio-circuit signal tracing.

erator, as shown in Fig. 4.2. The amplitude and waveform of the input signal are measured on the scope.

The scope probe is then moved to the input and output of each stage, in turn, until the final output is reached. The gain of each stage is measured as voltage on the scope. Also, it is possible to observe any *change in waveform* from that applied to the input. Stage gain and distortion (if any) are established quickly with a scope.

4.6 BASIC AUDIO TESTS

Most of the components described in the remaining chapters of this book contain audio amplifiers of some kind. For that reason, it is essential that you understand how to make basic audio-amplifier tests and measurements. The following paragraphs describe the basics of such procedures and are included here for those totally unfamiliar with audio work. Detailed procedures for test, measurement, and adjustment of specific audio components are given in the related chapter.

4.6.1 Frequency Response

The frequency response of an audio amplifier can be measured with a generator and a meter or scope. When a meter is used, the generator is tuned to various frequencies, and the resultant output response is measured at each frequency. The results are then plotted in the form of a graph or response curve as shown in Fig. 4.3.

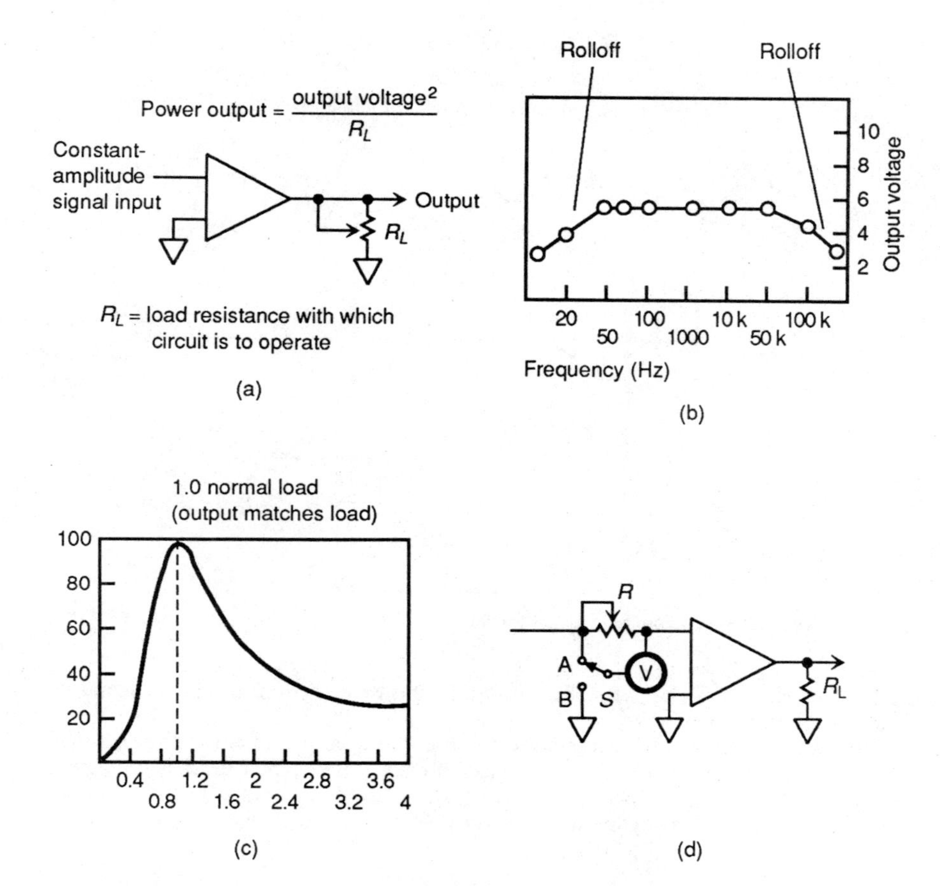

FIGURE 4.3 Basic audio-circuit test connections.

The procedure is essentially the same when a scope is used to measure audio-circuit frequency response. The scope gives the added benefit of a visual analysis of distortion, as discussed throughout the remainder of this chapter.

The basic procedure for measurement of frequency response (with either meter or scope) is to apply a *constant-amplitude* signal while monitoring the circuit output. The input signal is varied in frequency (but not in amplitude) across the entire operating range of the circuit. Any well-designed audio circuit should have a constant response from about 20 Hz to 20 kHz. The voltage output at various frequencies across the range is plotted on a graph as follows:

1. Connect the equipment as shown in Fig. 4.3*a*.

2. Initially, set the generator frequency to the low end of the range. Then set the generator amplitude to the desired input level.

3. In the absence of a realistic test input voltage, set the generator output to an arbitrary value. A simple method of finding a satisfactory input level is to monitor the circuit output (with the meter or scope) and increase the generator amplitude at the circuit's center frequency (or at 1 kHz) until the circuit is overdriven. This point is indicated when further increases in the generator output do not cause further increases in meter reading (or the output waveform peaks begin to flatten on the scope display). Set the generator output *just below* this point. Then return the meter or scope to monitor the generator voltage (at the circuit input) and measure the voltage. Keep the generator at this voltage throughout the test.

4. If the circuit is provided with any operating or adjustment controls (volume, loudness, gain, treble, balance, and so on), set these controls to some arbitrary point when making the initial frequency-response measurement. The response measurements can then be repeated at different control settings if desired.

5. Record the circuit-output voltage on the graph. Without changing the generator amplitude, increase the generator frequency by some fixed amount, and record the new circuit-output voltage. The amount of frequency increase between each measurement is an arbitrary matter. Use an increase of 10 Hz where rolloff occurs and 100 Hz at the middle frequencies.

6. Repeat the process, checking and recording the circuit-output voltage at each of the check points. For a typical audio amplifier, the curve resembles that shown in Fig. 4.3*b*, with a flat portion across the middle frequencies and a rolloff at each end. For audio amplifiers found in home-entertainment equipment, the low-end rolloff starts at about 20 Hz, while the high-end rolloff starts at about 18 kHz.

7. After the initial frequency-response check, the effects of operating or adjustment controls should be checked. Volume, loudness, and gain controls should have the *same effect* all across the frequency range. Treble and bass controls may have some effect at all frequencies. However, a treble control should have the *greatest effect* at the high end, whereas bass controls should have the greatest effect at the low end.

8. Note that generator output may vary with changes in frequency (a fact often overlooked in making a frequency-response test during troubleshooting). Even precision lab generators can vary in output with changes in frequency, thus producing considerable error. It is recommended that the generator output be monitored after each change in frequency (many generators have a built-in output meter). Then, if necessary, the generator-output amplitude can be reset to the correct value. It is more important that the generator amplitude *remain constant* rather than set at some specific value when making a frequency-response check.

4.6.2 Voltage Gain

Voltage gain in an audio amplifier is measured in the same way as frequency response. The ratio of output voltage to input voltage (at any given frequency or across the entire frequency range) is the voltage gain. Because the input voltage (generator output) is held constant for a frequency-response test, a voltage-gain curve should be identical to a frequency-response curve.

4.6.3 Power Output and Power Gain

The power output of an audio amplifier is found by noting the output voltage across the load resistance (Fig. 4.3*a*) at any frequency or across the entire frequency range. The power output is calculated as shown in Fig. 4.3*a*. For example, if the output voltage is 10 V across a load resistance of 50 Ω, the power output is $10^2/50$, or 2 W.

Never use a wire-wound component (or any component that has reactance) for the load resistance. Reactance changes with frequency, causing the load to change. Use a composition resistor or potentiometer for the load.

To find the power gain of an amplifier, it is necessary to find both the input and the output power. Input power is found in the same way as output power except that the input impedance must be known (or calculated). Calculating input impedance is not always practical in some amplifiers, especially in designs where input impedance depends on transistor gain. (The procedure for finding dynamic input impedance is described in Sec. 4.6.7.) With input power known (or estimated), the power gain is the ratio of output power to input power.

Input Sensitivity. In some audio-amplifier equipment, the input-sensitivity specification is used. Input-sensitivity specifications require a *minimum power output* with a given voltage input (such as 100-W output with a 1-V input).

4.6.4 Power Bandwidth

Many audio circuits include a power-bandwidth specification. Such specifications require that the audio amplifier deliver a given power output across a given frequency range. For example, a circuit may produce full power output up to 20 kHz even though the frequency response is flat up to 100 kHz. That is, voltage (without a load) remains constant up to 100 kHz, whereas power output (across a normal load) remains constant up to 20 kHz.

4.6.5 Load Sensitivity

An audio-amplifier circuit is sensitive to changes in load. This is particularly true of power amplifiers (including power-amplifier ICs). An amplifier produces maximum power when the output impedance is the same as the load impedance.

The test circuit for load-sensitivity measurement is the same as the circuit for frequency response (Fig. 4.3*a*) except that the load resistance is variable. (Again, never use a wire-wound load resistance. The reactance can result in considerable error.)

Measure the power output at various load-impedance and output-impedance ratios. That is, set the load resistance to various values (including a value equal to the true amplifier-output impedance) and note the voltage and/or power gain at each setting. Repeat the test at various frequencies.

Figure 4.3*c* shows a typical load-sensitivity response curve. Note that if the load is twice the output impedance (as indicated by a 2:1 ratio, or a normalized load-impedance of 2, in Fig. 4.3*c*), the output power is reduced to about 50 percent.

4.6.6 Dynamic Output Impedance

The load-sensitivity test (Sec. 4.6.5) can be reversed to find the dynamic output impedance of an audio circuit. The connections (Fig. 4.3*a*) and the procedures are the same, except that the load resistance is varied until *maximum output power* is found. Power is removed, the load resistance is disconnected from the circuit, and the resistance is measured with an ohmmeter. This resistance is equal to the dynamic output impedance of the circuit (but only at the measurement frequency). The test can be repeated across the entire frequency range if desired.

4.6.7 Dynamic Input Impedance

Use the circuit in Fig. 4.3*d* to find the dynamic input impedance of an audio amplifier. The test conditions are identical to those for frequency response, power output, and so on. Move switch *S* between points A and B, while adjusting resistance *R* until the voltage reading is the same in both positions of *S*. Disconnect *R* and measure the resistance of *R*. This resistance is equal to the dynamic impedance of the amplifier input.

Accuracy of the impedance measurement depends on the accuracy with which the resistance of *R* is measured. Again, a noninductive (not wire-wound) resistance must be used for *R*. The impedance found by this method applies only to the frequency used during the test.

4.6.8 Sine-Wave Analysis

All amplifiers are subject to possible distortion. That is, the output signal may not be identical to the input signal. Theoretically, the output should be identical to the input, except for amplitude.

Some troubleshooting techniques are based on analyzing the waveshape of signals passing through an amplifier to determine possible distortion. If distortion (or an abnormal amount of distortion) is present, the circuit is then checked further by the usual troubleshooting methods (localization, voltage measurement, and so on).

Amplifier distortion can be checked by sine-wave analysis. The procedures are the same as those used for signal tracing (Sec. 4.5.1 and Fig. 4.2). The primary concern in distortion analysis is deviation of the output (amplifier or stage) from the input waveform. If there is no change (except in amplitude), there is no distortion. If there is a change in waveform, the nature of the change often reveals the cause of distortion. For example, the presence of second or third harmonics can distort the fundamental signal, as discussed in Sec. 4.6.10.

In practical troubleshooting, analyzing sine waves to pinpoint amplifier problems that produce distortion is a difficult job that requires considerable experience. Unless distortion is severe, it may pass unnoticed.

Sine waves are best used where harmonic-distortion (Sec. 4.6.10) or intermodulation-distortion (Sec. 4.6.11) meters are combined with the scope for distortion analysis. If a scope is to be used alone, square waves (Sec. 4.6.9) provide the best basis for distortion analysis. (The reverse is true for frequency-response and power measurements.)

4.6.9 Square-Wave Analysis

The procedure for checking distortion with square waves is essentially the same as that used with sine waves. Distortion analysis is more effective with square waves because of the high odd-harmonic content in square waves (and because it is easier to see a deviation from a straight line with sharp corners than from a curving line).

Square waves are introduced into the circuit input, and the output is monitored with a scope, as shown in Fig. 4.4. The primary concern is deviation of the output waveform from the input waveform (which is also monitored on the scope). (If the scope has a dual-trace feature, the input and output can be monitored simultaneously.) If there is a change in waveform, the nature of the change often reveals the cause of the distortion.

The third, fifth, seventh, and ninth harmonics of a clean square wave are emphasized. If an amplifier passes a given frequency and produces a clean square

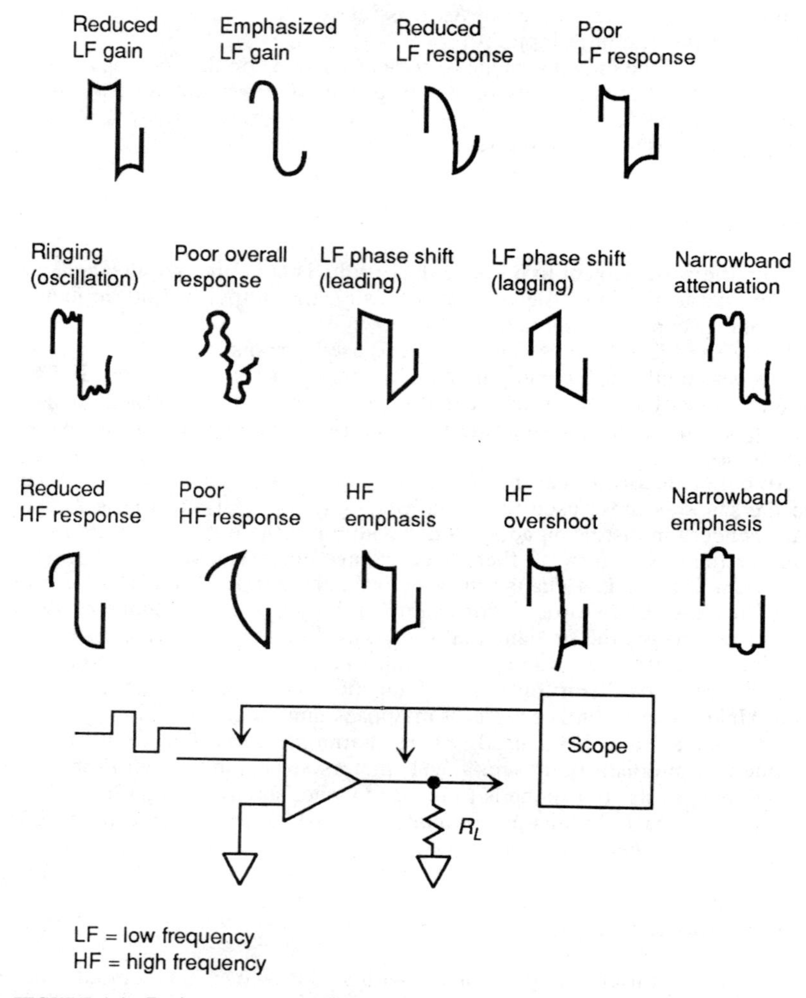

FIGURE 4.4 Basic square-wave distortion analysis.

wave output, it is fair to assume that the frequency response is good up to at least *9 times* the square-wave frequency.

4.6.10 Harmonic Distortion

No matter what amplifier circuit is used or how well the circuit is designed, there is the possibility of odd or even harmonics being present with the fundamental. These harmonics combine with the fundamental and produce distortion, as is the case when any two signals are combined. The effects of second- and third-harmonic distortion are shown in Fig. 4.5.

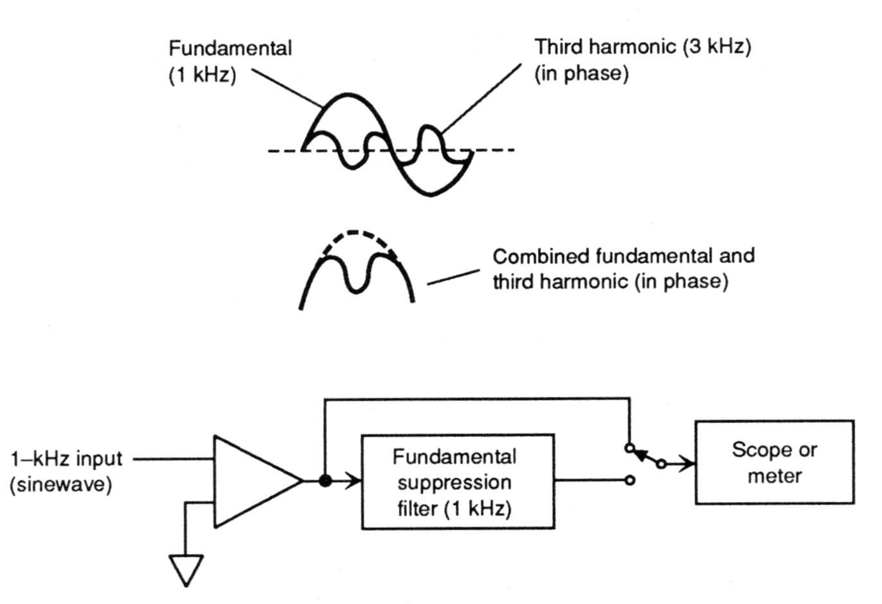

FIGURE 4.5 Basic harmonic-distortion analysis.

Commercial harmonic-distortion meters operate on the *fundamental-suppression* principle. A sine wave is applied to the amplifier input, and the output is measured on the scope, as shown in Fig. 4.5. The output is then applied through a filter that suppresses the fundamental frequency. Any output from the filter is then the result of harmonics.

The output is also displayed on the scope. (Some commercial harmonic-distortion meters use a built-in meter instead of, or in addition to, an external scope.) When the scope is used, the frequency of the filter-output signal is checked to determine harmonic content. For example, if the input is 1 kHz and the output (after filtering) is 3 kHz, *third-harmonic distortion* is indicated.

The percentage of harmonic distortion is also determined by this method. For example, if the output is 100 mV without the filter and 3 mV with the filter, a 3 percent harmonic distortion is indicated. (The *total harmonic distortion*, or THD, of a typical home-entertainment audio amplifier is 0.05 percent or less.)

On some commercial harmonic-distortion meters, the filter is tunable; therefore the amplifier can be tested over a wide range of fundamental frequencies. On other harmonic-distortion meters, the filter is fixed in frequency but can be detuned slightly to produce a sharp null.

4.6.11 Intermodulation Distortion

When two signals of different frequencies are mixed in an amplifier, there is a possibility that the lower-frequency signal will modulate the amplitude of the higher-frequency signal. This produces a form of distortion known as intermodulation distortion.

Commercial intermodulation-distortion meters consist of a signal generator and a high-pass filter, as shown in Fig. 4.6. The generator portion of the meter produces a high-frequency signal (usually about 7 kHz) modulated by a low-frequency signal (usually 60 Hz). The mixed signals are applied to the circuit input. The circuit output is connected through a high-pass filter to a scope. The

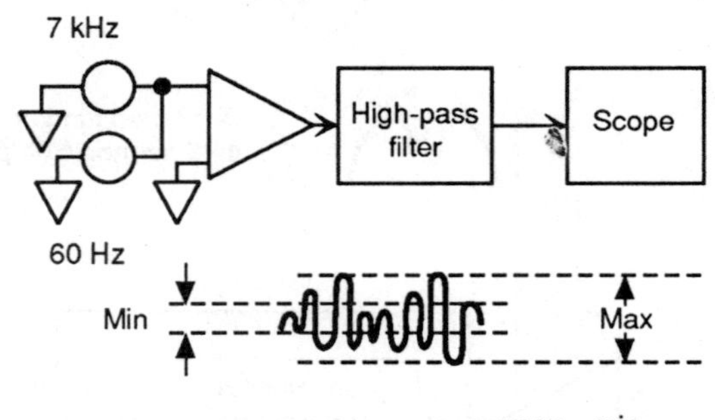

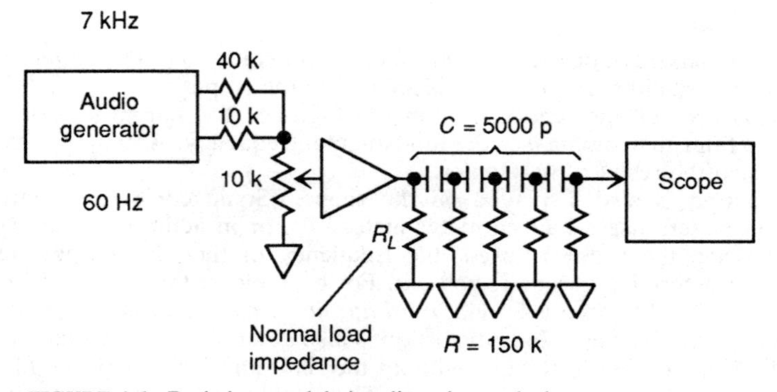

FIGURE 4.6 Basic intermodulation-distortion analysis.

high-pass filter removes the low-frequency (60 Hz) signal. The only signal appearing on the scope should be the high-frequency (7 kHz) signal. If any 60-Hz signal is present on the scope display, the 60-Hz signal is being passed through as modulation on the 7-kHz signal.

Figure 4.6 also shows an intermodulation test circuit that can be fabricated in the shop. Note that the high-pass filter is designed to pass signals that are about 200 Hz. The purpose of the 40- and 10-k resistors is to set the 60-Hz signal at *4 times* the amplitude of the 7-kHz signal. (Some audio generators provide a line-frequency output, at 60 Hz, that can be used as the low-frequency modulation source.)

If the shop circuit shown in Fig. 4.6 is used instead of a commercial meter, set the generator line-frequency (60 Hz) output to 2 V (if adjustable) or to some value that does not overdrive the amplifier being tested. Then set the generator output (7 kHz) to 2 V (or to the same value as the 60-Hz output).

Calculate the percentage of intermodulation distortion using the equation shown in Fig. 4.6. For example, if the maximum output (shown on the scope) is 1 V and the minimum is 0.9 V, the percentage of intermodulation distortion is approximately

$$\frac{1.0 - 0.9}{1.0 + 0.9} \approx 0.05 \qquad 0.05 \times 100 = 5 \text{ percent}$$

(The intermodulation distortion of a typical home-entertainment audio amplifier is 0.09 percent or less.)

4.6.12 Background Noise

If the scope is sufficiently sensitive, it can be used to check and measure the background-noise level of an amplifier as well as to check for the presence of hum, oscillation, and the like. The scope should be capable of a measurable deflection with about 1 mV (or less).

The basic procedure consists of measuring amplifier output with the volume or gain controls (if any) at maximum but without an input signal. The scope is superior to a meter for noise-level measurements because the frequency and nature of the noise (or other signal) are displayed visually.

The basic connections for measuring the level of background noise are shown in Fig. 4.7. The scope gain is increased until there is a noise or "hash" indication.

It is possible that a noise indication can be caused by pickup in the leads between the amplifier output and the scope. If in doubt, disconnect the leads from the amplifier but not from the scope.

If it is suspected that there is a 60-Hz line hum present in the amplifier output (picked up from the power supply or any other source), set the scope sync control (or whatever other control is required to synchronize the scope trace to the line frequency) to line. If a stationary signal pattern appears, the signal is the result of line hum.

If a signal appears that is not at the line frequency, the signal can be the result of oscillation in the amplifier or stray pickup. Short the amplifier input terminals. If the signal remains, suspect oscillation in the amplifier circuits.

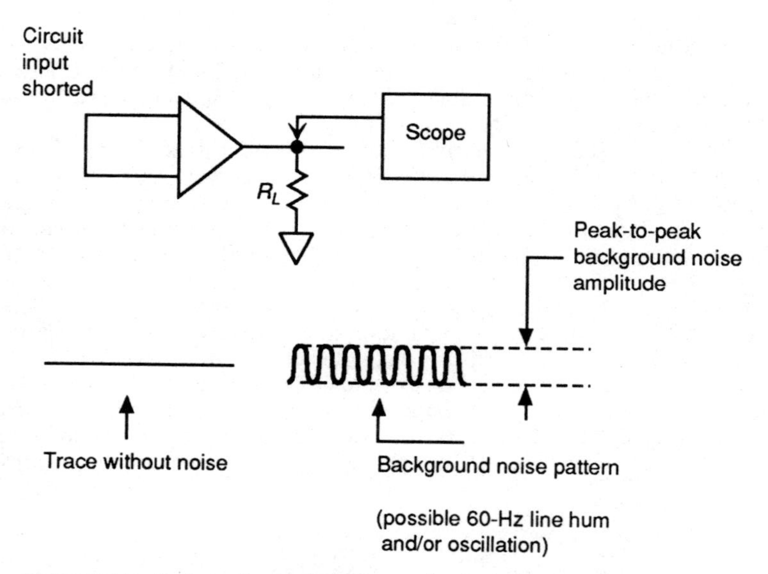

FIGURE 4.7 Basic audio-circuit background-noise test connections.

4.7 FEEDBACK PROBLEMS IN AUDIO CIRCUITS

Troubleshooting amplifiers without feedback is a relatively simple procedure. When the amplifier has feedback, the task is more difficult. Problems such as measurement of gain can be of particular concern.

For example, if you try opening the loop to make gain measurements, you usually find so much gain that the amplifier saturates and the measurements are meaningless. On the other hand, if you start making waveform measurements on a working closed-loop system, you often find the input and output signals are normal (or near normal), although many of the waveforms are distorted inside the loop. For this reason, feedback loops (especially internal-stage feedback loops) require special attention.

4.7.1 Typical Feedback-Amplifier Circuits

Figure 4.8*a* is the schematic of a basic feedback amplifier. Note the various waveforms around the circuit. These waveforms are similar to those that appear if the amplifier is used with sine waves.

Note that there is approximately 15 percent distortion inside the feedback loop (between Q_1 and Q_2) but only 0.5 percent distortion at the output. This is only slightly greater distortion than the input 0.3 percent. Open-loop gain for this circuit is about 4300; closed-loop gain is about 1000. The gain ratio (open loop to closed loop) of 4:1 is typical for a discrete-component amplifier.

4.7.2 Amplification of Signals in Feedback Circuits

Transistors in feedback amplifiers behave just like transistors in any other circuits. The transistors respond to all the same rules for gain and input-output im-

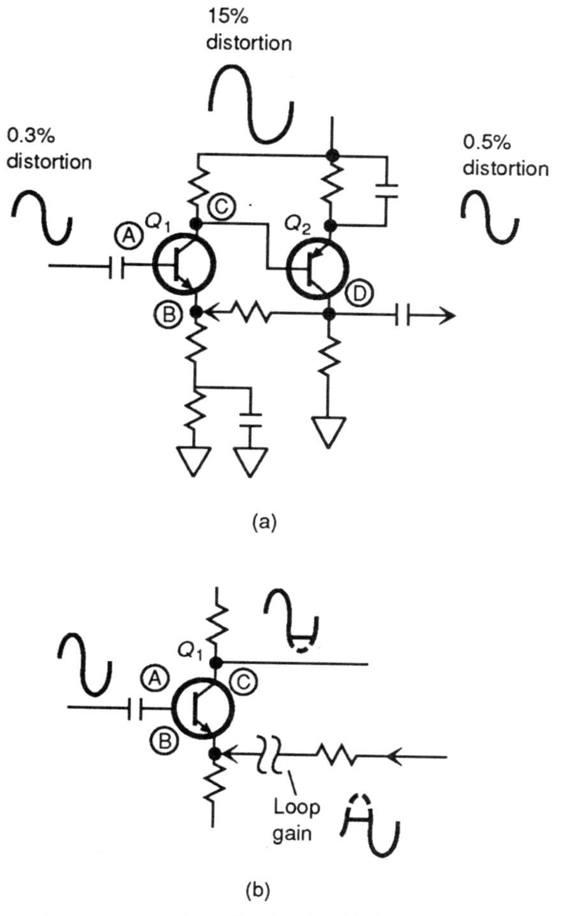

FIGURE 4.8 Basic audio circuit with feedback.

pedance. Each transistor amplifies the signal appearing between the emitter and base. It is here that the greatest difference between gain in feedback amplifiers and gain in nonfeedback amplifiers occurs.

4.7.3 Difference in Open-Loop and Closed-Loop Gain

Transistor Q_1 in Fig. 4.8 has a varying signal on both the emitter and base rather than on one element. In a nonfeedback amplifier, the signal usually varies at only one element, either the emitter or base. Because most feedback systems use *negative* feedback, the signals at both the base and emitter are in phase. The resultant gain is much less than when one of these elements is fixed (no feedback, open loop).

Assume that a perfect input signal is applied at point A (Fig. 4.8). If the amplifier is perfect (produces no distortion), the signal returning to B is also undistorted. Because the system uses negative feedback, the signal that travels around the loop a second time is undistorted as well. If the amplifier is not perfect (assume an extreme case of *clipping* distortion), the returning signal shows that effect of distortion, as illustrated in Fig. 4.8*b*.

To simplify the explanation, assume that the clipping is introduced in Q_1 and that Q_2 is perfect. Now the signals applied to the base and emitter of Q_1 are not identical. The resultant signal, applied at the control point of Q_1, is quite distorted. In effect, the distortion is a mirror image of the distortion introduced by Q_1. Transistor Q_1 then amplifies the distortion and adds the counterdistortion (produced by Q_1).

The final result is that there can be distortion inside the loop, but the distortion is counterbalanced by the feedback system. The output from Q_2 is undistorted (or relatively free from amplifier-induced distortion). The higher the amplification and the greater the feedback, the more effective this cancellation becomes and the lower the output distortion becomes. This last fact marks the basic difference in troubleshooting a feedback amplifier.

4.7.4 Causes of Distortion

There are three basic causes of distortion in any audio circuit: overdriving, operating the transistor at the wrong bias point, and the inherent nonlinearity of any solid-state device.

Overdriving. There are many causes for overdriving an amplifier (too much input signal, too much gain in the previous stage, and so on). The net result is that the output signal is clipped on one signal peak (when the transistor is driven into saturation) and also on the other peak (when the transistor is driven below cutoff).

Wrong Bias. Operating at the wrong bias point can also produce clipping but of only one peak. For example, if the input signal is 1 V and the transistor is biased at 1 V, the input swings from 0.5 to 1.5 V. Assume that the transistor saturates at any point where the base goes above 1.6 V and is cut off when the base goes below 0.4 V. No problem occurs when the bias is correct at 1 V.

Now assume that the bias point is shifted (because of component aging, transistor leakage, and so on) to 1.3 V. When the 1-V input signal is applied, the base swings from 0.8 to 1.8 V, and the transistor saturates when one peak goes from 1.6 to 1.8 V. (If the bias point is shifted down to 0.7 V, the base swings from 0.2 to 1.2 V, and the opposite peak is clipped as the transistor goes into cutoff.)

Even if the transistor is not overdriven, it is still possible to operate a transistor on a nonlinear portion of the input-output curve because of wrong bias. Some portions of the curve for all transistors are more linear than others (the output increases or decreases directly in proportion to input). For example, an increase is 10 percent at the input produces an increase of 10 percent at the output. Ideally, transistors are operated at the center of this curve. If the bias point is changed, the transistor can operate on a portion of the curve that is less linear than the desired point.

Inherent Nonlinearity. The inherent nonlinearity of any solid-state device (diode, transistor, etc.) can produce distortion even if a stage is not overdriven and is properly biased (the output never increases or decreases directly in proportion to the input). For example, an increase of 10 percent at the input produces an increase of 18 percent at the output. This is one of the main reasons for feedback in amplifiers where low distortion is required.

4.8 TROUBLESHOOTING AUDIO CIRCUITS WITH FEEDBACK

In any feedback circuit, the negative-feedback loop operates to minimize distortion in addition to stabilizing gain. As a result, the feedback-takeoff point has the least distortion of any point within the loop. From a practical troubleshooting standpoint, if the final output distortion and overall gain are within limits, all of the stages within the loop can be considered to be operating properly. Even if there is some abnormal gain in one or more of the stages, the overall feedback system has compensated for the problem. Of course, if the overall gain and/or distortion are not within the limits, the individual stages must be checked.

Most feedback-circuit problems can be pinpointed by waveform and voltage measurements, as discussed throughout this book. Give special attention to the following when troubleshooting any feedback-circuit or system.

4.8.1 Opening the Feedback Loop

Some troubleshooting literature recommends that the loop be opened and the circuits checked under no-feedback conditions. In some cases, *this can cause circuit damage*. Even if there is no damage, the technique is rarely effective. Open-loop gain is so great that some stage blocks or distorts badly. If the technique is used, as it must be for some circuits, remember that *distortion is increased* when the loop is open. A normally closed-loop circuit can show considerable distortion when operated as an open-loop circuit, even though the circuit is good. This also applies to IC amplifiers where there is a feedback loop outside the IC.

4.8.2 Measuring Gain

Take care when measuring the gain of amplifier stages in a feedback amplifier. For example, in Fig. 4.8*a*, if you measure the signal at the base of Q_1, the base-to-ground voltage is not the same as the input voltage.

To get the correct value of gain, connect the low side of the measuring device (meter or scope) to the emitter and the other lead (high side) to the base, as shown in Fig. 4.9*a*. In effect, measure the signal that appears across the base-emitter junction. This measurement includes the effect of the feedback signal.

As a general safety precaution, *never connect the ground lead of a meter or scope to the transistor base* unless the lead connects back to an insulated inner chassis on the meter or scope. Large *ground-loop currents* (between the measuring device and the equipment being serviced) can flow through the base-emitter junction and possibly burn out the transistor. (This can usually be eliminated by an *isolation transformer*.)

4.8.3 Low-Gain Problems

As we have noted, low gain in a feedback amplifier can also result in distortion. If gain is normal in a feedback amplifier, some distortion can be overcome. With

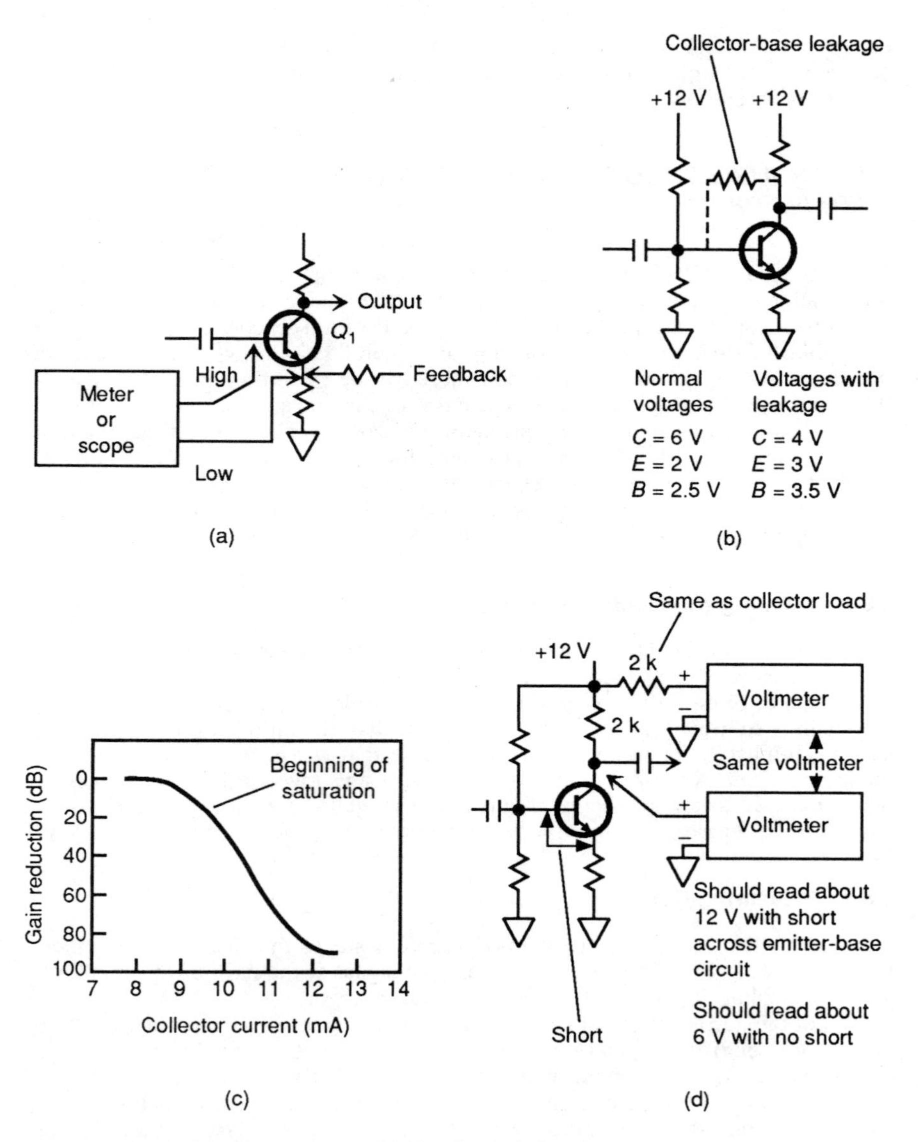

FIGURE 4.9 Basic audio-circuit troubleshooting techniques.

low gain, the feedback may not be able to bring the distortion within limits. Of course, low gain by itself is sufficient cause to troubleshoot a circuit (with or without feedback).

As an example, assume the classic failure pattern of a solid-state feedback amplifier that was working properly but that now has a decrease in output of about 10 percent. This indicates a general deterioration of performance rather than a major breakdown.

Remember that most feedback amplifiers have a very high open-loop gain that is set to some specific value by the ratio of resistor values (feedback-resistor value to load-resistor value). If the closed-loop gain is low, this usually means that the open-loop gain has fallen far enough so that the resistors no longer determine the gain. For example, if the beta or gain of Q_2 in Fig. 4.8*a* is lowered, the open-loop gain is lowered. Also, the lower beta lowers the input impedance of Q_2. In turn, the effective value of the load resistor for Q_1 is reduced (reducing overall gain further).

In troubleshooting such a situation, if waveforms indicate low gain and element voltages are normal, try replacing the transistors. Never overlook the possibility of open or badly leaking emitter-bypass capacitors. If the capacitors are open or leaking (acting as a resistance in parallel with the emitter resistor), there is considerable negative feedback and little gain. (Of course, a completely shorted emitter-bypass capacitor produces an abnormal voltage indication at the transistor emitter.)

4.8.4 Distortion Problems in Feedback Amplifiers

As discussed, distortion can be caused by improper bias, overdriving (too much gain), or underdriving (too little gain, preventing the feedback signal from countering the distortion). One problem often overlooked in a feedback amplifier with a pattern of distortion trouble is overdriving that results from transistor leakage. (The problem of transistor leakage is discussed further in Sec. 4.9.)

Generally, it is assumed that the collector-base leakage reduces gain because the leakage is in opposition to the signal-current flow. Although this is true in the case of a single stage, it may not be true when more than one feedback stage is involved.

Whenever there is collector-base leakage, the base assumes a voltage nearer to that of the collector (nearer than is the case without leakage). This increases both transistor forward bias and transistor current flow. An increase in the transistor current causes a reduction in input resistance (which may or may not cause a gain reduction, depending on where the transistor is located in the circuit).

If the feedback amplifier is direct coupled, the effects of feedback are increased. This is because the operating point (set by the base bias) of the following stage is changed, possibly resulting in distortion. For example, in Fig. 4.8*a*, the collector of Q_1 is connected directly to the base of Q_2. If Q_1 starts to leak (or if the normal collector-base leakage increases with age), the base of Q_2 (as well as the collector of Q_1) shifts the operating point of the circuit.

4.9 EFFECTS OF TRANSISTOR LEAKAGE ON AUDIO PERFORMANCE

When there is considerable leakage in a solid-state audio circuit, the gain is reduced to zero, and/or the signal waveforms are drastically distorted. Such a condition also produces abnormal waveforms and transistor voltages. These indications make troubleshooting relatively easy. The troubleshooting problem becomes really difficult when there is just enough leakage to reduce circuit gain but not enough to distort the waveform seriously (or to produce transistor voltages that are way off).

4.9.1 Collector-Base Leakage

Collector-base leakage is the most common form of transistor leakage and produces a classic condition of low gain (in a single stage). When there is any collector-base leakage, the transistor is forward biased, or the forward bias is increased, as shown in Fig. 4.9*b*.

Collector-base leakage has the same effect as a resistance between the collector and the base. The base assumes the same polarity as the collector (although at a lower value), and the transistor is forward biased. If leakage is sufficient, the forward bias can be enough to drive the transistor into or near saturation. When a transistor is operated at or near the saturation point, the gain is reduced (for a single stage), as shown in Fig. 4.9*c*.

4.9.2 Working with Known Transistor Voltages

If the normal transistor-element voltages are known (from the service literature or from previous readings taken when the circuit was operating properly), excessive transistor leakage can be spotted easily because all transistor voltages are off. For example, in Fig. 4.9*b*, the base and emitter are high and the collector is low (when measured in reference to ground).

4.9.3 Working with Unknown Transistor Voltages

If the normal operating voltages are not known, the transistor can appear to be good because all of the voltage relationships are normal. The collector-base junction is reverse biased (collector more positive than base for an NPN), and the emitter-base junction is forward biased (emitter less positive than base for an NPN).

A simple way to check transistor leakage is shown in Fig. 4.9*d*. Measure the collector voltage to ground. Then short the base to the emitter and remeasure the collector voltage.

If the transistor is not leaking, the base-emitter short turns the transistor off, and the collector voltage rises to the same value as the supply. If there is any leakage, a current path remains (through the emitter resistor, base-emitter short, collector-base leakage path, and collector resistor). There is some voltage drop across the collector resistor, and the collector voltage is at some value lower than the supply.

Note that most meters draw current, and the current passes through the collector resistor when you measure as shown in Fig. 4.9*d*. This can lead to some confusion, particularly if the meter draws heavy current (has a low ohms-per-volt rating), a fault common to meters not designed for solid-state work. To eliminate any doubt, connect the meter to the supply through a resistor with the same value as the collector resistor. The voltage drop, if any, should be the same as when the transistor collector is measured to ground. If the drop is much different (lower) when the collector is measured, the transistor is leaking.

For example, assume that in the circuit in Fig. 4.9*d* the collector measured 4 V with respect to ground. This means that there is an 8-V drop across the collector resistor and a collector current of 4 mA (8/2000 = 0.004). Normally, the collector is operated at about one-half the supply voltage (at about 6 V in this case). Note that simply because the collector is at 4 V instead of 6 V does not make the circuit faulty. Some circuits are designed that way.

In any event, the transistor should be checked for leakage with the emitter-base short test shown in Fig. 4.9*d*. Now assume that the collector voltage rises to 10.5 V when the base and emitter are shorted. This indicates that the transistor is cutting off but that there is still some current flow through the resistor, about 1 mA (2/2000 = 0.001).

A current flow of 1 mA is high for a meter. To confirm a leaking transistor, connect the same meter through a 2-k resistor (same as the collector load resistor) to the 12-V supply (preferably at the same point where the collector resistor connects to the power supply). Now assume that the indication is 11.7 V through the external resistor. This shows that there is some transistor leakage.

The amount of transistor leakage can be estimated as follows: 11.7 − 10.5 = 1.2-V drop, and 1.2/2000 = 0.0006, or 0.6 mA. However, from a practical troubleshooting standpoint, the presence of any current flow with the transistor supposedly cut off is sufficient cause to replace the transistor.

4.10 EXAMPLE OF AUDIO AMPLIFIER TROUBLESHOOTING

This step-by-step troubleshooting problem involves locating the defective part in a solid-state audio amplifier and then repairing the trouble. Figure 4.10 shows the schematic diagram of a discrete-component audio amplifier. (This is the type of amplifier circuit that is usually replaced as a package rather than being serviced. However, the circuits involved provide good examples of using schematics in basic audio troubleshooting.)

Regardless of the type of trouble symptom, the actual fault can eventually be traced to one or more of the circuit parts (transistors, ICs, diodes, capacitors,

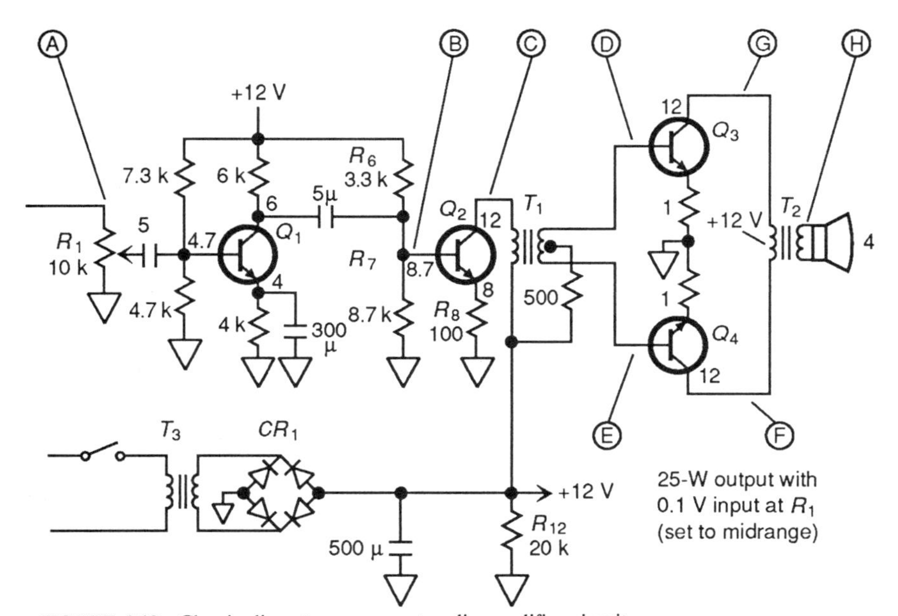

FIGURE 4.10 Classic discrete-component audio-amplifier circuit.

etc.). The waveform-voltage-resistance checks then indicate which branch within a circuit is at fault. You must locate the particular part that is causing the trouble in the branch.

This requires that you be able to read a schematic diagram. These diagrams show what is inside the blocks on a block diagram and provide the final picture of equipment operation. Quite often, you must troubleshoot audio equipment with only a schematic diagram. If you are fortunate, the diagram gives some voltages and waveforms.

4.10.1 Studying the Literature

It is essential that you study the service literature before starting into any troubleshooting job. In this example, our only "literature" is a schematic diagram similar to that shown in Fig. 4.10. There is no servicing block diagram. There is a somewhat cryptic note on the schematic stating that the output is 25 W across a 4-Ω load (loudspeaker) when a 0.1-V (100-mV) signal is introduced at the input (across R_1) and R_1 is at midrange. The test points shown in Fig. 4.10 are to be penciled in by you. The voltage information is incomplete, and there is no resistance information.

Using this fragmentary information (which is probably more than you will get in practical troubleshooting situations), you can pencil in the "logical" test points as shown in Fig. 4.10. The test points are logical because they show the input and output of each stage.

Using what you know, the input to be introduced at test point A should be 0.1 V at some audio frequency. You can connect an audio generator to test point A and set the generator to produce 0.1 V (100 mV) at a frequency of 1000 Hz (or some other frequency in the audio range) as shown in Fig. 4.11*a*. Under these conditions, the output voltage at test point H should be about 10 V. This voltage is found because the output is supposed to be 25 W across a 4-Ω load, or output voltage $= \sqrt{25 \times 4} = \sqrt{100} = 10$ V.

You have no idea which signals will appear at the other test points. However, with sine-wave signals introduced at A, the remaining test-point signals should be sine waves at the same frequency. Also there will probably be considerable voltage gain at points B and C, but you can only guess how much gain.

You can monitor each of the test points with an ac voltmeter, a dc voltmeter with a rectifier probe, or a scope. We use the scope because any really abnormal distortion at the test points appears on the scope display (as does the voltage).

Armed with this mass of information, you are now ready to begin the troubleshooting effort. You make notes as you go along, if that will help. Remember that each troubleshooting problem is always slightly different from the last. There is no surefire step-by-step procedure that fits every situation.

4.10.2 Determine the Symptoms

A trouble of "no audio output from the speaker" is reported to you, and you check the equipment yourself (never trust anyone else). You find that the symptom of no output is correct when the volume control R_1 is set at midrange (which is the normal operating position for R_1).

However, by rotating R_1 fully clockwise (for maximum volume) you note that

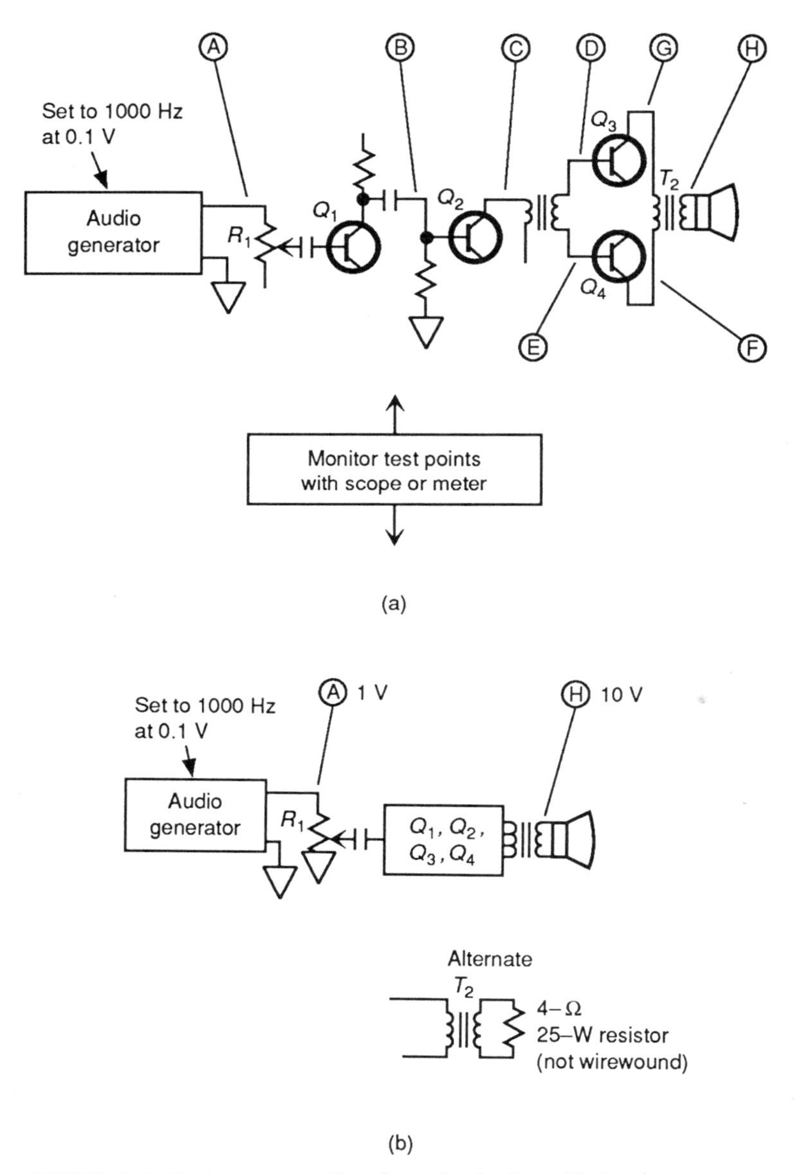

FIGURE 4.11 Basic test connections for audio-circuit troubleshooting.

there is a very weak tone from the speaker. The "no audio output" symptom is not exactly correct, but it is obvious that the amplifier is not operating properly. (With R_1 at nearly full on and a 0.1-V input, the output should be over 25 W and the tone would probably burst your eardrums.)

Reset R_1 to the normal midrange position and make your first decision.

You could decide that the audio generator is bad and that there is no signal at test point A. This is not logical. When R_1 is set for maximum volume, there is a weak tone from the speaker, telling you that there is an input signal present. If there is no tone whatsoever, you could say that the generator is *possibly bad* (or not properly connected). Just to satisfy yourself, connect the scope across the input terminals (test point A) and observe the input waveform.

You could decide that the power supply is defective. This is slightly more logical than the decision concerning the generator. If the power supply is bad, you will get a symptom of no output tone whatsoever when R_1 is set to maximum volume. Because there is a weak output, *possibly* caused by stray coupling around the defective stage or by a weak transistor, the power supply is working. Of course, the power supply could be producing a low voltage to one or more stages. To confirm this, check the voltage at each stage. Remember that you are trying to determine the symptoms, not isolate the trouble, at this point in troubleshooting.

You should decide to check the output of the circuit group. This is the first step in isolating the trouble to a circuit.

4.10.3 Isolating the Trouble to a Circuit

Once you have determined the symptoms, check the waveform at the output (test point H) of the circuit group, as shown in Fig. 4.11*b*. If the amplifier is operating properly, there should be a sine wave at H, with an amplitude of about 10 V when R_1 is set to midrange.

If the sine-wave is present but there is no tone from the speaker, suspect the speaker. If a replacement speaker is available, check by substitution. If no speaker is available, connect a 4-Ω resistor (as shown in Fig. 4.11*b*) across the secondary terminals of T_2 (to substitute as the load), and observe the sine wave at H. If the sine wave is correct with a substitute load, the speaker is at fault. *Do not operate the circuit without a load.* To do so could damage the transistors.

If the sine wave is absent at H with R_1 set to midrange, you can place a bad-output bracket at H. We have a good-input bracket at A and a bad-output bracket at H. Now it is time to make another decision.

You could decide that the output transformer T_2 *is defective.* This is possible but not logical. Remember that you are trying to isolate trouble to a single circuit by checking the input and output points. Thus far you know there is a normal input at test point A (the input to the circuit group) and an abnormal output at test point H (the output of the circuit group). The trouble is located somewhere between points A and H, somewhere within the four circuits preceding test point H, *possibly* in T_2. If the trouble is in T_2 and you find the trouble immediately, it is a lucky guess, not logical troubleshooting. Further testing must be done before you can say definitely that any one circuit is defective.

You could decide to use the half-split technique and make the next check at test point D or E. Assume that you monitored point D and found no signal present (with R_1 at midrange). This is definitely an abnormal condition, but proves very little. The problem could be associated with the circuits of Q_1, Q_2, or Q_3. Even the Q_4 circuit is not definitely eliminated (until you check at E). The fact that by chance you have selected a test point yielding an abnormal signal does not make your procedure correct.

You could decide to use the half-split technique and make the next check at test point B. This is not a bad choice, but there is a more logical choice. You must make a test at B sometime during the troubleshooting sequence. If the signal at B is abnormal, you have isolated the trouble to the Q_1 circuit (in one lucky jump). If the signal is normal at B, you are left with three possibly defective circuits (Q_2, Q_3, or Q_4).

You should decide to use the half-split technique and make the next check at test point C. This is the best choice because you have isolated the trouble to one-half of the equipment (or two circuits) in one jump.

If the signal at C is abnormal, the trouble is in Q_1 or Q_2. If the signal at C is normal, the trouble is isolated to Q_3 or Q_4. (Note that the primary winding of T_1 is considered as part of the Q_2 circuit, but the secondary of T_1 is part of the Q_3 and Q_4 circuits.)

The terms *normal* and *abnormal* applied to the signals at test points B and G are arbitrary and relative. The service literature does not tell you the signal amplitudes or the correct waveforms. However, it is reasonable to assume that all signals are sine waves (at least ac voltages at the frequency of the signal introduced in test point A).

It is also reasonable to assume that there is some voltage gain at each test point as you proceed along the signal path. With an input of 0.1 V, the service literature says that you can expect an output of 10 V. This is a voltage gain of 100. Most of the gain (at least half, probably more) is obtained in the Q_1 and Q_2 circuits (test points B and C) because the Q_3 and Q_4 circuits are essentially power amplifiers. Similarly, the circuit of Q_1 should show more voltage gain than the circuit of Q_2 because the emitter resistor of Q_1 is bypassed.

No matter what actual values may be present, the *difference in gain* should be greater between points A and B than between points B and C. Because the exact values are not available from our "service literature," you must ultimately isolate the trouble with voltage and resistance checks, component checks, and the like.

Now assume that you make the check at test point C and find the signal abnormal. If you are paying any attention to what you have read, you know that the next logical step is to monitor the signal at test point B. There is no reason to monitor any other test point under these conditions. The signals in the paths beyond point C are abnormal if point C is abnormal.

Now assume that you make the check at test point B and find the signal normal. You have now isolated the trouble to a circuit (the Q_2 circuit), and you have done so in three logical jumps (from test point H to C to B). Your next step is to locate the specific trouble in the Q_2 circuit.

4.10.4 Locating the Trouble

The first step in locating the specific trouble (after the trouble has been isolated to a circuit) is to perform an inspection using the senses. In our case, the transistor is good (probably) because there is no evidence of physical damage (look to see if this is true), and there is no evidence of overheating (touch the transistor; it should not be hot). There is no indication of burning (no characteristic burning smell), and there are no obvious physical defects. The inspection using the senses points to no outward sign of where the trouble is located. You must now rely on test procedures to locate the defective component.

Figure 4.12 shows both the physical relationship of the Q_2 circuit parts and the

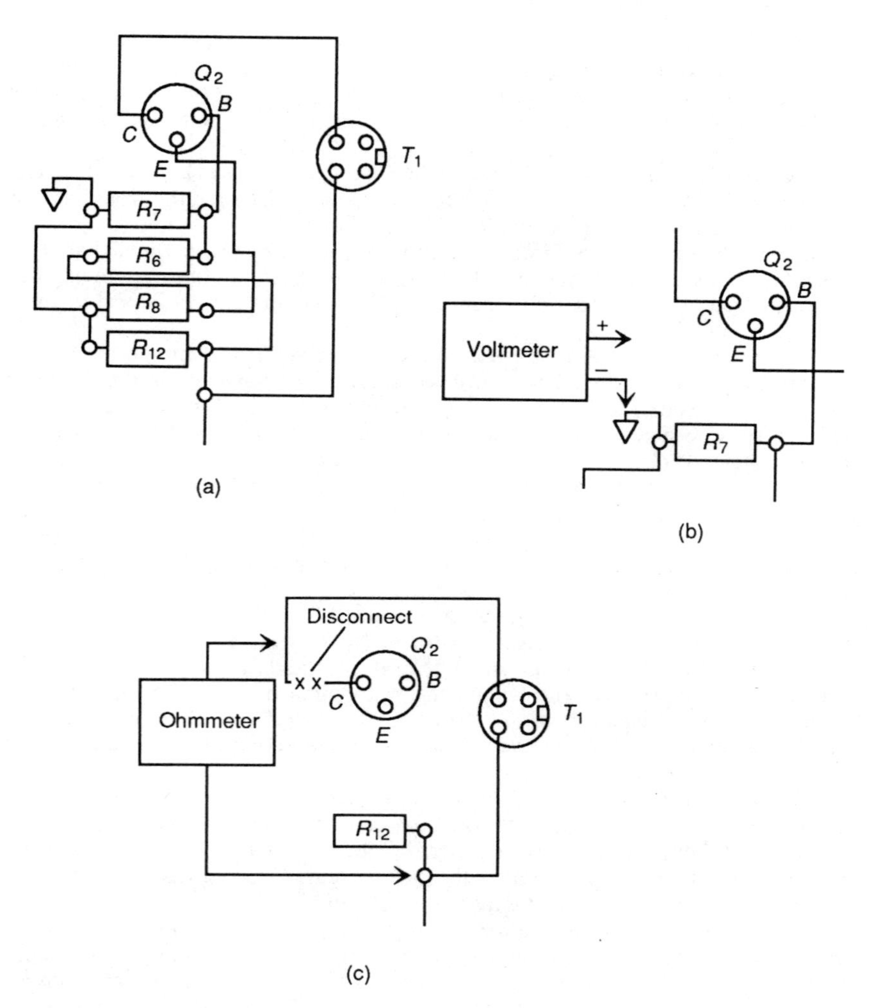

FIGURE 4.12 Point-to-point wiring of circuit parts.

point-to-point wiring. This illustration is similar to that provided in well-prepared service literature. Count yourself lucky if you have such data when troubleshooting audio equipment.

Note that Fig. 4.12 shows connections between the parts as *wires*. Present-day audio equipment generally uses *printed-circuit*, or PC, wiring, with a few wires connecting certain parts. However, the connections shown in Fig. 4.12 provide good examples of how point-to-point measurements are made during troubleshooting.

It is now time to make another decision concerning your next step in troubleshooting. *You could observe the waveform at the emitter of* Q_2. This is of little value. With a typical audio circuit, the emitter waveform is similar to the collector waveform (test point C) except that the emitter usually shows lower amplitude. Because there is a low-amplitude waveform (or no amplitude) at the collector, you will find nothing of value at the emitter.

You could decide that transformer T_1 *is defective*. To prove your assumption, you could measure the voltage at the collector of Q_2. The schematic (Fig. 4.10) shows that the voltage should be 12 V. If there is no voltage at the collector of Q_2, or the voltage is very low, T_1 is probably open. Your assumption is logical, but you are ahead of yourself. What if the voltage at the collector of Q_2 is correct? Then T_1 is probably good (at least the primary is not open).

The last sequence should teach you one important point: *Do not make hasty decisions with regard to faulty parts*. Make your decisions after you have gained enough information from conducting the proper tests and measurements. There is a more logical choice than faulting T_1 immediately.

You should check the voltages at each element of Q_2. Figure 4.12*b* shows the test connections for measuring the voltages. Note that the negative terminal of the meter is connected to the ground terminal on the PC board and that the positive terminal is connected to each of the transistor elements, in turn. (The power supply is positive with respect to ground, as indicated on the schematic.)

Figure 4.10 shows that the voltages should be +8, +8.7, and +12 V, respectively, for the emitter, base, and collector. If all voltages are normal, or nearly so, suspect Q_2. You can make an in-circuit test of Q_2 (as described in Sec. 4.9), or you can substitute a known-good transistor, whichever is most practical. (If the transistor fails the in-circuit test, you must try substitution.)

If the voltage at one or more of the Q_2 *elements is abnormal, you must now make resistance checks*. This is done by measuring the resistance from each element of the transistor to ground, using the resistance charts supplied in the service literature as a guide. But in this case, you have no resistance charts. Furthermore, the schematic does not give enough information for you to calculate the resistance to ground for each Q_2 element.

The emitter-to-ground resistance is a possible exception. The resistance reading should be 100 Ω because R_8 (a 100-Ω resistor) is connected to the emitter. Both the base and collector have several resistances in parallel. In any event, the correct resistance to ground for most transistor elements is only a wild guess.

Under these circumstances, continuity checks are your best bet. Let us examine each element of Q_2 in turn.

The collector of Q_2 is connected to the power supply through the primary of T_1. To check continuity in this line, disconnect the collector lead from the PC board, and make the connections as shown in Fig. 4.12*c*. Remember that you do not know the resistance of the T_1 primary winding, but you should have a continuity indication, probably a few ohms. If the ohmmeter shows an infinite resistance (with the meter set on one of the high-resistance scales), the T_1 primary is probably open. If the meter resistance is zero (on the lowest scale), the T_1 primary is probably shorted. Remember that you can skip this continuity measurement if the collector of Q_2 shows a normal voltage (about +12 V).

The emitter of Q_2 is connected to ground through R_8. To check continuity here, disconnect the emitter lead and make the connections as shown in Fig. 4.13*a*. The resistance should be about 100 Ω. Remember that all resistances have some tolerance, so the reading rarely (if ever) is exactly 100 Ω. Again, a high or infinite reading indicates an open, whereas a low or zero reading indicates a short. (From a practical standpoint, resistors usually do not short, but they do open. Also, the resistor leads can be shorted.)

The base of Q_2 is connected to the power supply through R_6. To check con-

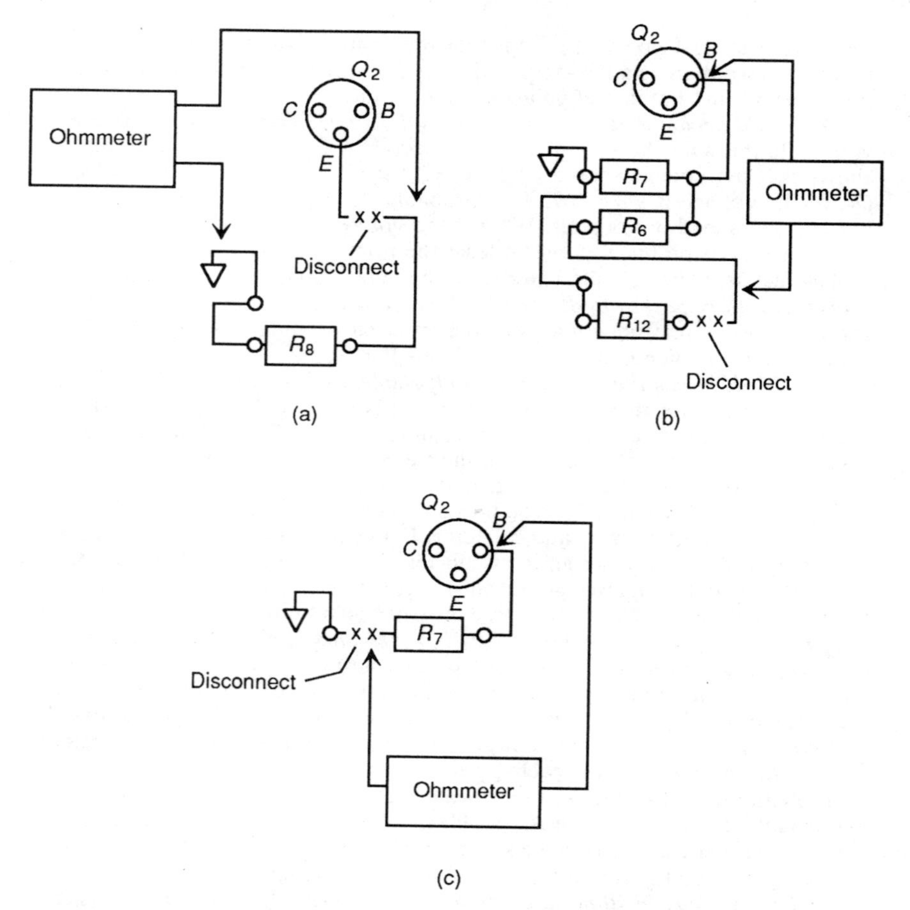

FIGURE 4.13 Additional point-to-point wiring of circuit parts.

tinuity on this line, disconnect the lead and make the test connections as shown in Fig. 4.13*b*. This removes any parallel resistance caused by R_7 or Q_2. The resistance of R_6 should be 3.3 k.

The base of Q_2 is also connected to ground through R_7. To check continuity in this line, disconnect the lead and make the test connections shown in Fig. 4.13*c*. This removes any parallel resistance caused by R_6 or Q_2. The resistance of R_7 should be 8.7 k.

Now assume that the continuity checks show that R_8 *is open*. Your next step is to make the necessary repairs and perform an operational check.

4.10.5 Repairs and Operational Check

After you have reviewed all data and are satisfied that you have located the specific cause of trouble, you should then repair the trouble. In this case you should replace emitter resistor R_8 with a known-good resistor. What is your next step?

You could turn on the power and make waveform and voltage measurements. Your first step after repairing the trouble should not be to turn on the power. Rather, you should first verify that the repair you have made is good. Because this trouble is an open emitter circuit, you should check out the emitter circuit (continuity and resistance measurements) to be sure that the repair is complete.

You should, before turning on the power, measure the resistance of R_8, *and check continuity from the emitter of* Q_2 *to ground.* If by some strange chance the emitter resistance is still abnormal, you can assume that the repair is not proper (or that you are incorrectly interpreting the resistance reading). Whatever the reason, you must take another look at your procedure.

Once proper resistance and continuity are established, you can turn on the power and make an operational check. What is the first thing you would do while making this check? *You could make voltage measurements at all elements of* Q_2. It is unnecessary to make these measurements at this time. You are now performing the operational check, not trying to isolate trouble to a defective branch of a circuit. You are fairly sure that the trouble is repaired; now you simply want to verify this fact.

You should check to be sure that all controls and switches, including those of the test equipment, are set for normal operation. Always make sure that all switches and controls are first set for normal operation when you are performing an operational check. If the volume control (R_1 in Fig. 4.10) is set for minimum volume, there will be no sound from the speaker and no waveforms available at the test points, even though the circuit is operating properly. Similarly, if the output control on the audio generator is accidentally set for minimum during troubleshooting, there will be no sound and no waveforms. Because these wrong control settings can be misinterpreted as equipment trouble, it is very important to set all controls for normal operation before attempting to perform the operational check.

CHAPTER 5

AUDIO AMPLIFIERS AND LOUDSPEAKERS

This chapter describes the overall function, user controls, operating procedures, installation, circuit theory, typical test and adjustment procedures, and step-by-step troubleshooting for state-of-the-art amplifiers and loudspeakers.

5.1 OVERALL DESCRIPTION

Figure 5.1 shows the front and back of a composite stereo amplifier. When this audio amplifier is used with a system (modular home entertainment, stereo system, etc.), the signal inputs to the amplifier are selected automatically by the system remote control. For nonsystem configurations, the signal inputs are selected manually by means of front-panel push buttons. In this book, we are concerned with operation of the amplifier in the stand-alone condition. Of course, the amplifier must have an audio input source (tuner, turntable, cassette deck, CD player, etc.) and must be used with loudspeakers.

5.1.1 Features

In addition to the basic amplifier functions, our amplifier has a number of features for both stand-alone and system operation. These include:

A *volume mute* feature that attenuates all audio passing through the amplifier.

A *speakers A/B* feature that provide for two sets of speakers that may be operated individually or turned off.

A *fluorescent power output and function display* that indicates the approximate power output for each channel and identifies the active input component. Note that when a system component is active, the fluorescent display is aux, except when that device is the turntable, in which case phono is displayed.

An *output protection with the auto reset* circuit that disconnects the speakers in the case of accidental short circuit and resets automatically when the amplifier is first turned off and then goes back on.

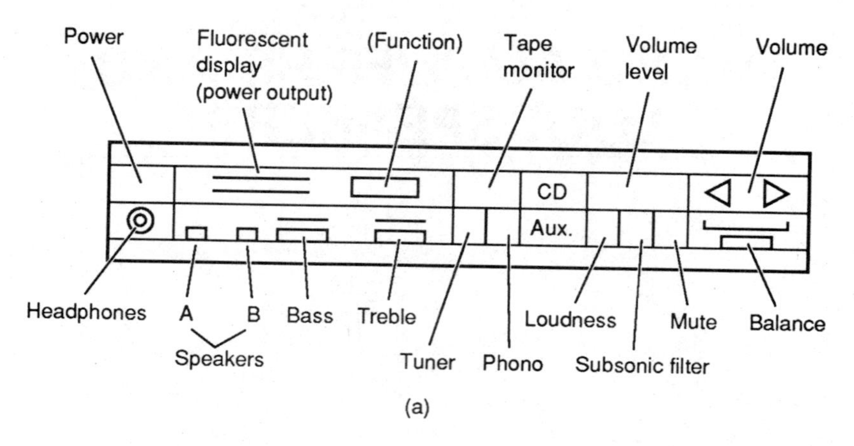

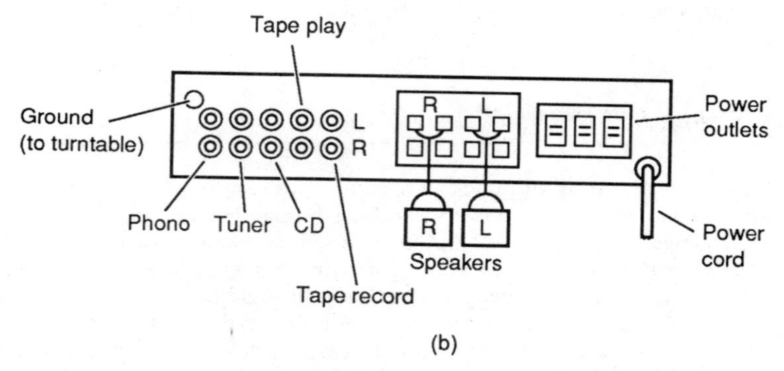

FIGURE 5.1 Operating controls, indicators, and connections for composite stereo amplifier.

An *electronic volume control with LED level indicators* that indicate the volume setting.

A *preset volume turn-on* circuit that presets the volume to a comfortable level at amplifier turn-on.

A *subsonic filter* that allows for virtual elimination of rumble caused by warped records by attenuating frequencies below 20 Hz.

A *loudness circuit* that automatically boosts bass and treble response at low listening levels to compensate for the low sensitivity of the human ear.

5.1.2 Operating Controls and Indicators

The following is a summary of the amplifier stand-alone front-panel operating controls and indicators, shown in Fig. 5.1*a*. When used in a system, many of these functions can be remotely controlled.

Push the power button to turn the amplifier on. Push power again to turn the amplifier off. Note that none of the other controls function when power is off. However, if the amplifier is used in a system, the control system remains active (the amplifier turns on when another system component is turned on). In stand-alone, the other components must be turned on separately.

The *fluorescent display* is divided into two sections. *Power output* indicators contain two rows of 10 light segments to indicate the approximate power output from each stereo channel. The *function indicator* displays the audio-input device selected (CD, phono, etc.). The function display *flashes* when the output-protection circuit is activated.

A row of eight LED segments indicate the volume setting. During muting, the segment farthest to the right that was on before muting remains on, to indicate where the volume level is set when muting is canceled.

Volume is automatically set to a comfortable level at turn-on. Pressing the left end of the volume bar decreases the volume; pressing the right end of the bar increases the volume. Increasing the volume setting during muting immediately cancels muting. The volume setting may be decreased while the amplifier is muted without canceling the muting function.

The balance slider control is moved left or right to equalize the volume of the stereo channels. When in doubt or as an initial setting, set balance to the mid-point.

Pressing the loudness button turns on a compensation circuit, which boosts both the bass and treble at low settings of the volume control. This compensates for the unequal sensitivity of the human ear to low, medium, and high frequencies at low volume levels. (As discussed in Sec. 5.5, the loudness circuit attenuates the midrange signals so that the high and low signals appear to be boosted.)

Pressing the subsonic filter button enables the subsonic filter, reducing the low-frequency rumble caused by warped records. The subsonic filter attenuates all frequencies below 20 Hz. Remember that the subsonic filter is used primarily to offset problems caused by warped records. If the filter must be used for all records, there is probably trouble in the turntable (possibly in the platter motor, Chap. 9).

Pushing the mute button reduces the volume to minimum. Note that when the amplifier is muted, only one of the fluorescent volume-indicator lights remains on. This light indicates the volume existing at the time muting is selected and the volume to which the amplifier is restored when muting is removed.

To increase treble response, move the treble control to the right. Move the treble control left to decrease treble response.

To increase bass response, move the bass control to the right. Move the bass control left to decrease bass response.

To select speaker operation, first set the speakers (A/B) control to A or B. Note that the amplifier provides for two sets of speakers, which may be operated one at a time or both may be turned off. (If only one set of speakers is connected to the amplifier, make certain to select the correct set.)

To select headphone operation, insert *stereo* headphones into the headphones jack. On this particular amplifier, using the headphones does not turn off the speakers. However, the speakers may be turned off without affecting the headphones. This is not always true for all amplifiers, so always check before you troubleshoot the circuits to solve a "no audio" trouble symptom.

Also note that the headphone jack is connected to the amplifier, no matter what loudspeaker configuration is used. If no headphones are used, a resistance is connected across the amplifier output. This is not always true for all amplifiers.

As a general precaution, make certain that loudspeakers and/or headphones are connected to an amplifier before turning on the power. *Solid-state and IC amplifiers should never be operated without a load.*

5.1.3 Typical Connections

Figure 5.1*b* shows the connection terminals on the amplifier back panel. Before connecting speakers to the amplifier (system or nonsystem) be sure to observe the precautions described in Sec. 5.1.4.

When making system connections, follow the service- and/or user-literature instructions. The following notes apply when using the amplifier in Fig. 5.1 in a nonsystem configuration.

Standard audio cables can be used for nonsystem signal connections to the amplifier input jacks. Most audio components have similar jacks, but if one component is different, adapters or special cables may be needed. Stereo cable pairs that have the left and right connectors marked (or color coded) are more convenient than two single-channel cables.

For power connections, it is usually more convenient to use the receptacles on the back of the amplifier for other components. Make sure not to exceed the wattage rating for these receptacles (300 W in the case of this amplifier). If the total wattage exceeds the limit, connect one or more components directly.

The tuner input is a high-level input designed for an AM/FM tuner output. However, the tuner input can be used for some other components (on many amplifiers).

The CD input is a high-level input designed for a CD-player output. However, the CD input can be used for some other components (on many amplifiers).

Note that on this amplifier, the tuner and CD inputs have the same impedance and can be interchanged. However, this is not usually the case. As discussed in Chap. 8, always use the CD input for a CD player. In any event, never connect a CD player to the phono input. This overdrives the amplifier and could damage the circuits (on most amplifiers).

The phono input is a low-level input, suitable for turntables that use a magnetic cartridge. If the turntable has a ground wire, attach the wire to the ground (GND) terminal.

The tape play input is a high-level input designed for a tape-deck output. However, the tape play input can be used for some other components (on many amplifiers).

The tape rec output is designed for use with a tape deck and produces a signal level (280 mV) suitable for recording on most decks. Note that the output from the tape rec jack depends on which input source (phono, tuner, or CD) is selected.

5.1.4 Speaker Installation

The following notes apply to most amplifier-speaker combinations. Compare these notes to the information found on service and user instructions.

Always try to match amplifier and speaker *impedances*. In some cases, separate terminals are provided for different impedances. In other cases, the amplifiers can be operated with speakers of different impedances, over a limited range.

For example, our amplifier can be used with speaker impedances of 6 to 16 Ω without damage or deterioration in sound. (Of course, an extreme mismatch reduces power output.)

It is generally best to use the speakers designed for a particular amplifier (although you will get an argument on this from audiophiles and stereo salespeople). Using the amplifier-speaker combination recommended by the manufacturer ensures a proper match of impedances, power-handling capabilities, and so on. However, using other speakers does not always mean a mismatch. If you are troubleshooting an existing system, it is reasonable to assume that the system was once satisfactory (the speakers match the amplifier and are properly connected). It is reasonable to assume this, but do not count on it.

As a guideline, the maximum input power for a speaker is twice the nominal input power (but always check the speaker specifications). Remember that you may be able to get by with speakers of lower power-handling capability when an amplifier is operated at low volume. However, the same speakers can be damaged (and will produce dreadful sound) if the amplifier is accidentally operated at full power. *Never connect or disconnect speakers when the amplifier is turned on*.

Do not connect more than one speaker to any pair of speaker terminals. (Here the term *speaker* refers to a loudspeaker unit or enclosure that may contain one or more individual speakers, such as a woofer and tweeter.)

Always observe polarity of the speaker leads, as well as the left-right positioning. No damage results from reversed polarity, but the sound is weird (sound has a directionless and diffused quality, and the spatial effects are lost).

Here is an old trick to tell if the polarity or speaker phase is good or bad. If the amplifier is used with an AM/FM tuner, play a *monaural source* (typical AM broadcast). If the speakers are connected in phase, the sound should seem to come from a point between the speakers (with balance at center). If the sound seems to come from a diffused source, rather than from a single point, the speakers are probably out of phase.

To correct an out-of-phase condition, make sure to reverse the connections *at only one speaker*. Many speaker wires are color-coded (typically, red for plus and black for minus, or one wire has a stripe).

Be careful of the *stranded wire* usually associated with speaker systems. Twist the bare ends of the wire tightly so that free strands cannot cause a short to adjacent terminals. Stray wire strands also find their way into fingertips or, even worse, under fingernails.

When working with stranded speaker wire, strip just enough insulation from the wires to allow the wires to seat firmly on the terminals. Note that both screw-type terminals and plug-in terminals are found on modern amplifier-speaker combinations.

As a guideline, do not use any wires smaller than no. 18 AWG for runs less than 30 ft or smaller than no. 16 AWG for runs longer than 30 ft. Try to avoid long runs if possible, and always use a good-quality speaker wire.

Note that some speakers have adjustment controls, particularly to adjust the high-frequency and/or midrange levels. Always check the setting of these controls *before* you start troubleshooting any speaker problems.

Place the speakers an equal distance from each side of the sound source (turntable, cassette deck, CD player). This is particularly true in home entertainment systems where a video monitor or receiver is involved. The sound is more realistic when the *picture is centered* in the field of sound.

Always check speaker placement when troubleshooting a speaker problem (such as a "my neighbor's stereo works better than mine, and I paid more" problem).

Stereo sound is best when the speakers are upright with the manufacturer's logo at the top and when the cabinets are positioned so that the tweeters are inside. In many cases, speakers are marked L and R. If speakers must be placed on their sides (try to avoid this), position the speakers so that the tweeters are toward the center of the listening area.

Place the speakers so that the tweeters are as close as possible to the *ear level of a seated* viewer or listener. In most (but not all) cases, the tweeter is located near the top of the speaker enclosure.

Try to keep each speaker at the same height, or at least at the same distance from the floor, walls, furniture, and so on. Unfortunately, there are some rooms where it is almost impossible to produce a good stereo sound, even though monophonic sound is no problem.

Remember that overall sound performance of any speaker system is influenced by variations in speaker placement and the acoustical qualities of the room (even though the customer will not believe it). That is a good reason for having a set of known-good speakers in the shop.

Finally, if you place the speakers in corners or with one side against a wall, you will get the strongest (but not necessarily the most accurate) *bass response*. If the speaker is adjustable, it may be necessary to correct this condition (lower the bass or increase the treble). Of course, there may be customers (such as the author) who want more bass than is available in the recording.

5.2 KEYBOARD AND DISPLAY CIRCUITS

Figure 5.2 shows the amplifier keyboard and display circuits. The front-panel control buttons are connected in a *key matrix*, consisting of two scan lines and six data-input lines, all interconnected to IC_{901}. The corresponding momentary-contact switch is closed when each front-panel button is pressed. IC_{901} detects the switch closure and performs the appropriate display operation, in addition to producing the corresponding function, as shown by the truth table in Fig. 5.2.

For example, when the CD button (Fig. 5.1*a*) is pressed, the scan A signal is applied to input 2 at pin 22 of IC_{901}. This causes IC_{901} to select the audio input from the CD player jacks (Fig. 5.1*b*) and produces a CD-function indication on the front-panel display FL_{901} through IC_{902}.

The amplifier has two front-panel display indicators, the volume-level display consisting of bar display D_{905} through D_{912} and fluorescent display FL_{901}. In turn, FL_{901} is divided into two sections, containing the function display and the L/R channel-output level indicators.

The volume-level display D_{905} through D_{912} is controlled directly by IC_{901}. At minimum volume, the volume-level display is not turned on. Pressing volume up causes IC_{901} to issue the volume 0 signal at pin 25. This turns on the LED at the left end of the display.

IC_{901} issues the volume 1 through volume 7 signals, pins 26 through 32, when the volume up button is pressed continuously. As volume is increased and the next segment turns on, the previous segment remains on. Pressing mute causes the segment that is on (farthest to the right) to remain on and all other segments

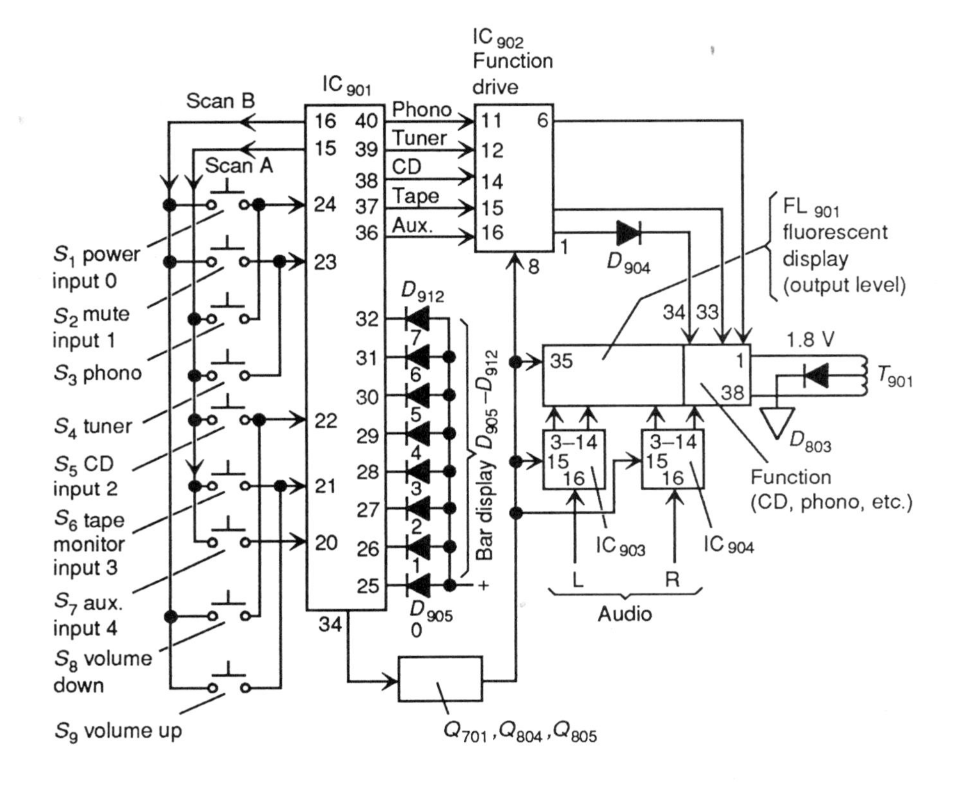

IC_{901} pin	Input	Scan A	Scan B
24	0	Phono	Power
23	1	Tuner	Mute
22	2	CD	Volume down
21	3	Tape	Volume up
20	4	Auxiliary	---

FIGURE 5.2 Amplifier keyboard and display circuits.

to turn off. This indicates to the user where the volume level is set when muting is canceled.

Function drive IC_{902} drives the function-display section of FL_{901}. IC_{901} supplies control information to IC_{902}. Pressing a device-select button directs IC_{901} to issue the appropriate function-control command. For example, pressing phono causes the phono signal to appear at pin 40 of IC_{901}. This phono signal is applied to IC_{902} and causes the phono display to appear at the right-hand side of FL_{901}.

The L and R channel audio output-level indicators of FL_{901} are driven by level-indicator drivers IC_{903} and IC_{904}. A portion of both the left and right audio outputs are rectified, and the positive portion of the audio is applied to IC_{903} and IC_{904} at pin 16. IC_{903} and IC_{904} contain nine level detectors, each with a separate output. As the audio output level to IC_{903} and IC_{904} increases, the audio output level indicators of FL_{901} are turned on, in sequence, by the signals at pins 3 through 14 of IC_{903} and IC_{904}.

5.2.1 Keyboard and Display Troubleshooting

If there is no function (input) display, first make sure that the power-supply voltages are available. Then press power and check for about 13 V at pin 8 of IC_{902} and pin 35 of FL_{901}. If it is missing, suspect Q_{701}, Q_{804}, and Q_{805}.

If power is available to FL_{901}, press aux and check that about 4.2 V appears at pin 36 of IC_{901}. If it does not, suspect IC_{901} or the aux switch circuit.

If pin 36 of FL_{901} is at 4.2 V, check that pin 1 of IC_{902} and pin 33 of FL_{901} are at about 12.6 V and that pin 34 of FL_{901} is at about 12.1 V. If pin 1 of IC_{902} is not at 12.6 V, suspect IC_{902}. If pin 34 is not at 12.1 V, suspect D_{904}.

If the input signals are available to FL_{901} but there is no function or input display, suspect FL_{901}. Before you remove the FL_{901} display from the amplifier (a tedious job), make certain that FL_{901} has operating power. Pins 38 and 1 receive about 1.8 V from T_{901}. If it does not, suspect T_{901} or D_{803}.

Note that the grid and anode function segment of FL_{901} is turned on (such as the aux display, phono display, etc.) with about 12 V from IC_{902}, while all other function segments are at 0 V (which is negative with respect to the cathode). This turns on the selected segment and turns off all other segments.

If there is no output power-level display, first make sure that the power-supply voltages are available. Then press power and check for about 13 V at pin 15 of IC_{903} and IC_{904}. If it is missing, suspect Q_{701}, Q_{804}, and Q_{805}.

If power is available to IC_{903} and IC_{904}, press aux and apply a signal to the auxiliary input. Check that audio is present at the speaker terminals. If not, refer to the troubleshooting notes in Secs. 5.4 through 5.6 (the audio-processing circuit troubleshooting sections).

If audio is present at the speaker terminals, check for rectified audio at pin 16 of IC_{903} and IC_{904}. If it is not there, trace back to the audio source from pin 16.

If audio is present at pin 16 of IC_{903} and IC_{904} but there is no output power-level display, suspect IC_{903} and IC_{904} and FL_{901}. Again, make sure that FL_{901} has about 1.8 V at pin 1. If it does not, suspect T_{901} or D_{803}.

5.3 POWER-PROTECTION CIRCUITS

Figure 5.3 shows the power-protection circuits. Our amplifier has a circuit that provides both long- and short-term ac power-loss protection. After a short-term power loss (less than 2 s), the amplifier returns to normal operation when power is reapplied. If an ac power loss is longer than 2 s, the amplifier remains off when power is reapplied (until power is pressed again).

With ac power applied and the −14-V supply operating, a high is applied to D_{913}, and about 0.7 V is applied to the base of Q_{901}. The 0.7 V is developed by a

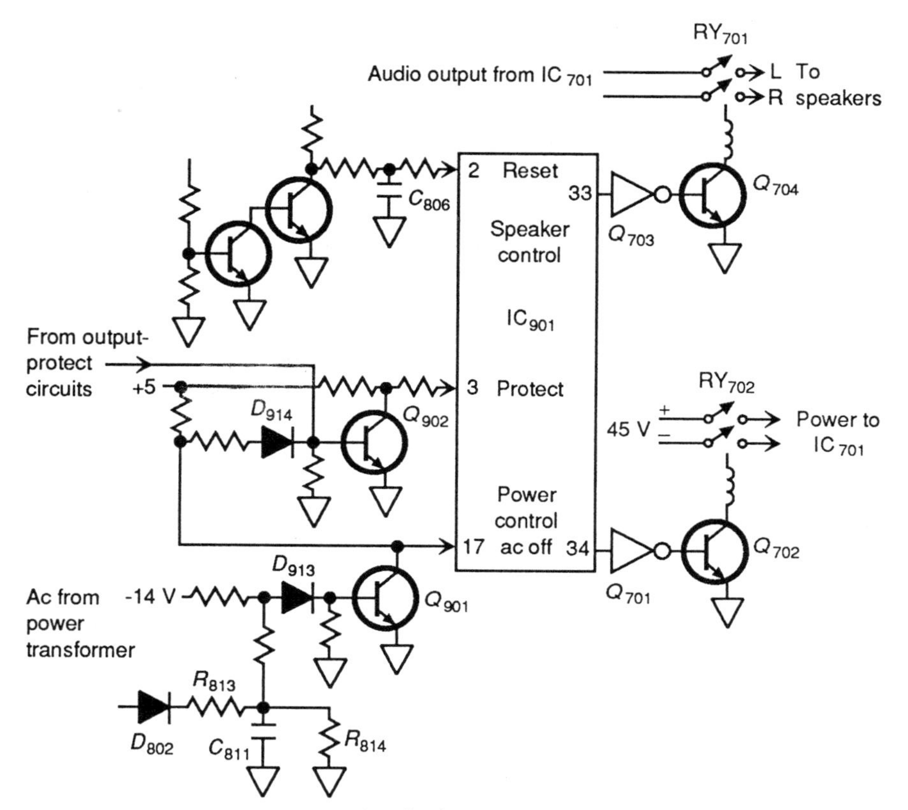

FIGURE 5.3 Amplifier power-protection circuits.

combination of the −14 V and about −16 V produced by D_{802} (which rectifies alternating current from the power transformer). The 0.7 V at the base of Q_{901} turns Q_{901} on and applies a low to pin 17 of IC_{901}. This low is also applied to the base of Q_{902} through D_{914}. The low keeps Q_{902} off and pin 3 of IC_{901} high during normal operation.

When the power button is pressed, pins 3 and 17 of IC_{901} are high and low, respectively, if ac power is present. Under these conditions, pin 34 of IC_{901} goes low. This low is inverted to a high by Q_{701}, causing Q_{702} to turn on and RY_{702} to be actuated. With RY_{702} actuated, power is applied to power amplifier IC_{701}.

Approximately 4 s after pin 34 of IC_{901} goes low, pin 33 goes low. This low is inverted to a high by Q_{703}, causing Q_{704} to turn on and RY_{701} to be actuated. With RY_{701} actuated, the output of IC_{701} is connected to the speakers.

D_{802}, R_{813}, R_{814}, and C_{811} form a 16-V dc supply. When ac power is removed, the +16-V supply decays much more rapidly than the +5- or −14-V supplies. (This is because C_{811} discharges rapidly through R_{814}, while the +5- and −14-V supply capacitors discharge slowly.)

The rapid decay of the +16-V supply reverse biases D_{913} and turns Q_{901} off. With Q_{901} off, pin 17 of IC_{901} goes high, while pin 3 of IC_{901} goes low. Under these conditions, pins 33 and 34 of IC_{901} go high, opening relays RY_{701} and

RY_{702}. With both RY_{701} and RY_{702} open, the output of power amplifier IC_{701} is disconnected from the speakers, and the power is removed from IC_{701}.

If ac power is reapplied in less than 2 s (before the +5- and −14-V capacitors discharge), a high is reapplied to pin 3 of IC_{901} as well as a low to pin 17 of IC_{901}. Under these conditions, both pins 33 and 34 of IC_{901} are made low, and the relays are energized to reapply power and reconnect IC_{701} to the speakers.

If ac power is reapplied after the 2-s limit (after the +5- and −14-V capacitors have discharged), IC_{901} is reset by discharge in the reset circuits.

When power is applied, C_{806} is charged and pin 2 of IC_{901} is high. If power is removed temporarily, the discharge of C_{806} keeps pin 2 of IC_{901} high for 2 s. When C_{806} discharges sufficiently, pin 2 of IC_{901} goes low (the reset condition) and all circuits within IC_{901} are reset. IC_{901} remains in reset (power standby) as long as the power cord is connected.

5.3.1 Power-Protection Troubleshooting

To check for proper power-loss operation, temporarily remove ac power (pull the cord) while monitoring pins 3 and 17 of IC_{901}. Check that pin 3 goes high while pin 17 goes low. Reapply power in less than 2 s, and check that pin 3 goes low while pin 17 goes high. If not, suspect Q_{901}, Q_{902}, D_{913}, D_{914}, and D_{802}.

Whenever there is a power turn-on problem, first check the reset circuits to be sure that IC_{901} has received a reset instruction. Pin 2 of IC_{901} should go low (to reset IC_{901}) when power is removed and then go high when power is reapplied. If it does not, suspect C_{806} and the associated circuit.

As a final check of the power-loss circuits, remove ac power for more than 2 s. Reapply power and check that the amplifier remains off but can be turned on when power is pressed.

5.4 DEVICE-SELECT OPERATION

Figure 5.4 shows the device- (audio input) select circuits. (Only the left channel is shown.) Our amplifier has provisions for both system and nonsystem control. In either case, the functions are controlled by input select IC_{602} under direction of IC_{901}.

Commands from the front-panel switches, or system control, are applied to IC_{901}, which generates a *device-select code* that is applied to IC_{602}. The code includes clock, data, and strobe information at pins 15, 16, and 13 of IC_{602}, respectively. The code is applied to decoder circuits within IC_{602} and causes corresponding IC_{602} switches to open or close.

Figure 5.4*b* shows a typical device select code. IC_{602} receives 14 bits of serial data from IC_{901}. The 14 bits are transmitted at the clock rate (4 MHz), or 1 bit per clock pulse. Once the 14 bits have been transmitted, the stobe signal is transmitted and instructs IC_{602} to perform the selection (close the appropriate switches).

As an example, if the phono (or turntable) audio is to be amplified and distributed to both channels, as well as to the cassette deck for tape recording, the amplifier front-panel phono button is pressed. This causes bits 1, 5, 7, 9, 11, and 13 to be selected (the data line goes high at those clock intervals or bits), and the corresponding IC_{602} switches are closed. The phono input is amplified by preamp IC_{401} (which also provides RIAA equalization, Sec. 5.5).

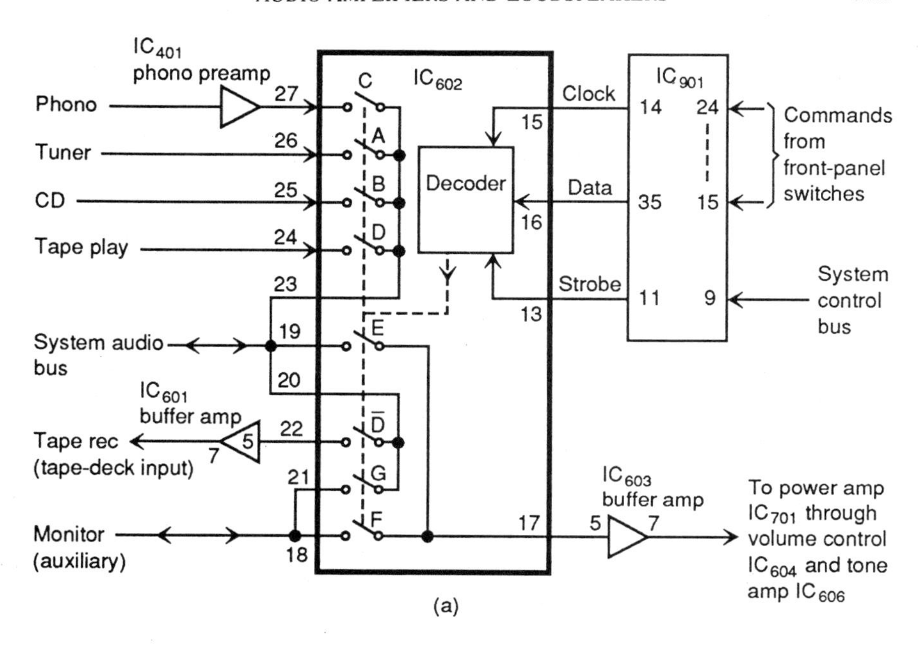

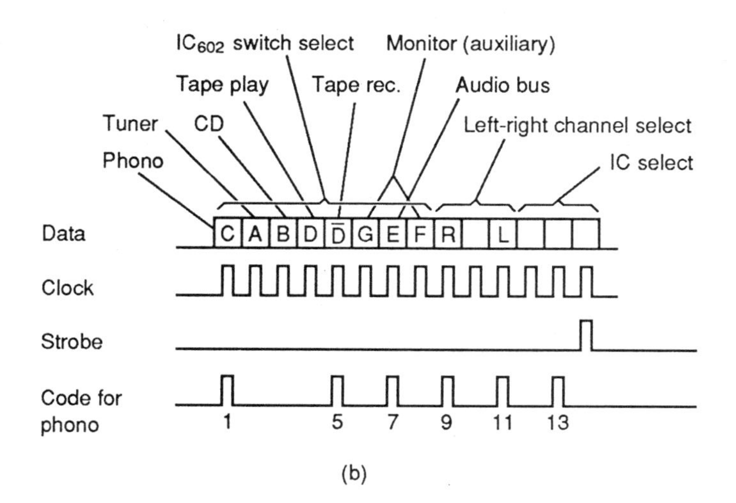

FIGURE 5.4 Amplifier device-select (audio input) circuits.

The audio at pin 27 of IC_{602} is distributed to pin 5 of buffer amp IC_{603} through switches C and E, and pins 23, 19, and 17 of IC_{602}. The audio is also available to the cassette deck input through switch $\overline{D}$, pins 20 and 22, and IC_{601}. The amplified audio output of IC_{603} is applied to the power amplifier IC_{701} (Sec. 5.6) through volume control IC_{604} (Sec. 5.5) and tone amplifier IC_{606} (Sec. 5.6).

5.4.1 Device-Select Troubleshooting

If you cannot select an audio input by remote or locally but other inputs are good, first check that the input is available from the rear-panel jacks. For example, if you cannot select the AM/FM tuner, check at pin 26 of IC_{602} for tuner audio.

If the tuner audio is missing, check the tuner audio cables as well as the wiring between the tuner jacks and pin 26. Of course, if you cannot receive audio from any input, the problem may be in the volume-control (Sec. 5.5) or audio-output (Sec. 5.6) circuits.

If only one input cannot be selected, first check that the corresponding command is received at IC_{901}. Remember that commands can be local (from front-panel switches, Fig. 5.2) or from the system (at pin 9 of IC_{901}, Fig. 5.4*a*). For example, if an auxiliary command is not completed, press aux, and check that pin 15 of IC_{901} is connected to pin 20, as shown in Fig. 5.2. If it is not, suspect wiring between the aux switch and IC_{901}.

If the command is received at IC_{901}, check that IC_{901} produces the corresponding command to IC_{602} at pins 11, 14, and 35 of IC_{901} and/or pins 13, 15, and 16 of IC_{602}. Monitor these lines with a scope or logic probe for the presence of clock, data, and strobe pulses. If any of the three lines shows a complete lack of pulses, suspect IC_{901}.

If all three lines show pulses but the corresponding input is not selected, suspect IC_{602}.

If there is no audio to the speakers or headphones but all input functions appear normal, suspect IC_{603}. Check for audio at pin 17 of IC_{602} and pin 7 of IC_{603}.

If all inputs can be selected but there is no audio to the tape recorder (cassette deck), suspect IC_{601}.

If only the phono input appears to be defective, suspect IC_{401}.

5.5 AMPLIFIER OPERATING AND ADJUSTMENT CONTROLS

Before we get into the controls for our particular amplifier, let us review typical amplifier controls. The most common operating controls for audio circuits used with music or voice reproduction equipment (hifi, stereo, public address, and the like) are the *volume*, or *loudness*, the *treble*, and the *bass* controls. The other most common audio control is the *gain* control.

The gain and volume controls are often confused, since both controls affect output of the amplifier circuit. A true gain control sets the *gain of one stage* in the amplifier. A true volume control sets the *level of the signal* passing through the amplifier, without affecting the gain of any or all stages. A gain control is usually part of a stage, whereas a volume control is usually found between stages or at the input to the first stage.

In addition to volume, bass, and treble controls, most stereo amplifiers have some form of *balance* control (so that both channels of audio can be balanced). Also, most hifi-stereo systems have a form of *playback equalization* (for tape and phono playback).

5.5.1 Volume Control

As shown in Fig. 5.5*a*, the basic volume control is a variable resistor or potentiometer connected as a voltage divider. The voltage output (or signal level) depends on the volume-control setting.

If the audio circuit is to be used with voice or music, the volume control is usually of the *audio taper* type where the voltage output is not linear throughout the setting range. (The resistance element is not uniform.) This produces a nonlinear voltage output to compensate for the human ear's nonlinear response to sound intensity. (The human ear has difficulty in hearing low-frequency sounds at low levels and responds mainly to the high-frequency components.)

If the audio circuit is not used with voice or music, the volume control is usually

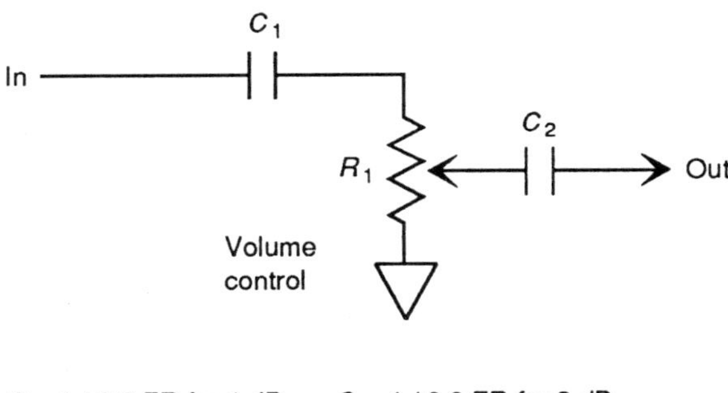

$C = 1 / 3.2FR$ for 1 dB $\quad C = 1 / 6.2FR$ for 3 dB

C in F, F in Hz, R in Ω

(a)

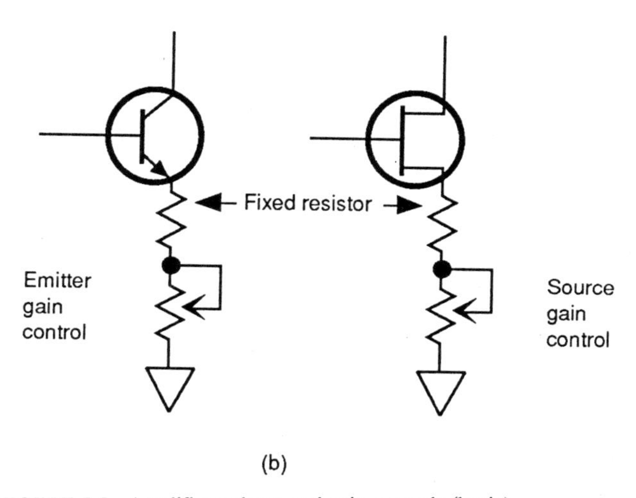

(b)

FIGURE 5.5 Amplifier volume and gain controls (basic).

of the linear type (unless there is some special circuit requirement). With such a control, the actual voltage or signal is directly proportional to the control setting.

No matter what type of volume control is used, the control should be isolated from the circuit elements. If a volume control is part of the circuit (such as a collector or base resistance), any change in volume setting can change impedance, gain, or bias.

The simplest method for isolating a volume control is to use coupling capacitors (as shown in Fig. 5.5*a*). However, the capacitors create a low-frequency response problem. As in the case of coupling capacitors (Chap. 3), capacitor C_1 forms a high-pass *RC* filter with the volume control R_1. Capacitor C_2 forms another high-pass filter with the input resistance of the following stage.

The volume control should be located at a low-signal-level point in the amplifier circuit. The most common locations for a volume control are at the amplifier input stage or between the first and second stages.

When a volume control is located at the input, the control resistance forms the input impedance (approximately). Volume controls are available in standard resistance values. Select the standard resistance value nearest the desired impedance.

When a volume control is located between stages, the resistance value should be selected to match the output impedance of the previous stage. Use the nearest standard value to produce the least signal loss.

Very little current is required for a volume control that is isolated as shown in Fig. 5.5*a*, so the power rating (in watts) required is quite low. Usually, a 1- or 2-W rating is more than enough for any volume control used in solid-state and IC circuits. Wire-wound potentiometers should not be used for any audio application. The inductance produced by a wire-wound potentiometer can reduce the frequency response of the circuit.

Figure 5.5*a* shows the equations for low-frequency cutoff versus *RC*-value relationships of typical audio volume controls.

5.5.2 Gain Control

As shown in Fig. 5.5*b*, the basic gain control is a variable resistance or potentiometer, serving as one resistance element in the amplifier circuit. Any of the three resistors (base, emitter, or collector) could be used as the gain control, since stage gain is related to each resistance value (all other factors remaining constant).

The emitter resistance is the most logical choice for a gain control. If the collector resistance is variable, the output impedance of the stage changes as the gain setting is changed. A variable base resistance produces a variable input impedance. A variable emitter resistance has minimum effect on the input or output impedance of the stage but directly affects both current and voltage gain.

With all other factors remaining constant, a decrease in emitter resistance raises both current and voltage gain. An increase in emitter resistance lowers stage gain.

The resistance value of an emitter gain control should be chosen on the same basis as the emitter resistor, except that the desired value should be the approximate midpoint of the control stage. For example, if a 500-Ω fixed resistor is normally used (or if 500 Ω is the calculated value for proper stage gain, bias stability, and so on), the variable gain control should be 1000 Ω.

In practical applications, it is usually desirable to connect an emitter gain control in series with a fixed resistance. If the gain control is set to minimum resistance value (0 Ω), there is still some emitter resistance to provide gain stabilization and prevent thermal runaway. As a guideline, the series resistance should be no less than one-twentieth of the collector resistor value. This provides a maximum stage gain of 20.

If the gain control must provide for reduction of the stage voltage gain from some nominal point down to unity, the maximum value of the control should equal the collector resistance.

If reduction to unity current gain is desired, the maximum value of the control should equal the input (base) resistance.

An audio-taper potentiometer should not be used as a gain control unless there is some special circuit requirement. The potentiometer should be of the *noninductive composition* type. The wattage of an emitter gain control should be the same as for an emitter resistor.

Use of a gain control in a power amplifier should be avoided. If a gain control must be used in power amplifiers, the control should be at the input stage where emitter current is minimum.

5.5.3 Electronic (IC) Volume Control

In present design, audio amplifiers are often provided with electronic volume controls. Such controls are usually in IC form and are (in turn) under the control of a microprocessor. Figure 5.6*a* shows the electronic volume-control circuits of our amplifier (Fig. 5.1). Note that only the left stereo channel is shown.

The amplifier volume up, volume down, muting, and loudness functions are controlled by IC_{604} (sometimes called a *volume control*, other times called an *attenuator*). Volume is adjusted in 40 steps, including full muting. (Note that most electronic volume controls provide for adjustment of the volume in steps, rather than continuous adjustment.) Volume control IC_{604} is under control of microprocessor IC_{901}.

Commands from front-panel switches (or system control) are applied to IC_{901}, which generates a *volume-control code* that is applied to IC_{604}. The code includes clock, data, and strobe information at pins 10, 11, and 12 of IC_{604}, respectively. The code is applied to decoder circuits within IC_{604} and causes IC_{604} switches and attenuators to be selected.

Figure 5.6*b* shows a typical volume-control code. IC_{604} receives 20 bits of serial data from IC_{901}. The 20 bits are transmitted at the clock rate (4 MHz), or 1 bit per clock pulse. Once the 20 bits are transmitted, the strobe signal is transmitted and instructs IC_{604} to produce the correct amount of attenuation.

As an example, assume that the volume is already at −10 dB, without the loudness function, and it is desired to turn the loudness function on, with a volume of −8 dB. The amplifier front-panel loudness button is pressed (once), and the volume up button is held until the desired −8 dB is obtained. Under these conditions, bits 1, 2, 3, 8, 9, and 20 are selected (data line high).

Audio from circuits ahead of IC_{604} is applied to 10-dB attenuators in IC_{604} through pin 5. The output of the 10-dB attenuators is applied to 2-dB attenuators through buffer amplifier IC_{605}. The output of the 2-dB attenuators is applied to the output circuits (Sec. 5.6) through a mute switch in IC_{604}.

The 10-dB attenuator in IC_{604} has eight steps, 0 to −70 dB, while the 2-dB

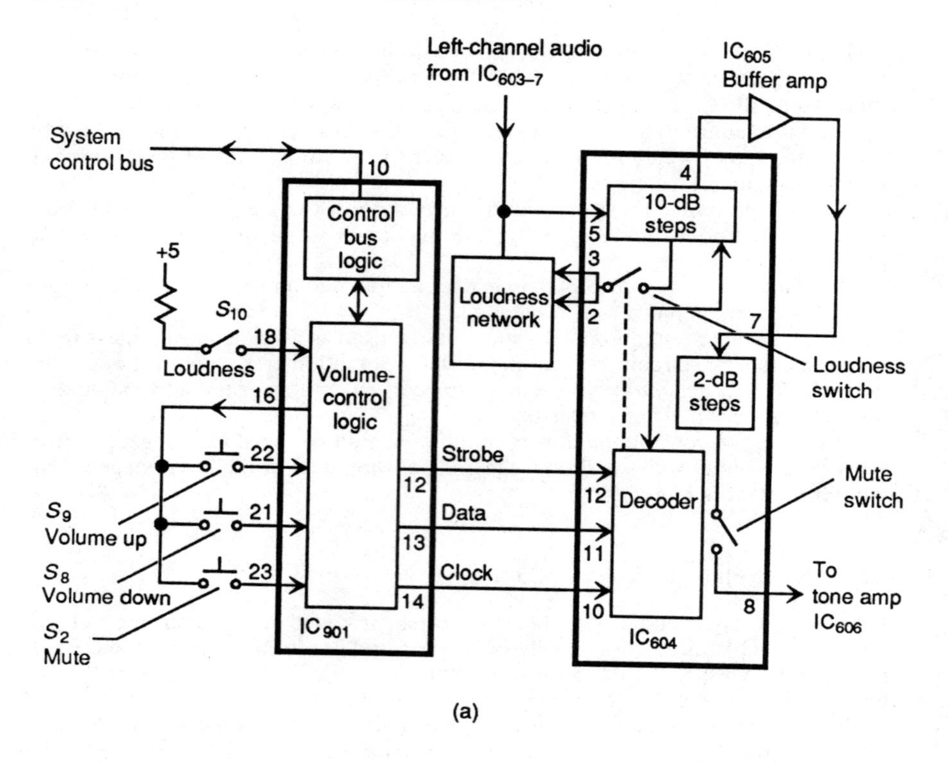

(a)

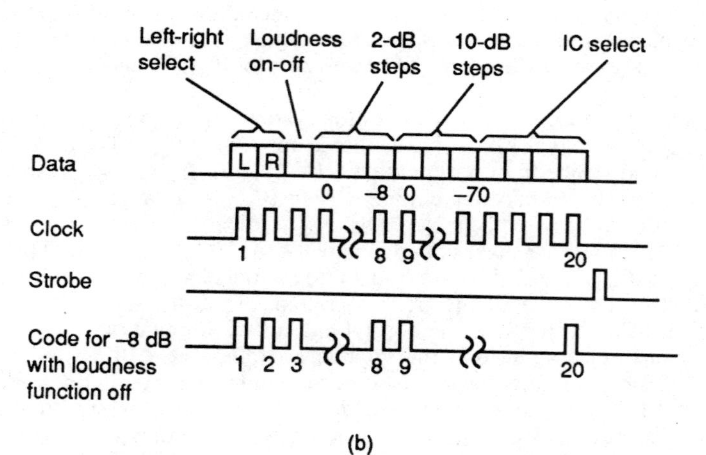

(b)

FIGURE 5.6 Amplifier electronic volume-control circuits.

attenuator has six steps, 0 to −8 dB. With the volume-control setting at minimum volume (−78 dB attenuation), continuously pressing (or holding) the volume up button causes the 2-dB attenuator to step up from −8 to 0 dB, in 2-dB steps.

Once the 2-dB attenuator reaches 0 dB, the attenuator resets to −8 dB. At the same time, the 10-dB attenuator switches to −68 dB. This results in a 2-dB step, from −70 to −68 dB.

At low-volume settings, pressing the loudness button causes the loudness switch in IC_{604} to close, connecting a loudness network between pins 2, 3, and 5 of IC_{604}. The loudness circuit attenuates the midrange audio frequencies and passes the bass and treble frequencies. This has the effect of supplying (or reinforcing) *positive* feedback to the audio at pin 5 of IC_{604} (at high and low frequencies).

Pressing the mute button opens the mute switch in IC_{604} to interrupt the audio. However, this does not affect the attenuators. When mute is pressed again, the audio is restored at the same level of attenuation (unless the attenuation is changed during mute).

5.5.4 Electronic Volume-Control Troubleshooting

If the volume control does not operate, check for audio at pin 5 of IC_{604}. If there is none, check the device-select circuits (Sec. 5.4). Remember that the electronic volume control is in the audio path between the amplifier input and output. So before you condemn the volume circuits, check for audio at IC_{604}.

If there is audio at pin 5 of IC_{604}, check for audio at pin 4 of IC_{604} while operating volume up and volume down buttons. The audio volume should increase in 10-dB steps at pin 4 each time you press the corresponding button. (If you hold the volume buttons, the audio should increase or decrease steadily in 10-dB steps.)

If there is no audio at pin 4 of IC_{604}, with audio at pin 5, suspect IC_{604}. If there is audio at pin 4 but there is no change when the volume buttons are pressed, monitor strobe (pin 12), data (pin 13), and clock (pin 14) outputs from IC_{901} while holding the volume buttons.

Although you probably cannot decode the information on the control lines, the presence of pulse activity on the lines usually indicates that IC_{901} is good. If any one of the lines shows no activity with the volume buttons operated, suspect IC_{901}. On the other hand, if there is pulse activity on all three lines but there is no change in the volume at pin 4 of IC_{604} (with the volume buttons held), suspect IC_{604}.

If there is audio at pin 4 of IC_{604} and the volume changes, check for audio at pin 7 of IC_{604}. If it is absent, suspect IC_{605}.

If there is audio at pin 4 of IC_{604}, check for audio at pin 8, and make sure that the audio changes in 2-dB steps when the volume buttons are operated. If there is no audio, suspect IC_{604}. If there is audio but there is no change with the volume buttons operated, suspect either IC_{901} or IC_{604}. Note that IC_{604} is the most likely suspect if the audio is good at pin 7, but it is possible that IC_{901} is not generating the correct code to produce 2-dB changes in volume.

If there is audio at pin 8 and the audio changes in 2-dB steps when the volume buttons are operated, press mute and check that audio is cut off at pin 8. If not, suspect the mute switch circuit, IC_{901}, or IC_{604}. You can also check for pulse activity on the data code outputs from IC_{901} (but you probably cannot decipher the code). Press mute again, and check that audio is restored at pin 8 and is at the same level (if you have not pressed the volume buttons during the mute condition). If it is not, suspect the mute circuit, IC_{901}, or IC_{604}.

Note that the *loudness network* is part of the electronic volume control (although there are external components). Also note that a failure symptom for the loudness functions is usually difficult to define. This is because the loudness function attenuates the midrange signals so that the ear hears what appears to be the same level across the audio range. (In some amplifiers, the loudness function boosts the treble and bass, but this is rare.)

Unfortunately, all ears are not the same, and not all loudness networks define midrange at the same frequencies. Usually, the customer complains that "there is no difference when I play the tape or recording with the loudness function on or off."

To troubleshoot such a symptom, apply an audio signal (say between 7 and 10 kHz) with loudness off. Then press loudness and check for drop of about 20 dB in level at pin 4 of IC_{604}. Repeat the test at 50 Hz and 20 kHz. There should be substantially no change in level (at pin 4 of IC_{604}) at the low and high ends of the audio range (unless a boost circuit is used) even though there is a change at the midrange.

No matter what type of loudness circuit is used, if there is no change in audio level at any frequency when the loudness function is switched in and out, there is a problem in the loudness circuit. Start by checking the loudness switch circuit, IC_{901}, IC_{604}, and the network connected at pins 2 and 3 of IC_{604}.

5.6 AMPLIFIER OUTPUT CIRCUITS

Before we get into the output circuits for our particular amplifier, let us review typical tone, balance, and playback-equalization functions usually associated with present-day amplifiers.

5.6.1 Tone and Balance

Tone (treble and bass) controls are found in most hifi amplifier systems. Balance controls are used in stereo amplifiers to balance the gain of both channels.

A *treble control* provides a means of adjusting the high-frequency response of an audio amplifier. Such adjustment may be necessary because of variation in response of the human ear or to correct the frequency response of a particular recording.

A *bass control* provides a means of adjusting the low-frequency response of an audio amplifier. Such adjustment may be necessary because of variation in response of the human ear, which does not respond as well to low-frequency sounds (at low levels) as to high-frequency sounds at the same level. Also, coupling capacitors present high reactance to low-frequency signals. Both of these conditions require that the low-frequency signals be boosted (in relation to high-frequency signals).

There are many circuit arrangements for tone controls. Some involve the use of adjustable feedback (mainly in treble controls). Other circuits involve bypassing the coupling capacitors with adjustable reactances (mainly in bass controls). However, the most common tone controls are *RC* filters using audio-taper potentiometers as the adjustable *R* portion of the filter.

In present design, tone-control networks are often used with IC amplifiers. One advantage to the IC-amplifier and tone-control design is that any insertion

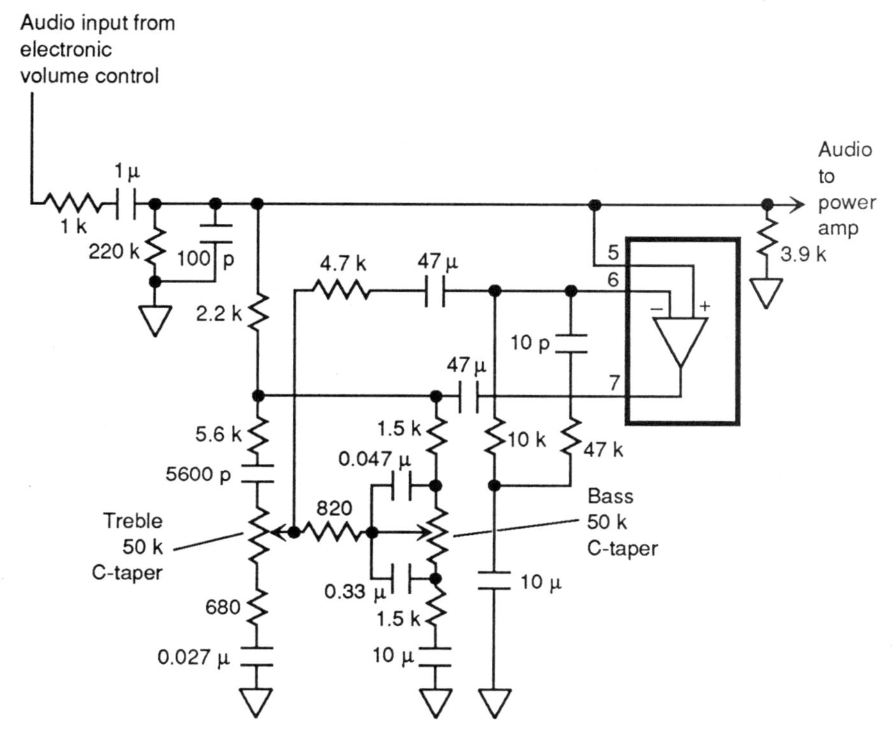

FIGURE 5.7 One channel of a tone-control network (IC).

loss presented by the circuit can be eliminated or minimized as desired. Figure 5.7 shows one channel of a tone-control network using an IC amplifier. Note that the circuit does not provide for volume control. This is because an electronic volume-control circuit (such as shown in Fig. 5.6*a*) is used. The tone-control network of Fig. 5.7 is connected between the electronic volume control and the output or power amplifier, as described in Sec. 5.6.3.

5.6.2 Playback Equalization

Many playback-equalization circuits are found in modern audio amplifiers. Most involve the use of frequency-selective feedback between stages or from the output to the input of an amplifier (typically an IC amplifier).

The feedback network usually consists of resistances and capacitances that form a feedback circuit. At any given frequency, the amount of feedback (and thus the frequency response) is set by selection of the appropriate *RC* combinations. As frequency increases, the capacitor reactance decreases, resulting in a change of feedback (and a corresponding change in frequency response).

Basic Playback-Equalization Network. Figure 5.8*a* shows a basic playback-equalization circuit where the voltage gain of two stages (with feedback) is about equal to the feedback-circuit impedance divided by the source impedance. (In

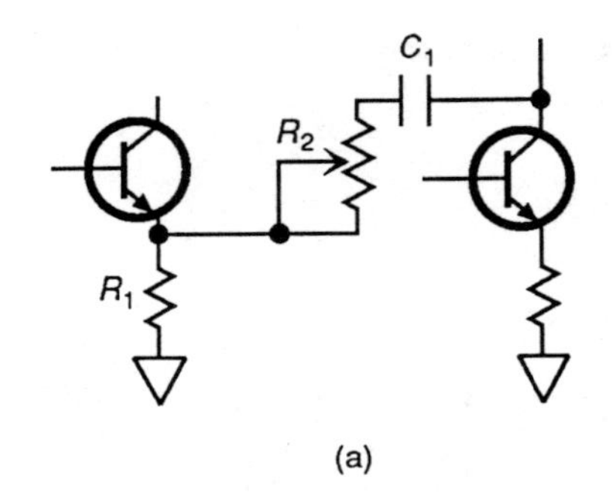

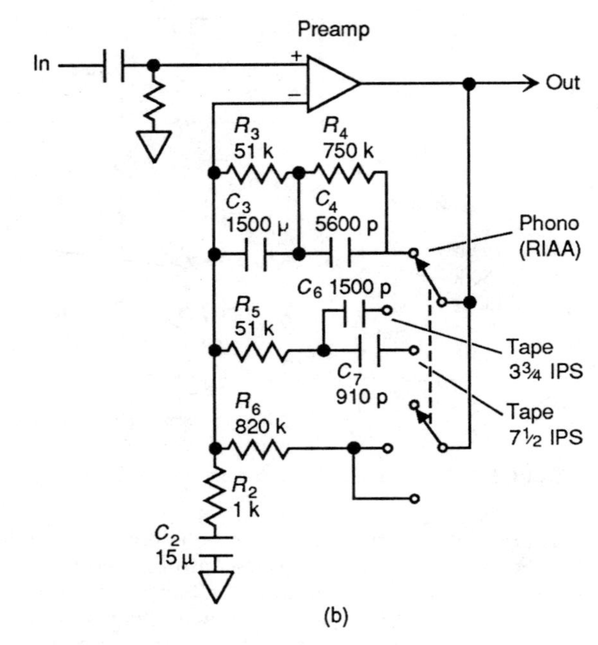

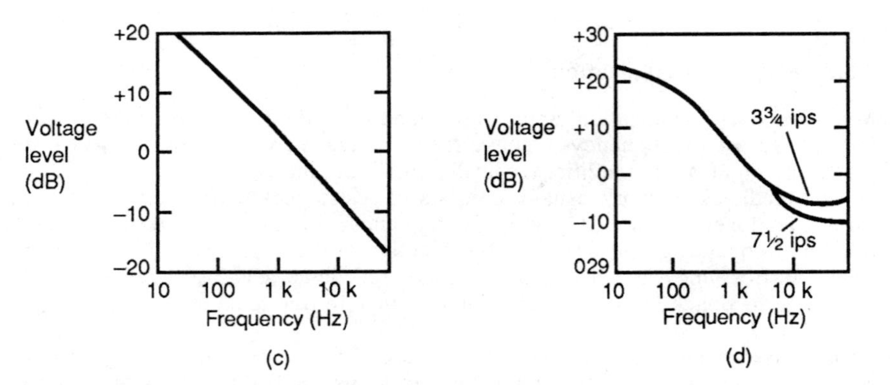

FIGURE 5.8 Amplifier playback-equalization circuits.

this case, the source impedance is the emitter-resistance R_1 value.) The feedback impedance is the vector sum of the R_2 resistance value and the C_1 reactance value. The voltage gain of the two stages can be set to any desired level for any given frequency with this simple feedback circuit.

IC Amplifier Equalization Network. Figure 5.8*b* shows an equalization network that uses feedback between the input and output of an IC amplifier (the preamp of a stereo system in this case). The closed-loop (with feedback voltage) gain of the preamp is set by the ratio of the feedback network to resistor R_2.

The feedback for the phonograph (RIAA) equalization network is composed of C_3, C_4, R_3, and R_4, and the tape network (NAB) is composed of R_5, C_6, and C_7. Note that CD players normally do not require playback equalization.

RIAA Playback Equalization. Figure 5.8*c* shows the standard RIAA equalization curve for phonograph use. The recording curve is the inverse of the playback curve, so addition of the two curves produces a flat frequency-versus-amplitude response. In phonograph recording, the high frequencies are emphasized to reduce effects of noise and the low inertia of the cutting stylus. The low frequencies are attenuated to prevent large excursions of the cutting stylus. It is the job of the frequency-selective feedback network to accomplish the addition of the recording and playback responses.

It is impossible to have the playback network be the exact inverse of the recording compensation since each recording system is slightly different. However, optional guidelines can be applied. A typical audio range is from 20 Hz to 20 kHz, so there is a rolloff at both the low and high ends.

In the circuit in Fig. 5.8*b*, the linear rolloff is produced by dividing the playback network into three sections. The R_2C_2 section sets the 10-Hz point at 3 dB down from the 20-Hz point, the R_4C_4 section covers frequencies up to about 1 kHz, and the R_3C_3 network covers higher frequencies.

NAB Playback Equalization. Figure 5.8*d* shows the standard NAB equalization curves for tape use. Again, the recording curve is the inverse of the playback curve, so addition of the two produces a flat response. Likewise, the high frequencies are emphasized and the lows are attenuated. However, unlike the phono playback, tape playback tends to flatten out after about 3 to 4 kHz.

A different response is required for different tape speeds. Figure 5.8*d* shows the *playback* response curves for both 3¾ and 7½ inches per second (ips). Up to about 1 kHz, the curves are almost identical. Because there is only one frequency breakpoint (where the curve must start to flatten) for each tape speed, a simple *RC* compensation network is sufficient (instead of the multisecond network used for phono playback).

The breakpoint for 3¾ ips occurs at about 1.85 kHz, whereas the breakpoint for 7½ ips is at about 3.2 kHz. (The reactance of C_6 equals the resistance of R_5 at 1.85 kHz, whereas the reactance of C_7 equals the resistance of R_5 at 3.2 kHz.)

Note that the accuracy of both the RIAA and NAB compensation is only as good as the components used. Typically, resistors and capacitors with a 1 percent (or better) tolerance are used in discrete-component playback networks. Also, it may be necessary to trim the value to get an exact (or near exact) performance curve for truly good hifi performance. (Remember this if you must replace any component in any playback-equalization network.)

Because C_6 and C_7 block the dc path for the IC preamp feedback input, R_6 is added when the phono-tape switch is in either tape position. Although R_6 limits the preamp boost, since R_6 is across the tape compensation network, the circuit does provide about 15 dB of boost.

5.6.3 IC Amplifier Output Circuits

Figure 5.9 shows the amplifier output circuits. The power amplifiers for both channels are contained within one hybrid integrated circuit IC_{701}. Because of the power rating of 50 W, the IC is mounted on a heat sink.

The tone amplifiers that provide for treble and bass adjustment are also contained within a single integrated circuit IC_{606}. Both left and right channels use a subsonic filter consisting essentially of C_{618} and subsonic filter switch S_{602}. The subsonic filter attenuates frequencies below 20 Hz to reduce rumble caused by warped records or defective turntables. The audio output is monitored by protection circuits, under control of IC_{901}, as described in Sec. 5.7.

Audio from electronic volume control IC_{604} (Sec. 5.5) is applied to the noninverting inputs of IC_{606}. Bass control R_{634} and treble control R_{635} are connected between the output and inverting input of IC_{606}. Decreasing the bass or treble negative feedback has the effect of boosting the bass or treble, and vice versa.

The output of IC_{606} is coupled to the balance control and subsonic filter. Balance control R_{637} is connected between the right- and left-channel audio, with the wiper connected to ground. Moving the wiper changes the impedance of both channels simultaneously. R_{637} is usually set to provide equal signal levels in both channels.

With subsonic filter switch S_{602} in the off position, C_{618} is out of the circuit, and there is no attenuation of low-frequency signals. With S_{602} in the on position, C_{618} acts as a high-pass filter, attenuating all frequencies below 20 Hz. The output of the subsonic filter is applied to the noninverting input of IC_{701}.

An *RC* network is connected between the output of IC_{701} and the inverting input of IC_{701} to prevent oscillation. The audio output from IC_{701} is applied through R_{716}, which acts as the sensing resistor for the protection circuits (Sec. 5.7). The audio from R_{716} is applied to the speaker terminals and headphone jack.

5.6.4 Amplifier Output Troubleshooting

Before troubleshooting the audio output circuits, check that there is audio at pins 3 and 5 of IC_{606}. If there isn't, check the device-select circuits (Sec. 5.4) and electronic volume control circuits (Sec. 5.5).

Next, set the bass, treble, and balance controls to midrange. This simple act of faith has been known to cure a "no audio or weak audio in one channel" symptom.

If there is a total loss of audio in one or both channels, check the speaker switches first. Many "no audio" problems are solved instantly by setting a speaker switch to on.

Our amplifier has *separate speaker switches* in each channel, and there are two sets of outputs (A and B) for each channel. Of course, not all amplifiers have such a configuration, but the speaker switches should be checked first, in all cases.

Once you are sure that all switches and controls are properly set, check for audio at pins 1 and 7 of IC_{606}. If it is absent but there is audio at pins 3 and 5,

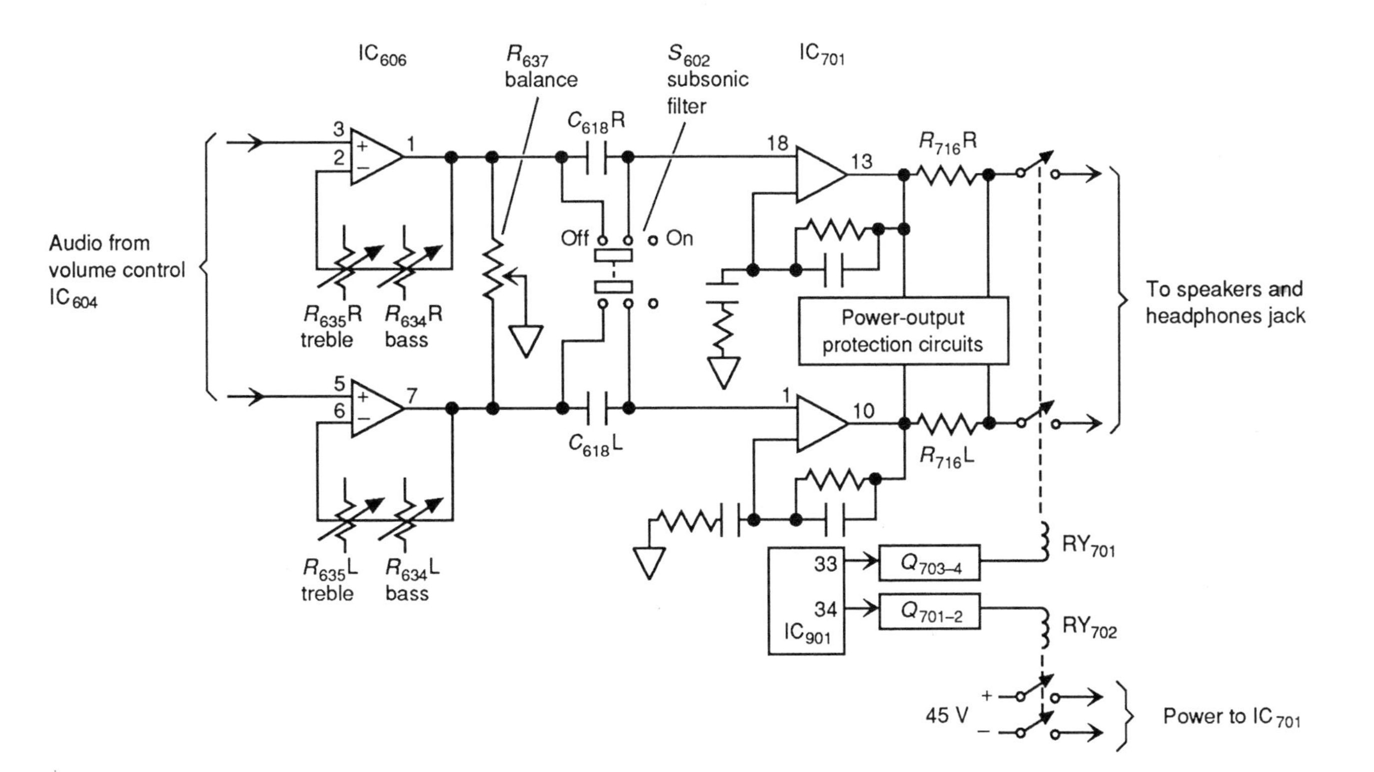

FIGURE 5.9 Amplifier output circuits.

suspect IC_{606}. Also check the bass and treble controls since they are in the negative-feedback path of IC_{606}.

If there is audio at pins 1 and 7 of IC_{606}, check for audio (at about the same level) at pins 1 and 18 of IC_{701}. If the audio level at IC_{701} is substantially different from the level at IC_{606}, trace the audio path through the subsonic filter.

The subsonic filter switch S_{602} setting should have little or no effect on the audio level, except at very low frequencies (below about 20 Hz). If you notice a drastic change in audio level at different settings of S_{602}, from about 1 kHz and up, look for problems in the subsonic filter circuit (such as leakage in C_{618}).

If there is audio at pins 1 and 18 of IC_{701} but not at pins 10 and 13, suspect IC_{701}. Before you pull IC_{701} (heat sink and all), make sure that the 45-V supply is applied to various IC_{701} terminals.

The 45-V (both plus and minus) supply is applied through RY_{702}. In turn, RY_{702} is turned on through Q_{701} and Q_{702} when pin 34 of IC_{901} goes low (when the amplifier power switch is pressed). Most of the other ICs receive operating power when the power cord is plugged in (whether the power button is pressed or not).

Because of the heavy current drain and high heat dissipation, power amplifier IC_{701} is turned on only during play. This is typical for most amplifiers where the final power stage is a single IC in the 40- to 50-W range. If the 45-V supply is absent at IC_{701}, suspect RY_{702}, Q_{701}, Q_{702}, or IC_{901} (check pin 34 of IC_{901} for a low).

If there is audio at pins 10 and 13 of IC_{701} but the audio does not reach the speakers or headphones, suspect RY_{701}, Q_{703}, or Q_{704}. Also check for a low at pin 33 of IC_{901}. Pin 33 should go low at the same time as pin 34.

Remember that the power-output protection circuits described in Sec. 5.7 are designed to cut off the audio circuit in the event of an overload. Defective protection circuits can cut off the audio, *even without an overload*.

5.7 OUTPUT-PROTECTION CIRCUITS

Figure 5.10 shows the output-protection circuits. The *overload-protection* circuit prevents damage to IC_{701} when a low-impedance or shorted speaker is connected. The *midpoint-potential* protection circuit prevents damage to the speakers in case of a defective IC_{701} and is turned on when a dc potential (usually called *dc offset*) is present at the output of IC_{701}. The *thermal-protection* circuit prevents damage to IC_{701} from excessive heat.

5.7.1 Thermal Protection

Thermal-protection switch S_{703} is mounted on the IC_{701} heat sink and is normally closed. This keeps D_{707} reverse biased and Q_{902} off. With Q_{902} not conducting, pin 3 of IC_{901} is high, and pin 33 of IC_{901} remains low to keep the speakers connected to the IC_{701} output.

If the temperature of the IC_{701} heat sink rises to 100°C, S_{703} opens, forward biasing D_{707}. This turns Q_{902} on and produces a low at pin 3 of IC_{901}. Under these conditions, pin 33 of IC_{901} goes high to disconnect the speakers from IC_{701}. Si-

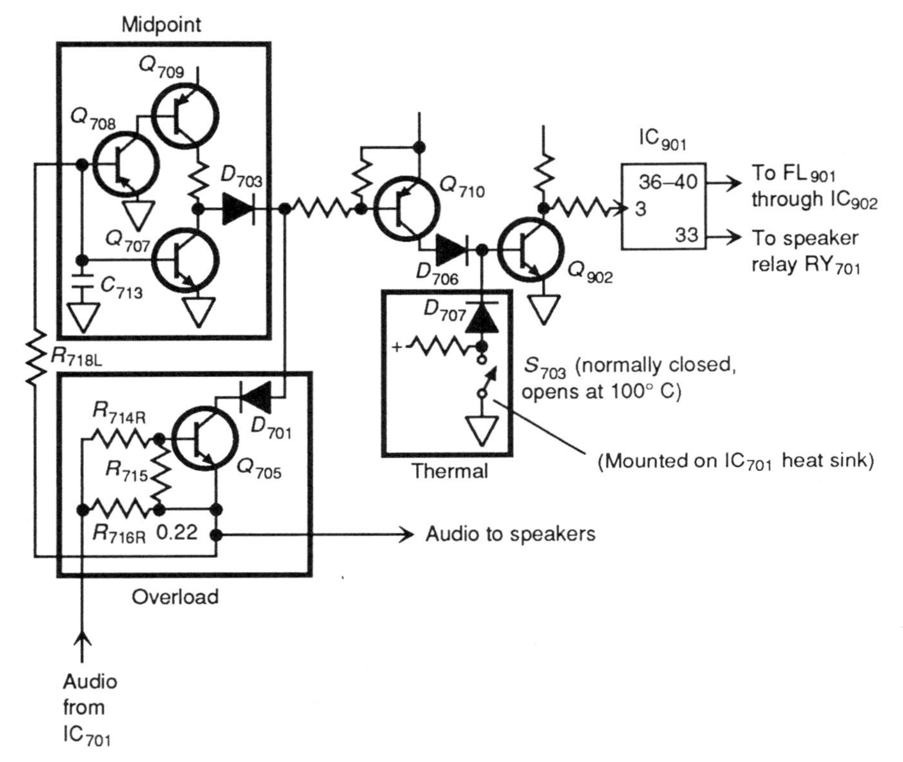

FIGURE 5.10 Amplifier output-protection circuits.

multaneously, pins 36 through 40 of IC_{901} are pulsed to flash the function-display portion of FL_{901}.

5.7.2 Midpoint-Potential Protection

The midpoint-potential protection circuit functions by monitoring the dc output from IC_{701}. In theory, there should be no dc output from IC_{701} to the speakers. (Excessive direct current can damage the speaker coils.) However, as a practical matter, there may be as much as ± 1.7 V at the IC_{701} output without damage to the speakers. The midpoint-potential protection circuit (called the *dc offset protection* circuit in some amplifiers) is turned on if the 1.7-V value is exceeded.

If there is any dc output from IC_{701} to the speakers, this potential causes C_{713} to charge through R_{718}L and R. C_{713} charges to the *average value* of the speaker voltage. During normal operation, with the dc output from IC_{701} less than ± 1.7 V, the midpoint-protection circuit is turned off.

If the average charge across C_{713} increases above ± 1.7 V, Q_{707} is turned on, forward biasing D_{703}. This applies a low to the base of Q_{710}, turning Q_{710} on and forward biasing D_{706}. This applies a high to the base of Q_{902}, turning Q_{902} on, and causes pin 3 of IC_{901} to go low. IC_{901} then produces a high at pin 33 to disconnect the speakers (and to pulse the FL_{901} function display) as discussed.

5.7.3 Overload Protection

The overload-protection circuit is the same for both channels, so only the right channel is covered here. Audio output from IC_{701} to the speakers is applied through R_{716R}, a 0.22-Ω resistor. This resistance is much smaller than the speaker load impedance (typically, 6 to 16 Ω).

During normal operation, the voltage across R_{716R} is very small. If a shorted or very low-impedance speaker is connected, excessive output current flows through R_{716R}, and the voltage across R_{716R} increases sharply.

Resistors R_{714R} and R_{715R} are connected as a voltage divider across R_{716R} to the base of Q_{705R}. As the current through R_{716R} increases (because of a short or low-impedance load), the voltage applied to Q_{705R} increases, turning Q_{705R} on. This forward biases D_{701}, turns on Q_{710}, forward biases D_{706}, turns on Q_{902}, and applies a low to pin 3 of IC_{901} to disconnect the speakers and pulse the FL_{901} display.

5.7.4 Output-Protection Troubleshooting

The front-panel function display FL_{901} should flash on and off, and the speakers should be disconnected when any one of the following occurs: FL_{701} becomes overheated (the IC_{701} heat sink reaches 100°C), the constant (no audio) dc voltage applied to the speakers exceeds ±1.7 V, or the speaker output line is shorted (or is at any impedance below that of the speakers).

Except for the low-impedance output, these conditions are difficult to simulate, making the circuits difficult to check. Also, if you do succeed in simulating any one of these conditions and the protection circuits are not functioning properly, you can damage the equipment (for example, burn out the speaker coil and/or overheat IC_{701}).

If you must check the circuits, try shorting the speaker lines (either L or R or both) to ground temporarily (*very* temporarily). Check that the function portion of FL_{901} flashes on and off and that the speakers are disconnected. If not, temporarily short pin 3 of IC_{901} to ground, and check for a flashing display with the speakers disconnected.

If the display flashes, and the speakers are cut with pin 3 of IC_{901} shorted but not when the speaker lines are shorted, suspect Q_{705}, Q_{707}, Q_{708}, Q_{709}, Q_{710}, and Q_{902}.

If the display does not flash and speakers are not disconnected with pin 3 of IC_{901} shorted, suspect IC_{901}.

You can also check that the anode of D_{707} is at ground (unless the IC_{701} heat sink is at 100°C or higher). If it is not, suspect that S_{703} is open.

You can also check the bases of Q_{707} and Q_{708}. Both bases should be 0 V (ideally) but may be at some potential less than ±1.7 V, without triggering the protection circuits. If the bases are at some value in excess of ±1.7 V, pin 3 of IC_{901} should go low, and the display should flash. If not, suspect Q_{707} through Q_{710} or Q_{902}.

5.8 TYPICAL TESTING AND ADJUSTMENTS

The procedures described in Chap. 4 are sufficient to test most present-day amplifiers. However, those procedures should be compared to those found in the service literature for the specific amplifier being serviced. Likewise, the service-literature adjustment procedures should be followed. (When all else fails, follow instructions.)

Note that the amplifier described in this chapter does not have any internal adjustment controls. This is true for many present-day IC audio amplifiers.

As a minimum, most amplifiers have bass, treble, and balance controls. In the absence of specific testing and adjustment procedures in the service literature, some points to consider when testing any audio amplifier follow.

5.8.1 Amplifier Tests

Operate the volume control for a midrange volume level or as specified in the service literature, and set the bass, treble, and balance controls to their midrange. Next, apply a 1-kHz sine-wave signal to each of the various inputs (phono, turner, CD, tape play, auxiliary, etc.).

Adjust the balance control until the outputs across the speakers (and/or tape record) are identical. If the amplifier has a front-panel level indicator, adjust the balance control until both channels show the same level indication.

If the balance control must be set far from the midrange to get equal output (with an identical signal at both inputs), there is a severe mismatch condition. This can be the result of problems in the balance circuits but is not limited to the balance network. For example, the problem can be a mismatch in IC_{701} (Fig. 5.9) if the problem is evident at all inputs of the amplifier. In such cases, it is necessary to replace IC_{701} as a package even though one channel may be good.

Of course, if there is a mismatch at only one input or one output of the amplifier, the problem can be pinned down easily. For example, if there is a mismatch at only the phono input, suspect IC_{401} (the phono preamp shown in Fig. 5.4). On the other hand, if the mismatch appears at only the tape record output, suspect IC_{601} (tape record buffer, Fig. 5.4).

If both channels produce essentially the same signal (with an identical signal at both inputs and with the balance control at midrange), the next step is to test the range of both the bass and treble controls.

Typically, the bass and treble controls have a ±8- to ±10-dB range, at some specific frequency. For example, the bass control of the amplifier described in this chapter has a ±8-dB range at 100 Hz, whereas the treble control has a ±10-dB range at 10 kHz. 50 Hz and 20 kHz are also common bass and treble frequencies.

Set both the bass and treble controls to midrange. Apply a 100-Hz signal to the inputs of both channels. Set the balance control so that the outputs of both channels are identical.

Keep the treble and balance controls at midrange. Vary the bass control from one extreme to the other. Note that the output of each channel varies about 8 dB above and below the output existing at the bass midrange setting.

Return both the bass and treble controls to midrange, and apply a 10-kHz signal to the inputs of both channels. Leave the balance control set so that both outputs are identical.

Keep the bass and balance controls at midrange. Vary the treble control from one extreme to the other. Note that the output of each channel varies about 8 dB above and below the output existing at the treble midrange setting.

If the amplifier passes the tests described here, it is reasonable to assume that the amplifier is functioning normally, and no troubleshooting is required.

5.8.2 Loudspeaker Tests

In addition to checking amplifier characteristics, the loudspeakers should also be checked. This brings up some problems. Although it is possible to test a loudspeaker for such characteristics as *sound pressure level* (SPL) under laboratory conditions, the most practical test is "by ear." Unfortunately, you and the customer have different ears, so the results are uncertain (at best).

To further complicate the speaker problem, some speakers are adjustable. For example, the speakers shown in Fig. 5.11 have volume controls (called *pads*, or potentiometers with more than one segment) in both the midrange and tweeters. Although these are not usually customer adjustment controls, the pads are often adjusted to some arbitrary setting "to match the customer's ear."

From a test or troubleshooting standpoint, make sure that the speaker adjustments (if any) can control volume at the corresponding speaker and that the controls are smooth (no abrupt changes in volume as the control is adjusted).

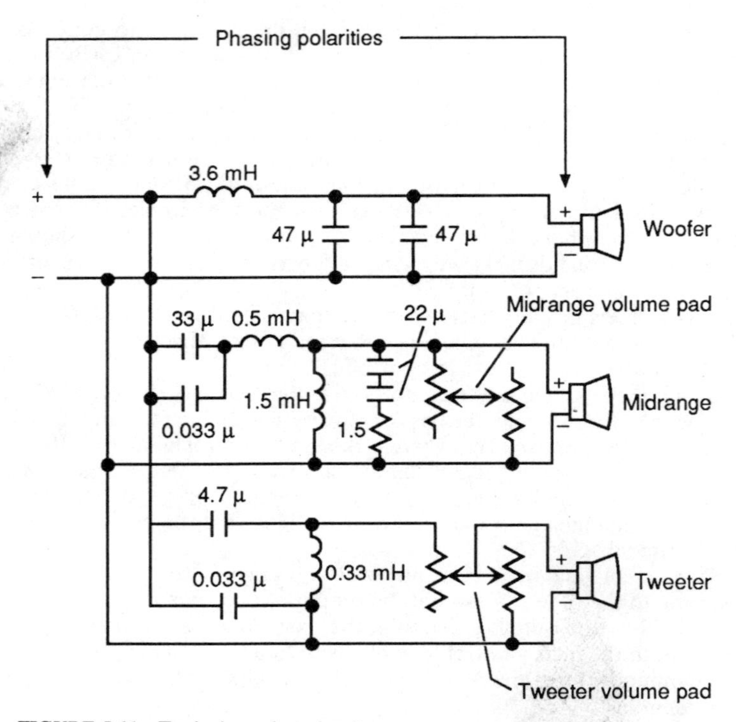

FIGURE 5.11 Typical speaker circuits.

5.9 PRELIMINARY TROUBLESHOOTING

Before you plunge into any amplifier circuit, here are some rather obvious but often overlooked checks that may cure mysterious problems.

If the amplifier does not turn on (either manually or by remote), make sure that the power cord is secure in the receptacle and that the receptacle has power.

If the amplifier operates manually but not by remote, make sure that the control cables are secure in the correct jacks. If all control cables are good, try replacing the remote-unit batteries.

If there is no sound, check that the speaker switches are set properly. Then check the input selector. It is possible that the wrong audio-source component has been selected or the component is defective. Try playing the amplifier with another source. It is also possible that the audio cables between the source and amplifier, or the speaker cables, are disconnected (even though the control cables are good). If the front-panel output power-level display is flashing, with no sound, the output-protection circuit has tripped.

If there is hum, the turntable may not be properly connected to the amplifier ground terminal (Fig. 5.1), or the audio-cable shields may be broken. It is also possible that the audio-cable connector may not be properly seated in the corresponding jack. If the cables, connectors, and ground terminals all appear to be good, check the installation notes in Sec. 5.1.4.

If the left or right speaker (only) is dead, temporarily reverse the left and right speaker leads. If the same speaker remains dead, the speaker is at fault. If the problem moves to the other speaker, one channel of the amplifier or input audio source is at fault. Temporarily reverse the left and right audio-source cables to the amplifier input. If the same speaker remains dead, the amplifier is probably at fault. If the other speaker goes dead, suspect the audio-source component.

CHAPTER 6
AM/FM TUNER AUDIO

This chapter describes the overall function, user controls, operating procedures, installation, circuit theory, typical test and adjustment procedures, and step-by-step troubleshooting for state-of-the-art AM/FM tuners found in audio systems.

6.1 OVERALL DESCRIPTION

Figure 6.1 shows the front and back of a composite AM/FM tuner. When this tuner is used with a system (modular home entertainment, stereo system, etc.), the AM or FM stations are selected automatically by the system remote control. For nonsystem configurations, the signal inputs are selected manually by means of front-panel push buttons. In this book, we are concerned with operation of the tuner in the stand-alone condition. Of course, the tuner must have an amplifier and loudspeakers.

6.1.1 Features

Quartz synthesized tuning system: Exact tuning is assured by the accuracy of the internal crystal oscillator and phase-locked-loop (PLL) circuit. These circuits select a preset frequency for the station chosen and fine tune for the best reception.

Signal strength indicator: The front-panel segmented light bar shows the strength of the signal tuned in.

Automatic stereo/mono switching: If the FM stereo broadcast signal is too weak for optimum reception, the tuner automatically switches from stereo to mono operating mode.

Sixteen-station random tuning memory: The tuner contains memory locations for 16 different stations (AM or FM), which may be programmed at random (via the front panel). The user may then select any one of these stations by entering the station preset number. The tuner remembers the preset stations indefinitely as long as power is connected. In case power is disconnected, the memory is retained for several days.

Preset channel scanning: If preset stations are programmed into the tuner memory, the user can scan up or down through the 16 preset stations to select the desired AM or FM station.

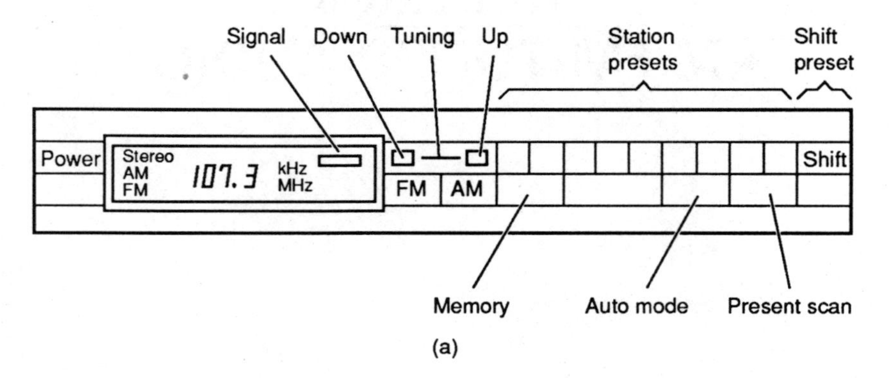

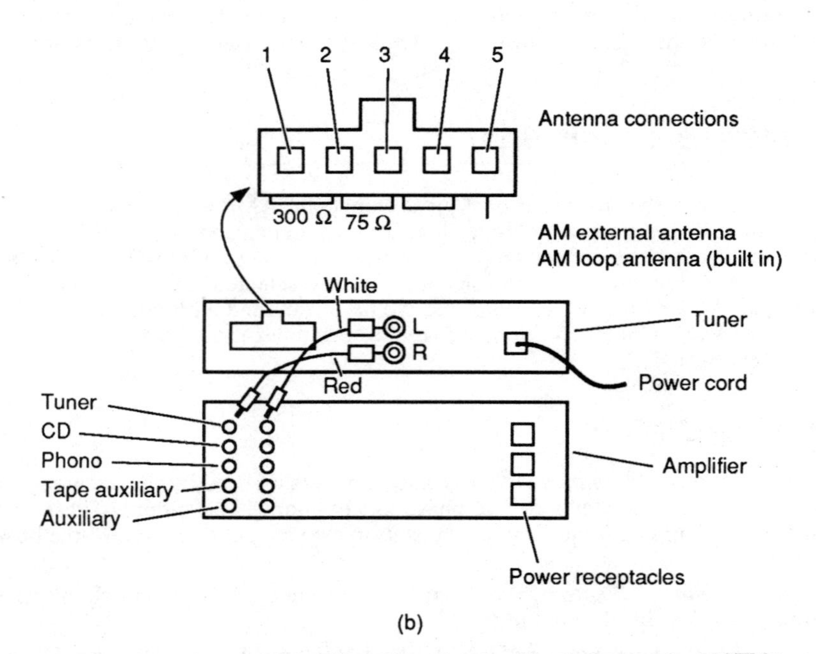

FIGURE 6.1 Operating controls, indicators, and connections for composite AM/FM tuner.

Last channel memory: The tuner remembers the last station selected and returns to that station when the tuner is again turned on. The last channel memory is retained as long as power is connected or, in case of power failure, for several days.

Search tuning (front panel): In the FM mode, while the tuner is in automatic mode, the user can search for a particular station by pressing the front-panel up/down controls. The tuning system automatically starts scanning at 200-kHz increments until a station is detected. At that time, the scanning function stops and waits for the user to enter an additional command.

6.1.2 Operating Controls and Indicators

The following is a summary of the tuner stand-alone front-panel operating controls and indicators, shown in Fig. 6.1*a*. When used in a system, many of these functions can be controlled remotely.

Press power to turn the tuner on and off. None of the other controls function when power is off.

Press the up side of the tuning bar to tune the next-higher frequency. Press the down side of the tuning bar to select the next-lower frequency.

When AM is selected, press tuning up/down momentarily to change the tuning frequency by 10 kHz. Hold the tuning bar to change the tuning frequency rapidly to the desired station. Audio is muted while the tuning up/down bar is pressed.

When FM is selected, the auto mode button determines the method of tuning in the FM band. With auto mode on, pressing the tuning up/down bar causes the tuner to seek the next-higher or next-lower frequency. With auto mode off, the tuner operates as in the AM mode, except that each station interval is 0.2 MHz (instead of 10 kHz). Also, the tuner output is mono audio, even if the station tuned in is broadcasting in stereo.

Preset stations may be selected using one of the preset buttons in conjunction with the shift button. Press shift to toggle between station presets 1 through 8 and 9 through 16. An indicator turns on, showing that the preset buttons are assigned to 1 through 8 or 9 through 16.

Press FM to change the tuning mode from AM to FM. It is not necessary to press FM when using the preset buttons after the presets are programmed.

Press AM to change the tuning mode from FM to AM. It is not necessary to press AM when using the preset buttons after the presets are programmed.

It is necessary to use the memory button when programming the preset memory. First select AM or FM with the corresponding buttons. Then tune in the desired station with the tuning up/down button. (The auto mode function must be on to tune FM stereo stations.) Press memory. The display should show the word *memory* for 5 s. Press the desired preset button within this 5-s interval to program the tuned station into memory. Repeat the programmed process for each preset position.

Pressing preset scan causes the preset stations to be scanned. Each station is auditioned for 3 s before scanning continues. Release preset scan when the desired station is reached.

The fluorescent display indicates the frequency tuned, the band selected, and the signal strength of the tuned station. The preset position is also displayed prior to the tuned frequency when the preset buttons are used.

6.1.3 Connections

Figure 6.1*b* shows typical connections for the tuner. External connections between the tuner and amplifier are made from the back of the tuner. A pin cord is supplied with the tuner (and sometimes the amplifier) for connections between left and right stereo outputs, and the corresponding inputs on the amplifier. Although the connections are very simple, certain precautions must be observed for all tuners.

Look for any color coding the the pin cord. Typically, red is used for the right channel, while white is used for the left channel (but do not always count on it). Also, look for any ground terminals or leads that are part of the pin cord.

The amplifier described in Chap. 5 has ac power receptacles for a number of external components. It may be necessary to use a standard polarized wall outlet for the tuner if the amplifier is not provided with power outlets.

Do not connect the tuner to the CD, aux, phono, or tape play inputs of an amplifier. Instead, always use the tuner input (or whatever the tuner input is called on the amplifier).

Generally, there is no damage if you connect the tuner to the wrong input. (A possible exception is the phono input.) However, the tuner output does not match the other audio components. For example, the typical 500- to 600-mV output of a tuner is too high in comparison to the typical 5-mV output of a turntable and too low for the typical 2-V output of a CD player. You can try a really novel approach and use the input specified in the service literature.

Figure 6.1*b* also shows the antenna connections for a typical tuner. Note that five terminals are involved. In our tuner, terminals 1 and 2 are for a 300-Ω FM antenna lead-in (or balun, TV/FM splitter, or other balanced FM signal source). Note that some tuners are provided with an FM dipole for indoor reception.

Terminals 2 and 3 are for a 75-Ω unbalanced or coax lead-in (center wire on terminal 2, shield on terminal 3). Terminals 3 and 4 are for a built-in AM loop antenna. Terminals 3 and 5 are for an external AM antenna. Use terminal 3 for ground and terminal 5 for the single lead of the external AM antenna. Do not remove the loop from terminals 3 and 4 when using the external antenna.

On some older tuners, the *power cord is used as the antenna*. However, this is not the case with present-day AM/FM tuners.

6.2 RELATIONSHIP OF TUNER CIRCUITS

Figure 6.2 shows the relationship of the tuner circuits. IC_{301} is both an amplifier and multiplex decoder and is used in the audio path for AM and FM. Once audio reaches pin 2 of IC_{301}, whether from the AM or FM section, IC_{301} produces corresponding audio at pins 6 and 7. IC_{301} has only one adjustment control. This is the multiplex VCO adjust potentiometer R_{305} connected at pin 15.

6.2.1 FM Signal Processing

FM broadcast signals are applied to FM tuner package MD_{101}. The IF output of MD_{101} is applied to the FM amplifier and detector IC_{201} through amplifiers Q_{101} and Q_{102} and ceramic filters MF_{201} and MF_{202}. This amplifier-filter combination removes any amplitude modulation and passes only signals of the desired frequency.

IC_{201} has three adjustments: muting-level adjustment R_{202}, S-curve null-point adjustment T_{201}, and FM-distortion adjustment T_{202}.

The audio output from the FM section is applied to multiplex decoder IC_{301} through Q_{201} and Q_{202}. IC_{301} functions as a mono amplifier when there is no FM stereo present or when the stereo signal is too weak to produce proper FM.

6.2.2 AM Signal Processing

AM broadcast signals are applied to AM buffer-amplifier Q_{151} through an RF circuit tuned by one section of D_{153}. The output of Q_{151} is applied to AM IF

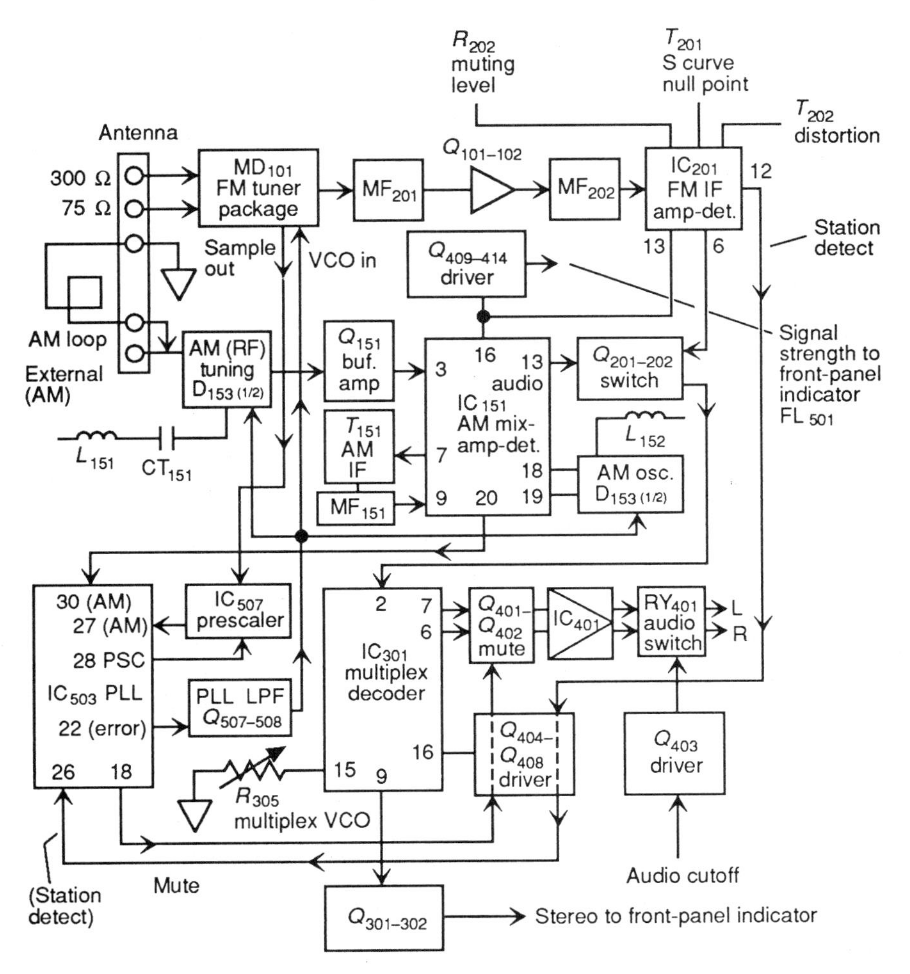

FIGURE 6.2 Relationship of tuner circuits.

amplifier-detector IC_{151}. The mixer output of IC_{151} is tuned by AM IF adjustment T_{151} and applied to the IF portion of IC_{151} through ceramic filter MF_{151}.

The audio output from the detector of IC_{151} is applied to multiplex decoder IC_{301} through Q_{201} and Q_{202}. Note that IC_{301} functions as a mono amplifier when the AM mode is selected. Since there is no stereo signal present in the AM mode, IC_{301} shifts to mono operation, just as when there is no FM stereo or the stereo is weak.

6.2.3 FM Tuning

The FM section is tuned to the desired frequency (and locked to that frequency) by signals applied to MD_{101}. The VCO signal (a variable dc voltage, sometimes called the *error voltage*) from PLL microprocessor IC_{503} is applied to MD_{101} through Q_{507} and Q_{508}. The IC_{501} error voltage shifts the MD_{101} oscillator as necessary to tune across the FM broadcast band or to fine tune MD_{101} at a selected station.

The frequency produced by MD_{101} is sampled and applied to the FM input of IC_{503} through prescaler IC_{507}. (Operation of the prescaler is described in Sec. 6.4.) The sampled signal serves to complete the FM tuning loop. For example, if MD_{101} drifts from the frequency commanded by IC_{503}, the error voltage from IC_{503} changes the MD_{101} oscillator as necessary to bring MD_{101} back on frequency.

A station-detect signal is produced at pin 12 of IC_{201}. This signal is used to tell IC_{503} that an FM station has been located and that the station has sufficient strength to produce good FM operation.

In between FM stations, or when the FM station is weak, pin 12 of IC_{201} goes high, turning on Q_{406}. This causes pin 26 of IC_{503} to go low and causes the audio to be muted.

When there is an FM station of sufficient strength, pin 12 of IC_{201} goes low, turning Q_{406} off and causing pin 26 of IC_{503} to go high. Under these conditions, the audio is unmuted, and IC_{503} fine tunes the FM section for best reception of the FM station. The muting level is set by R_{202}.

6.2.4 AM Tuning

The AM section is also tuned to the desired frequency (and locked to that frequency) by signals from IC_{503} applied to the two sections of diode D_{153}. This is the same error voltage applied to the VCO of FM tuner MD_{101}. The IC_{503} error voltage shifts the RF and oscillator circuits as necessary to tune across the AM broadcast band.

The oscillator circuit of IC_{151} is adjusted by L_{152}, while the RF input circuit is tuned by tracking adjustments L_{151} and CT_{151}.

The frequency produced by IC_{151} is sampled and applied to the AM input of IC_{503}. The sampled signal serves to complete the AM tuning loop. For example, if the AM section drifts from the frequency commanded by IC_{503}, the error voltage from IC_{503} changes both circuits controlled by D_{153} as necessary to bring the AM section back on frequency.

6.2.5 Output Display

The outputs of both IC_{201} (FM) and IC_{151} (AM) are applied to front-panel fluorescent display FL_{501} through Q_{409} through Q_{414}. These outputs indicate the relative signal strength of AM or FM broadcast signals. Also note that the FM multiplex decoder IC_{301} senses an FM stereo signal of sufficient strength. IC_{301} applies a signal to the stereo display of FL_{501}.

6.3 KEYBOARD AND DISPLAY CIRCUITS

Figure 6.3 shows the keyboard and display circuits. IC_{503} provides the PLL synthesized quartz tuning as well as an interface between the front-panel keyboard

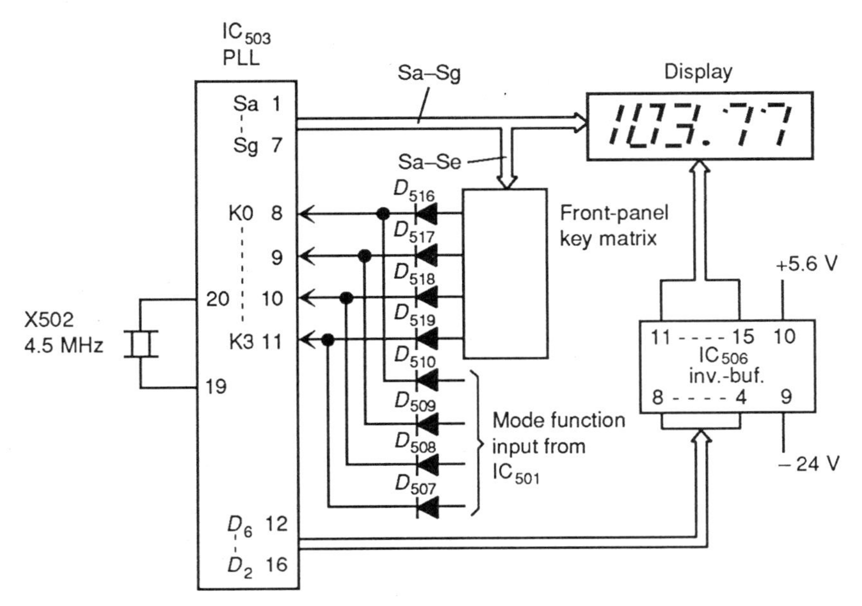

FIGURE 6.3 Tuner keyboard and display circuits.

and display circuits and the tuner. All of these functions are under control of 4.5-MHz clock pulses developed by X_{502} connected at pins 19 and 20 of IC_{503}.

IC_{503} generates *segment-drive* pulses Sa and Sg at pins 1 through 7. These pulses are applied to the front-panel fluorescent display (Sec. 6.9). Portions of the pulses (Sa and Se) are also applied to the front-panel keyboard matrix. IC_{503} also generates *digit-drive* pulses D_2 through D_6 at pins 12 through 16. These pulses are applied to the display through IC_{506}.

Mode-selection scanning pulses from the front-panel keyboard (developed by the Sa and Se drive pulses) are returned to IC_{503} through D_{516} through D_{519}. Mode-selection inputs from the remote are applied through IC_{501} and D_{507} through D_{510}. No matter what the source, IC_{503} responds to the mode-select input pulses K_0 through K_3 at pins 8 through 11 by instructing the tuner circuit to perform the selection function (power on, channel change, etc.). Simultaneously, the selected functions shown on the front-panel display by means of the Sa and Sg and D_2 and D_6 drive signals.

6.3.1 Keyboard and Display Troubleshooting

If the tuner fails to respond to a command, either from the front panel or remote, make the following checks. Confirm the presence of Sa and Sg pulses at IC_{503} and the front-panel display. If any of the pulses are missing, suspect IC_{501}. If all of the pulses are missing, suspect X_{502}. Check for 4.5-MHz clock pulses at pin 19 of IC_{503}.

If the Sa and Sg pulses are all present, check for pulses at the K_0 and K_3 inputs of IC_{503} when front-panel keyboard buttons are pressed. If the K_0 and K_3 pulses are absent or abnormal, suspect the keyboard matrix switches and/or D_{516} and

D_{519}. If the K_0 and K_3 pulses are present at the correct input of IC_{503} but the tuner does not respond, suspect IC_{503}.

If the tuner responds to the front-panel commands but not to remote commands, suspect IC_{501} and/or D_{507} through D_{510}. *If the tuner fails to respond to an on-off command*, refer to Sec. 6.5. *If the tuner fails to respond properly to a channel-change* command, refer to Sec. 6.4. *If the tuner fails to respond properly to an AM/FM select command*, refer to Sec. 6.6. *If the tuner responds properly to commands but the response is not shown on the front-panel display*, refer to Sec. 6.9.

6.4 FREQUENCY SYNTHESIS TUNING

Frequency synthesis, or FS, tuning provides for convenient push-button or preset AM and FM station selection, with automatic station search or scan, and automatic fine-tune (AFT) capability.

The key element in any FS system is the PLL that controls the variable-frequency oscillator (VFO) and/or RF tuning, as required for station selection and fine tuning. Note that the PLLs used in the AM/FM tuners of audio systems are essentially the same as those used in the FS tuners of TV sets and VCRs.

6.4.1 Elements of an FS System

Figure 6.4 shows the elements of an FS system. *Phase locked loop* (Fig. 6.4*a*) is the term used to designate a frequency-comparison circuit in which the output of a VFO is compared in frequency and phase to the output of a very stable (usually crystal controlled) fixed-frequency reference oscillator. Should a deviation occur between the two compared frequencies or should there be any phase difference between the two oscillator signals, the PLL detects the degree of frequency or phase error and automatically compensates by tuning the VFO up or down in frequency or phase until both oscillators are locked to the same frequency and phase.

The accuracy and frequency stability of a PLL circuit depends on the accuracy and frequency stability of the reference oscillator (and on the crystal that controls the reference oscillator). No matter what reference oscillator is used, the variable-frequency oscillator of most PLL circuits is a VCO where frequency is controlled by an error voltage.

6.4.2 AM Section

Figure 6.4*b* shows how the PLL principles are applied to the AM section of a typical audio-system tuner. The 1-kHz reference oscillator in Fig. 6.4*a* is replaced by a reference signal, obtained by dividing down the PLL IC_{503} clock (4.5 MHz). This reference signal is applied to a phase comparator within IC_{503}. The other input to the phase comparator is a sample of the AM signal at pin 30 of IC_{503} (taken from pin 20 of IC_{151}).

The output of the phase comparator is an *error signal* or *tuning-correction voltage* applied through low-pass filter Q_{507} and Q_{508} to the AM tuning circuits. The filter acts as a buffer between the comparator and tuning circuits.

Note that the AM sample is applied to the phase comparator through a *pro-*

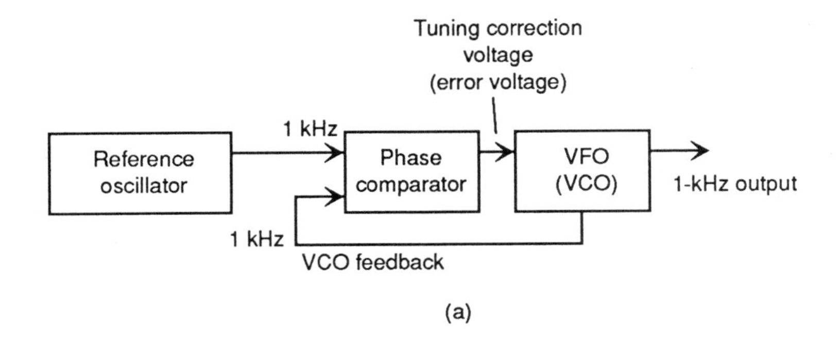

(a)

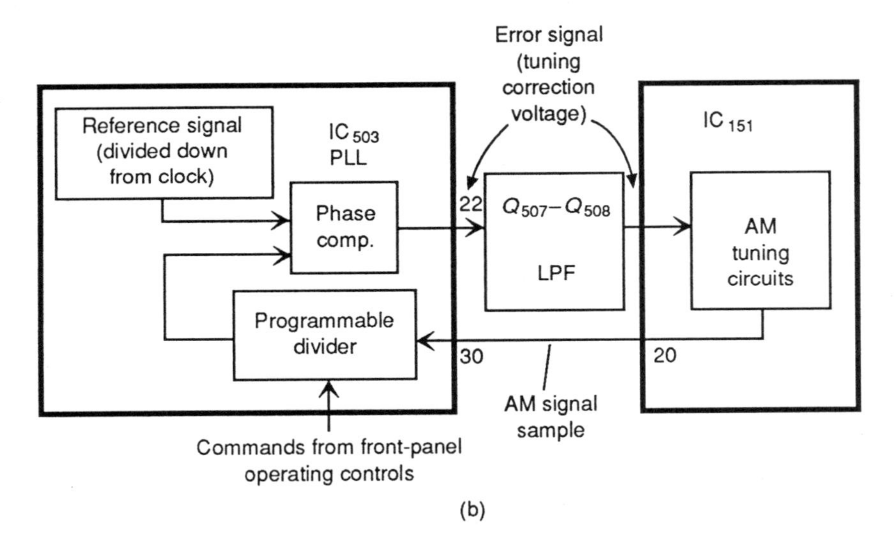

(b)

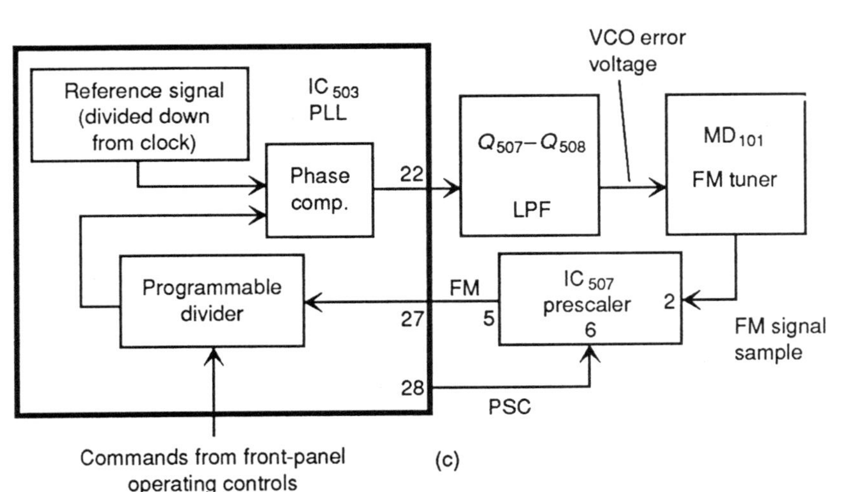

(c)

FIGURE 6.4 Elements of an FS system. (*a*) PLL; (*b*) AM; (*c*) FM.

grammable divider or *counter*. The division ratio of the programmable divider is set by commands from the front-panel operating controls (tuning up/down, preset scan, etc.). In effect, the divider is programmed to divide the AM sample by a specific number.

The variable-divider function makes possible many AM local-oscillator frequencies. An AM frequency change is done by varying the division ratio with front-panel commands. This produces an error signal that shifts the tuning circuits until the AM signal (after division by the programmable divider) equals the reference-signal frequency, and the tuning loop is locked at the desired frequency.

6.4.3 FM Section

Figure 6.4*c* shows the PLL circuits for the FM section of a typical audio-system tuner. This circuit is similar to the PLLs found in TV sets and VCRs, in that a *prescaler* is used. The system in Fig. 6.4*c* is generally called an *extended PLL* and holds the variable-oscillator frequency to some harmonic or subharmonic of the reference oscillator (but with a fixed phase relationship between the reference and variable signals).

The PLL in Fig. 6.4*c* uses a form of *pulse-swallow control*, or PSC, that allows the division ratio of the programmable divider to be changed in small steps. As in the case of AM, the division ratio of the programmable divider is determined by commands applied to IC_{503} from the front-panel controls.

The PSC system uses a very high-speed prescaler IC_{507}, also with a variable division ratio. The division ratio of the prescaler is determined by the PSC signal at pin 28 of IC_{503} and can be altered as required to produce subtle changes in frequency needed for optimum station tuning (fine tuning) in the FM mode.

The PSC signal at pin 28 of IC_{503} is a series of pulses. As the number of pulses increases, the division ratio of the prescaler also increases. When a given FM station or frequency is selected by the front-panel controls, the number of pulses on the PSC line is set by circuits in IC_{503} as necessary for each station or frequency.

The overall division ratio for a specific FM station or frequency is the prescaler division ratio, multiplied by the programmable-divider division ratio. The result of division at any FM station or frequency is a fixed output to IC_{503} when the FM tuner is set to the desired frequency.

6.4.4 FS Troubleshooting

A failure in the PLL IC or in the FS circuits controlled by the PLL can cause many trouble symptoms. Unfortunately, a failure in other circuits can cause the same symptoms. For example, assume that you operate the tuning up/down buttons and see that the front-panel frequency display varies accordingly, but the AM or FM section does not tune across the corresponding band (no stations of any kind are tuned in). This can be caused by a PLL failure or by a failure of the commands to reach the PLL IC_{503} (even though the commands are displayed), which creates special troubleshooting problems for AM/FM tuners with PLL.

Common FS-Tuning Failure Symptoms. The most common symptoms for failure of the PLL are a combination of *no stations received* (on both AM and FM) and *noisy audio* (audio not muted when you tune across the broadcast band). Of course, not all tuners have both auto and mono modes, as does our tuner, and not all muting circuits operate in exactly the same way. However, the following approach can be applied to the basic PLL problem without regard to the exact method of PLL tuning.

Operating Control Checks. First, make certain that the operating controls are properly set for a particular PLL function. For example, on our tuner, you must be in the auto mode (auto mode button pressed, auto mode indicator on) before the PLL seeks FM stations as you tune across the FM band with the tuning up/down button.

If you have selected mono (auto mode indicator off) or the circuits have gone into mono because of a failure, the PLL tunes across the FM band in 200-kHz increments whether the stations are present or not. Always look for some similar function on the tuner you are servicing.

When you are certain that the controls are set properly and that there is a true malfunction, the next step in troubleshooting PLLs is to isolate the problem to the tuning circuit or the PLL circuits. The two basic approaches are discussed in the following sections.

Error Voltage Checks. The first approach to troubleshooting PPLs is to apply a frequency-change command to the PLL and see if the error voltage and/or sample voltage changes accordingly. For example, to check the AM section of our tuner (Fig. 6.2), press the tuning up/down buttons and see if the voltage at pin 22 of IC_{503} changes as the frequency display changes. If not, suspect IC_{503} or the circuits between the front-panel tuning up/down buttons (key matrix) and IC_{503} (Fig. 6.3).

If the error voltage at pin 22 of IC_{503} changes, check that the frequency of the signal at pin 30 of IC_{503} (or pin 20 of IC_{151}) also changes as the frequency display changes. If not, suspect the tuning circuit D_{153}, Q_{151}, and/or low-pass filter Q_{507} and Q_{508} (Fig. 6.2).

Next, press the tuning up/down buttons while in the FM mode and see if the error voltage at the VCO input of MD_{101} changes. If not, suspect IC_{503} and/or IC_{507}. You can also check that the PSC pulses at pin 28 of IC_{503} change, but this is usually more difficult to monitor.

If the error voltage applied to the VCO input of MD_{101} changes, check that the frequency of the sample-frequency output of MD_{101} and pin 27 of IC_{503} change as the front-panel frequency display changes. If not, suspect MD_{101}. If the frequency does change at MD_{101} and IC_{507}, but not at pin 27 of IC_{503}, suspect IC_{507}. (It is also possible that IC_{507} is not receiving proper PSC pulses from pin 28 of IC_{503}.)

Substitute Tuning Voltage. An alternative technique for troubleshooting PLL and FS circuits is to apply a substitute tuning voltage, or error voltage, to the tuning circuits and see if the circuits respond by producing the correct frequency. Although this sounds simple, here are some considerations.

First, you must make certain that the substitute tuning voltage is in the same range as the error voltage. For example, the error voltage in our tuner varies from about 1 to 20 V (at the tuning circuits). You can cover this range with a typical shop-type variable dc supply. However, the shop supply can possibly load the tuning circuit with unwanted impedance, reactance, and so on.

Remember that if you apply a lower voltage, the circuits will not respond properly. If you apply a voltage higher than the tuning-circuit range, the circuits can be damaged.

Although a number of tuners use circuits similar to our tuner (the MD_{101} package is quite common in present-day audio-system tuners), the tuning circuits are not the same for all tuners. Some tuners combine the AM and FM functions in a single package. Of course, if you are lucky, you can find the error-voltage range in the service literature, often in the adjustment chapter.

Station-detect. If the tuner has a station-detect function (most tuners do, at least in the FM section), you can use this feature together with a substitute tuning voltage to isolate PLL problems. Simply vary the substitute tuning voltage across the range and see if stations are detected. For example, in our tuner (Fig. 6.2), you can check at pin 12 of IC_{201} and/or pin 26 of IC_{503} for a change of status each time a station is tuned in and out. Pin 26 of IC_{503} should go high and pin 12 of IC_{201} should go low each time an FM station of sufficient strength is tuned in. The status of the pins should reverse when the station is tuned out (in between stations).

6.5 AUDIO OUTPUT-SELECT CIRCUITS

Figure 6.5 shows the audio output-select circuits. These circuits select the AM or FM audio for the input to multiplex decoder-amplifier IC_{301}.

When FM is selected, FM B+ is applied to Q_{202}, turning Q_{202} on. Since there is no B+ applied to Q_{201}, the AM audio does not pass. However, FM audio present at pin 6 of IC_{201} is passed to pin 2 of IC_{301} through Q_{202}, C_{301}, and R_{301}.

When AM is selected, Q_{201} is turned on, Q_{202} is turned off (blocking the FM audio), and AM audio present at pin 13 of IC_{151} is passed to pin 2 of IC_{301} through Q_{201}, C_{301}, and R_{301}.

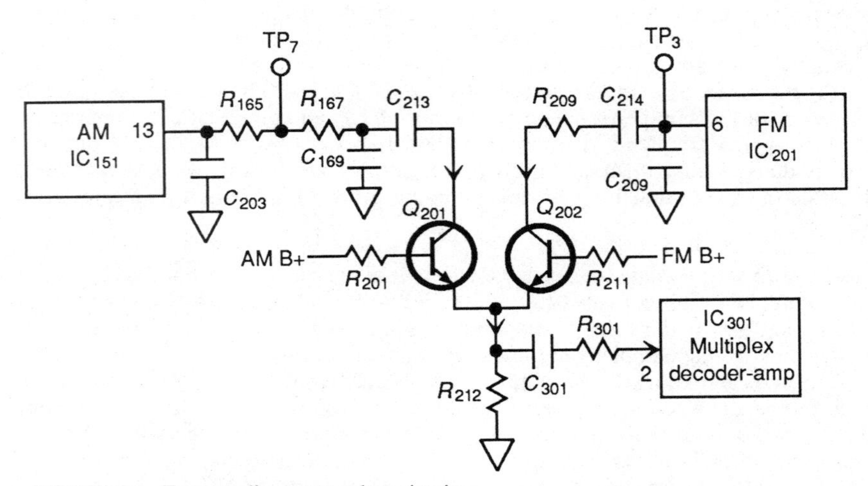

FIGURE 6.5 Tuner audio output-select circuits.

6.5.1 Audio Output-Select Troubleshooting

If you get AM audio but no FM audio, it is fair to assume that the circuits from R_{212} through to pin 2 of IC_{301} (and beyond to the audio-output terminals) are good. The first place to trace FM audio is at pin 6 of IC_{201} (TP_3) and at the emitter of Q_{202}.

If there is no audio at pin 6 of IC_{201} (or TP_3), suspect the FM circuits. Check the FM PLL circuits as described in Sec. 6.4. If necessary, go through the FM adjustments described in the service literature (or as covered in Sec. 6.9).

If there is audio at pin 6 of IC_{201} (or TP_3) but not at the emitter of Q_{202}, suspect C_{209}, C_{214}, R_{209}, R_{211}, and Q_{202}. Make certain that Q_{202} is turned on by FM B+. (The base of Q_{202} should be about 3 or 4 V.)

If there is FM audio at pin 2 of IC_{301} but no audio at the tuner-output terminals, suspect the muting circuits (Sec. 6.7) or possibly the audio-control circuits (Sec. 6.6).

If you get FM audio but no AM audio, it is fair to assume that the circuits from R_{212} through to pin 2 of IC_{301} (and beyond to the audio-output terminals) are good. The first place to trace AM audio is at pin 13 of IC_{151}, TP_7, and at the emitter of Q_{201}.

If there is no audio at pin 13 of IC_{151} or TP_7, suspect the AM circuits. Check the AM PLL circuits as described in Sec. 6.4. If necessary, go through the AM adjustments described in the service literature (or as covered in Sec. 6.9).

If there is audio at pin 13 of IC_{151} but not at TP_7, suspect C_{203} and R_{165}.

If there is audio at TP_7, suspect C_{169}, C_{213}, R_{167}, R_{212}, and Q_{201} if the audio does not appear at the emitter of Q_{201}. Make certain that Q_{201} is turned on by AM B+. (The base of Q_{201} should be about 3 or 4 V.)

If there is AM audio at pin 2 of IC_{301} but no audio at the tuner-output terminals, suspect the muting circuits (Sec. 6.7) or possibly the audio-control circuits (Sec. 6.6).

6.6 AUDIO-CONTROL CIRCUITS

Figure 6.6 shows the audio-control circuits. IC_{501} provides interface between the tuner circuits and the remote or system control. The remote-system input is at pin 39, while the output is at pin 37 of IC_{501}. The output signal is buffered by Q_{506} and then applied to the input of driver Q_{506}. The remote-system line (also called a control bus) is tied to +5.6 V through a pull-up resistor. Q_{506} pulls the line low and lets the line return to high (to send data).

During the period when the tuner is not receiving or transmitting data (from the remote or to the system control), the signal on the remote-system line is at a high (about 3 to 4 V). Zener ZD_{501} helps reduce the possibility of damage to the system in the event of static discharge or excessive voltage (resulting from the wrong equipment being applied to the remote-system line).

Remote commands are passed to PLL IC_{503} through the remote-system line and IC_{501}. In turn, IC_{503} sends status back to the system control through IC_{501} and the line. This status information is sent to the system control whether the information represents a change caused by the remote or front-panel buttons.

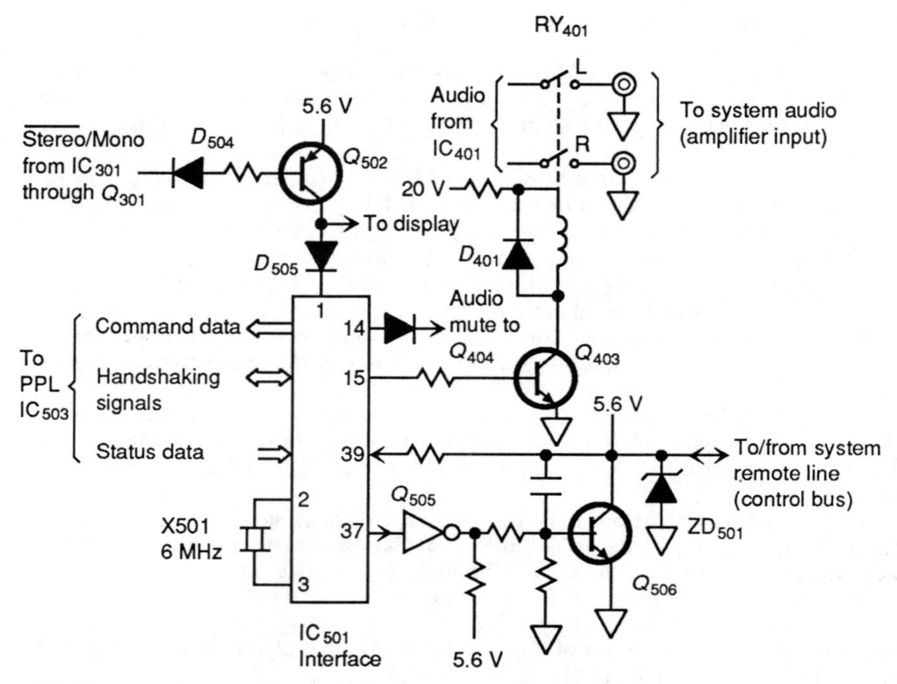

FIGURE 6.6 Tuner audio-control circuits.

6.6.1 Stereo/Mono-Select Circuits

A stereo/mono logic signal from the FM stereo decoder IC_{301} is routed to Q_{502} through Q_{301} and D_{504}. During stereo reception, the line from IC_{301} is low, pulling the base of Q_{502} low. This turns Q_{502} on and applies about 5.2 V to the stereo indicator of the front-panel display.

If mono is selected or if the system automatically switches to mono, the anode of D_{504} goes high, turning Q_{502} off. This turns off the front-panel stereo display. The low also reverse biases D_{505}, applying a low to pin 1 of IC_{501}. The stereo/mono signal applied to pin 1 of IC_{501} is passed to the system control through IC_{501} and the remote-system line as part of the status transmission.

6.6.2 Audio-Muting Circuits

An audio-muting signal is applied to the muting circuits (Sec. 6.7) from pin 14 of IC_{501} through D_{403}. This mute signal is high during periods when the PLL tuning system is selecting stations, during preset-scan conditions, between station changes, and during the time when the tuner is on but is not being used as the audio source (say in an audio system where the tape deck or CD player is being used). As discussed in Sec. 6.7, the high mutes audio from the tuner to the audio-system amplifier.

6.6.3 Audio-Control

Audio from the tuner is applied to the audio system through relay RY_{401}. When the tuner is off, pin 15 of IC_{501} is low. This turns off Q_{403}, deenergizes RY_{401}, and opens the path between the tuner audio output and the audio system line (both R and L). When the tuner audio output is selected (either by the remote or front-panel buttons), pin 15 of IC_{501} goes high. This turns on Q_{403}, energizes RY_{401}, and connects the tuner audio output to the audio system line. Note that when a different component is selected to provide the audio, the tuner remains energized, but pin 15 of IC_{501} stays low, preventing tuner audio from reaching the audio system.

6.6.4 Audio-Control Troubleshooting

If the tuner does not respond to front-panel commands, suspect the keyboard matrix circuits of IC_{503} (Sec. 6.3).

If the tuner operates properly from the front panel but not the remote, first check that other components can be operated from the remote. (The problem may be with the remote unit or with the system control functions.)

Next check that there are commands being applied to pin 39 of IC_{501} when the remote is operated. If not, suspect the remote or system-control functions. If there are commands at pin 39 of IC_{501} but there is no tuner response or the response is not correct, suspect IC_{501} or X_{501} (which is plug-in on our tuner). Check for 6-MHz clock signals at pins 2 and 3 of IC_{501}.

If the tuner responds properly to commands but does not transmit status information to system control, check for status signals at pin 37 of IC_{501}. If they are absent, suspect IC_{501}. If commands are present at pin 37 of IC_{501} but no status information is applied to system control, suspect Q_{505}, Q_{506}, ZD_{501}, or the remote-system line.

If the tuner appears to be operating properly but there is no tuner audio on the audio-system lines, first check that pin 14 of IC_{501} goes high during power-up, channel changes, preset-scan, or when the tuner is not selected as the audio component but then goes low at all other times when the tuner is on. If not, suspect IC_{501} and/or D_{403}. Then check the muting circuits as discussed in Sec. 6.7.

If the tuner operating circuits and muting circuits are good but there is no audio, check for a high at pin 15 of IC_{501}. If it is missing, suspect IC_{501}. If pin 15 is high but there is no audio, suspect Q_{403}, RY_{401}, or D_{401}.

6.7 AUDIO-OUTPUT AND MUTING CIRCUITS

Figure 6.7 shows the audio-output and muting circuits. Note that there are four signals used to mute audio output from the tuner. Similarly, the effect of these muting signals is changed, depending on the operating mode selected. For example, when an FM signal of sufficient strength is tuned in during auto mode, pin 12 of IC_{201} goes low. This low turns Q_{406} off and Q_{407} on. With Q_{407} on, Q_{404}, Q_{408}, Q_{401} and Q_{402} are turned off to unmute the audio.

When the FM signal is weak or between FM stations, pin 12 of IC_{201} goes high. The high is applied to Q_{406} and Q_{407} (connected as a Schmitt trigger). In

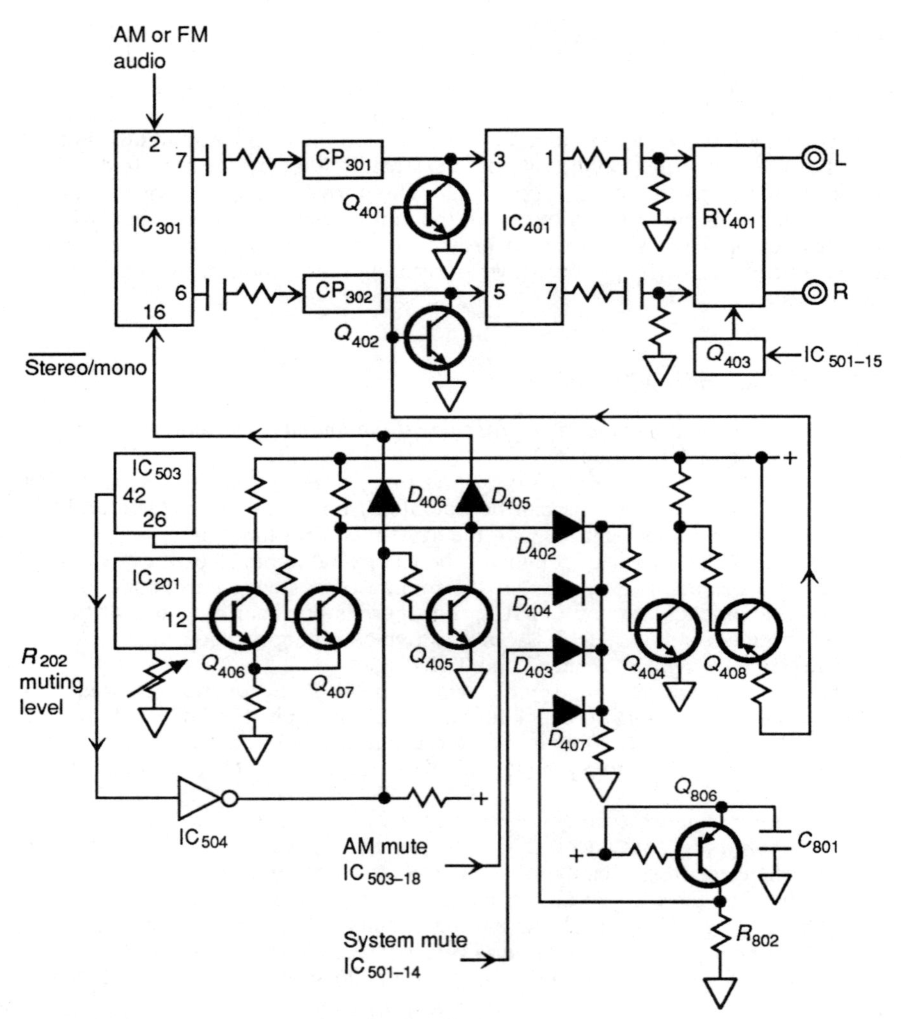

FIGURE 6.7 Tuner audio-output and muting circuits.

auto mode, the collector of Q_{407} goes high, turning Q_{404} (through D_{402}) and Q_{408} on. With Q_{408} on, B+ is applied to Q_{401} and Q_{402}, turning both Q_{401} and Q_{402} on. This grounds both the left and right audio lines and mutes the audio.

In the mono mode (auto mode indicator off), pin 42 of IC_{503} goes low. This low is inverted by IC_{504}, turning Q_{405} on and turning Q_{404}, Q_{408}, Q_{401}, and Q_{402} off. This removes the ground from the left- and right-channel audio lines and unmutes the audio. The high from IC_{504} also tells IC_{301} to operate as a mono amplifier (same audio signal to both channels).

In auto mode, pin 42 of IC_{503} goes high. This high is inverted by IC_{504}, and applied to pin 16 of IC_{301} through D_{406}, telling IC_{301} to operate as an FM stereo multiplexer (left- and right-channel audio). The low from IC_{504} has no effect on Q_{405}.

The audio can be muted (or unmuted) by signals from pin 18 of IC_{503} (for AM operation) and pin 14 of IC_{501} (when the tuner is used as part of a system).

The audio can also be muted by a temporary power interruption. During normal operation C_{801} charges to the B + level of about 13 V. If the B+ drops because of a momentary power interruption, Q_{806} turns on, and C_{801} discharges through Q_{806} and R_{802}. The temporary high across R_{802} turns Q_{404}, Q_{408}, Q_{401}, and Q_{402} on to mute the audio (temporarily, until C_{801} discharges).

6.7.1 Audio-Output and Muting Troubleshooting

If there is no automatic search (auto mode) operation, first make sure that you are in FM mode. (There is no AM auto mode.) Try pressing FM and auto mode and make sure that the FM and auto mode indicators are on (which might just cure the problem).

Next make sure that the front-panel stereo indicator is on when an FM station is tuned. It is possible that the station signal is not sufficient for good FM operation. If the stereo indicator is on, check that pin 12 of IC_{201} is low and pin 26 of IC_{503} is high. If pin 12 of IC_{201} is not low, suspect IC_{201}. If pin 12 of IC_{201} is low and pin 26 of IC_{503} is not high, and the FM tuning does not step when the stereo indicator turns on, suspect IC_{503}. Also check the frequency-synthesis circuits (Sec. 6.4).

If there is no audio muting or the audio is muted at all times, troubleshooting the circuits may prove difficult. As with most AM/FM tuners, our tuner is muted between stations or when stations are too weak to provide good operation. The muting function can also be removed so that the audio is not muted under any conditions. So, if the tuner never mutes in any mode, there is an obvious problem in the muting circuits. On the other hand, if the muting circuits are defective and mute the audio under all conditions (or under the wrong conditions), this can lead you to believe there are no stations of sufficient strength to unmute the audio or that the circuits ahead of the muting function are defective.

The most practical approach to any muting-circuit problem is to make some preliminary isolation steps. Start by checking for audio at pins 2, 6, and 7 of IC_{301}. If there is no audio at pin 2, check the front-end circuits (Sec. 6.5).

If there is audio at pin 2 but not at pins 6 and 7 of IC_{301}, suspect IC_{301}. If there is audio at pins 6 and 7 of IC_{301}, trace the audio through to the left- and right-channel audio output terminals. If the audio drops off at pins 3 and 5 of IC_{401}, it is possible that Q_{401} and Q_{402} are turned on by a muting signal from Q_{408} and Q_{404}. Also, if audio is available at pins 1 and 7 of IC_{401} but not at the tuner audio output terminals, it is possible that RY_{401} is turned off by a low at pin 15 of IC_{501} (Sec. 6.7).

If the audio path between IC_{301} and the audio output terminals is good, press auto mode and check that pin 42 of IC_{503} goes high (and that the auto mode indicator turns on). If not, suspect IC_{503} or the circuits between the auto mode switch and IC_{503}.

Next check that pin 16 of IC_{301} is low (in auto mode). If pin 16 of IC_{301} is high, IC_{301} operates as a mono amplifier rather than an FM stereo multiplex/decoder. However, IC_{301} should pass audio in either mode (auto or mono). Check the base of Q_{404}. If the base is low (0 V), the audio should be unmuted. If it is not, suspect

Q_{404} and Q_{408}. If the base is high (about 0.6 V), the audio should be muted. If it is not, suspect Q_{404}, Q_{408}, Q_{401}, and Q_{402}.

Note that in auto mode, the base of Q_{404} should go high only when an FM station of sufficient strength is tuned in. If this is not the case, try correcting the problem by adjustment of the muting level R_{202}, as described in the service literature (or Sec. 6.9.4). If this does not cure the problem, check that pin 12 of IC_{201} goes high and low as you tune across the FM band. If it does not, suspect IC_{201}. If pin 12 of IC_{201} changes status as FM stations are tuned in and out but the base of Q_{404} does not change status, suspect Q_{406} and Q_{407}.

It is also possible that Q_{405} has not been turned off or on because of a defect. (Q_{405} should be on only in mono mode.) If Q_{405} is on, Q_{406} and Q_{407} and the signal at pin 12 of IC_{201} have no effect on the base of Q_{404}. However, Q_{404} can be turned on to mute the audio by a high from pin 18 of IC_{503} (AM mute), by a high from pin 14 of IC_{501} (system mute), or a high across R_{802} (temporary power-loss mute).

6.8 FRONT-PANEL DISPLAY CIRCUITS

Figure 6.8 shows the front-panel fluorescent display circuits. Note that the segments of the display are turned on in sequence, in a time-sharing mode. However, the time-share rate is high enough so that the segments appear to be on constantly.

Segments Sa through Sg are applied directly to FL_{501} from IC_{503}, as described in Sec. 6.3. Segments D_6 through D_2 are applied to FL_{501} through buffers within

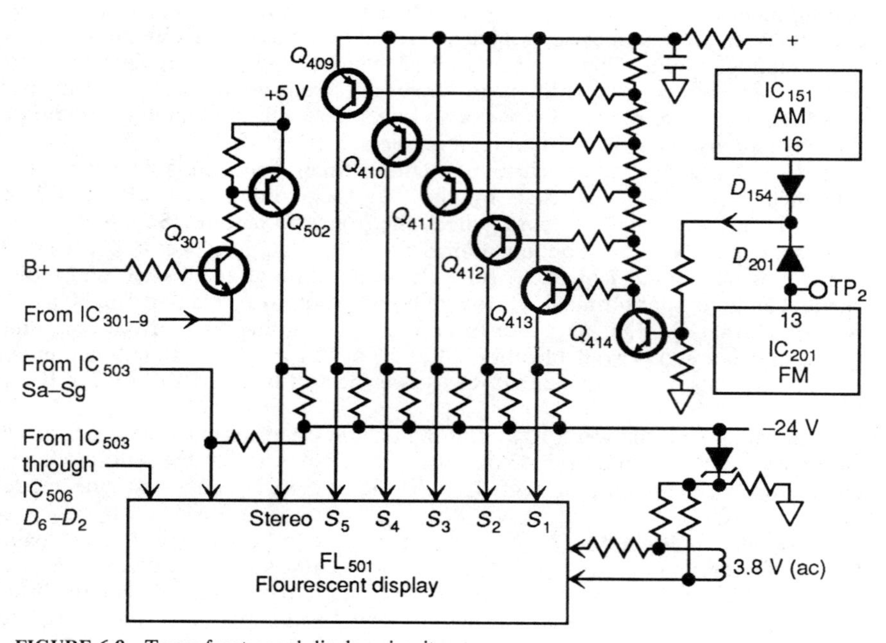

FIGURE 6.8 Tuner front-panel display circuits.

IC_{506}. The S and D segments produce all displays (frequency, operating mode, etc.) except for stereo indication and relative signal strength.

When an FM stereo station (of sufficient strength) is received, pin 9 of IC_{301} goes low. This turns Q_{301} and Q_{502} on, allowing B+ to be applied to the stereo segment of FL_{501} (turning on the stereo display).

A sample of the AM and FM audio signal is applied to Q_{414}. Variations in AM/FM signal strength cause Q_{409} through Q_{413} to turn on, supplying B+ to segments S_1 through S_5. This provides a display of relative signal strength.

6.8.1 Front-Panel Display Troubleshooting

If there is no stereo indication on the front-panel display but stereo is available at the tuner audio output terminals, check for a low at pin 9 of IC_{301}. If the low is available but the stereo display is not on, suspect Q_{301} and Q_{502} or FL_{501}. If pin 9 of IC_{301} is not low (with stereo available) suspect IC_{301}.

If there is no signal-strength indication on the front-panel display but audio is available at the tuner audio output terminals, check for signals at the base of Q_{414} at pin 16 of IC_{151} (AM) and pin 13 of IC_{201} (FM).

If the signals are missing during AM but available during FM, suspect IC_{151} and D_{154}. If signals are missing during FM only, suspect IC_{201} and D_{201}. If signals are available at the base of Q_{414} but not at FL_{501}, suspect Q_{409} through Q_{414} or FL_{501}.

6.9 TYPICAL TESTING AND ADJUSTMENTS

This section describes the test and adjustment procedures for a typical AM/FM tuner, using our tuner an an example.

6.9.1 FM Stereo Generator

An FM stereo generator is essential for troubleshooting the FM portion of an AM/FM tuner. This is because an FM generator simulates the very complex modulation system used by FM stereo broadcast stations. Without an FM stereo generator, you are totally dependent on the constantly changing signals from such stations, making it impossible to adjust the FM portion of the tuner or to measure frequency response after adjustment.

FM Stereo Modulation System. Before we describe the characteristics of an FM stereo generator, let us review the basic FM stereo modulation system (which is quite complex when compared to that of the AM broadcast system). It is essential that you understand the FM system to troubleshoot the FM portion of any AM/FM tuner.

Figure 6.9*a* shows the composite audio modulating signal used for FM stereo. This FM system permits stereo tuners and receivers to separate audio into left and right channels and permits mono FM tuners to combine left- and right-channel audio into a single output.

Figure 6.9*b* shows the block diagram of a FM-stereo modulator and transmit-

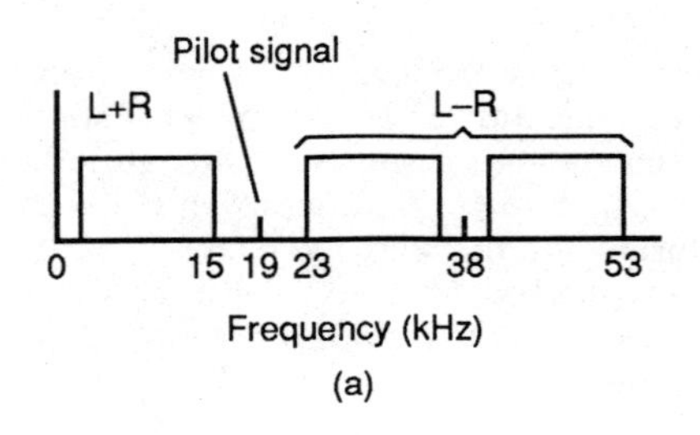

(a)

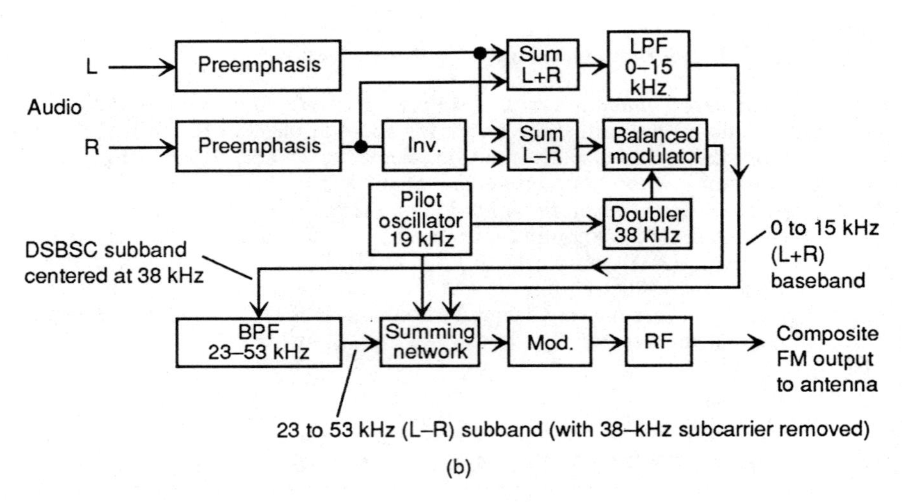

(b)

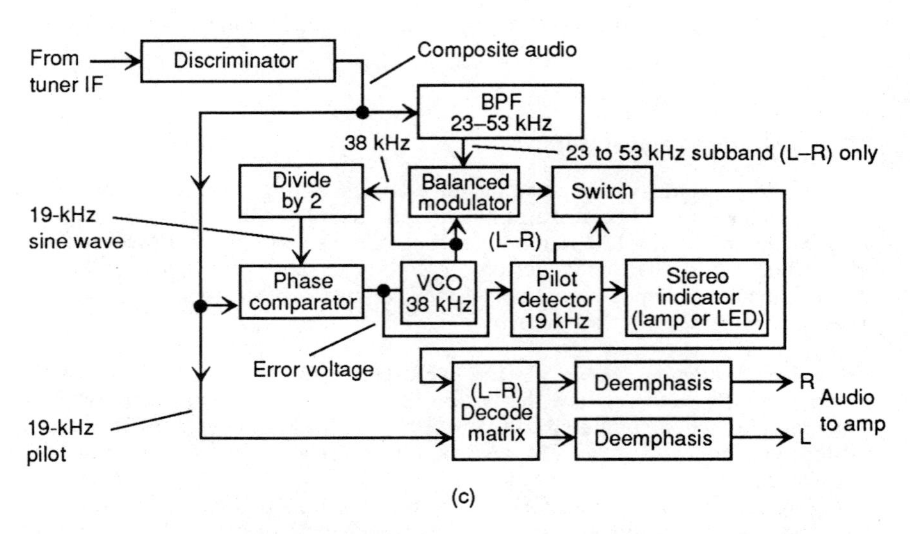

(c)

FIGURE 6.9 FM stereo modulation system.

ter. Left- and right-channel audio signals are applied through preemphasis networks to a summing network that adds the two signals. A low-pass filter limits this signal to the 0- to 15-kHz audio band, which is the maximum authorized for FM broadcast service. This (L + R) signal contains both left- and right-channel audio in a 0- to 15-kHz baseband.

The left-channel audio is applied to another summing network, along with the right-channel audio, which is inverted. The summing network effectively subtracts the two signals. The 0- to 15-kHz (L − R) signal is fed to a balanced modulator along with a 38-kHz sine wave. The balanced modulator produces a double sideband suppressed carrier (DSBSC) subband, centered around 38 kHz. A bandpass filter limits the signal to the 23- to 53-kHz range (±15 kHz of the 38-kHz carrier). The resulting left − right signal is a 23- to 53-kHz subband, with the 38-kHz subcarrier fully suppressed.

When FM stereo is broadcast, a low-level 19-kHz pilot signal is transmitted simultaneously. The pilot signal is generated by a stable, crystal-controlled oscillator operating at 19 kHz. The 19-kHz pilot oscillator output is also applied to a frequency doubler, providing the 38-kHz carrier for the balanced modulator.

The 0- to 15-kHz baseband (L + R), 23- to 53-kHz subband (L − R), and 19-kHz pilot signal are applied to a summing network, resulting in a composite audio signal consisting of the three components shown in Fig. 6.9*a*. The composite audio signal is applied to the modulator, which FM-modulates the RF carrier of the transmitter. For a fully-modulated RF carrier, the (L + R) signal accounts for 45 percent, the (L − R) signal for 45 percent, and the pilot signal for 10 percent.

Stereo tuners and receivers decode the signal shown in Fig. 6.9*a* and then separate the audio into original left and right channels. Since mono FM tuners have only a 0- to 15-kHz audio response, the 19-kHz pilot signal and the 23- to 53-kHz (L − R) subband are rejected. However, the (L + R) baseband signal is accepted and combines left- and right-channel audio to produce mono audio.

Figure 6.9*c* shows the FM section of a typical AM/FM tuner, from the discriminator to the audio amplifier inputs. The discriminator output is a composite audio signal similar to that shown in Fig. 6.9*a*.

A PLL locks the sine-wave output of the 38-kHz VCO in phase with the received 19-kHz pilot signal as follows. A divide-by-2 circuit converts the 38-kHz VCO output to a 19-kHz sine wave, which is one of the inputs to a phase comparator. The other input is the received 19-kHz pilot signal. The phase comparator produces an error voltage to lock the VCO in-phase with the 19-kHz pilot signal.

The composite audio signal from the discriminator is applied through a 23- to 53-kHz bandpass filter, which blocks the 0- to 15-kHz (L + R) baseband and 19-kHz pilot signals, while passing only the 23- to 53-kHz subband (L − R) signal. The 23- to 53-kHz subband signal is applied to a balanced demodulator. The other input to the demodulator is a 38-kHz carrier from the VCO. The VCO carrier is locked in phase to the 19-kHz pilot signal by the PLL.

During stereo broadcast, with the 19-kHz pilot signal present, the error voltage from the phase comparator approaches zero as phase-lock is achieved. This error voltage is applied to a 19-kHz pilot detector and to the 38-kHz VCO. When the error voltage drops below the threshold of the 19-kHz pilot detector, the stereo indicator is turned on, as is the switch that couples the (L − R) signal from the balanced demodulator to the L − R decode matrix.

The decode matrix separates the L and R components of the (L + R) and (L − R) signals into independent left- and right-channel outputs. When the (L − R) signal is absent, such as during mono reception, the error signal from the phase comparator does not approach zero, and the 19-kHz pilot detector does not actuate the switch or turn on the stereo indicator. As a result, the (L + R) input is not separated into L and R components within the matrix. Instead, both outputs from the matrix are identical (L + R) signals.

As shown in Fig. 6.9*b*, both the L and R audio signals are subjected to *preemphasis* before reaching the FM modulator circuits. Likewise, as shown in Fig. 6.9*c*, the L and R outputs from the decode matrix are applied to audio amplifiers through *deemphasis* networks. The use of preemphasis and deemphasis is mostly to improve the S/N ratio.

Preemphasis increases the highs, while deemphasis increases the lows. Frequency modulation, or FM, is usually a form of phase modulation, or PM. The noise-modulation characteristic of PM is not flat but increases with noise frequency. So, with deemphasis in the tuner, the high-frequency noise is reduced to the same level as the low-frequency noise, thus improving the S/N ratio.

Remember that noise modulation is essentially an internal modulation of constant level. By preemphasizing the external audio of the transmitter, an overall flat audio response is possible. This compensates for the deemphasis characteristics of the receiver but does not affect noise modulation.

The *RC* time constants of the coupling components in a preemphasis or deemphasis network determine the center frequency and are specified in microseconds (μs); 75 μs is standard for FM broadcasts in the United States, and 50 μs is used in some other countries. The standard rolloff rate for both preemphasis and deemphasis is 6 dB per octave.

FM Stereo Generator Features. A typical FM stereo generator produces a stereo multiplex FM-modulated RF carrier signal that conforms to FCC regulations and duplicates the type of signal radiated by an FM stereo broadcast transmitter. External or internal modulation audio is converted to a composite audio signal containing a (L + R) 0- to 15-kHz baseband, and (L − R) 23- to 53-kHz subband. Generally, the modulating signal (composite audio) is also available for injection directly into audio and stereo decoder circuits.

On most generators, the composite audio signal is continuously adjustable, and a modulation meter is calibrated to measure the rms value of composite audio. The RF output may be internally or externally modulated. In either case, modulation is continuously adjustable up to 75 kHz. A calibrated modulation meter reads FM deviation in kilohertz.

For external modulation, independent left and right input jacks permit stereo modulation through 75- or 50-μs preemphasis networks, or with no preemphasis, on most generators. The 50-Hz to 15-kHz audio input bandwidth equals that of an FM broadcast transmiter, thus permitting full audio-range frequency response test of tuners.

The typical internal-modulation frequency is 1 kHz, which can be selected in one of five combinations: left-channel only, right-channel only, (L + R) baseband, (L − R) subband, and left and right (line) with 1-kHz signal applied to left channel and 50- to 60-Hz line-voltage signal applied to the right channel. These internal-modulation combinations permit complete testing of stereo decoder circuits, including channel-balance and channel-separation characteristics (if required).

During either internal or external modulation, a highly stable 19-kHz pilot signal is generated and combined with the composite audio. The pilot signal may be switched off when desired, such as for the testing operation of a pilot detector circuit (if any).

Some FM stereo generators have a detachable telescoping antenna to simulate an FM stereo broadcast transmitter. On most generators, the RF output may be turned off whenever desired, thus generating the composite audio only.

6.9.2 FM IF Adjustments (Preliminary)

Figure 6.10*a* shows the test and adjustment points. The purpose of this procedure is to set the IF circuits of the FM section of an AM/FM tuner. The procedure can be used at any time but is of most value when performed before any extensive troubleshooting. (The procedure just might cure a number of problems.) A preferred procedure is described in Sec. 6.9.3.

With the sweep generator connected to TP_2, adjust the sweep generator for 10.7 MHz with a 200-kHz sweep width. Adjust the generator output amplitude for a weak signal that produces noise patterns as shown on the waveforms in Fig. 6.10*a*.

Using a nonmetallic alignment tool, adjust the IF tuning (IFT) of the MD_{101} FM tuner so the IF waveform (S curve) in Fig. 6.10*a* is maximum.

Move the sweep generator to TP_3, but leave the frequency and output as previously set. If necessary, reduce the scope gain. Alternately, adjust T_{201} for a symmetrical S curve and T_{202} for linearity between the positive and negative peaks of the IF waveform.

6.9.3 FM IF Adjustments (Preferred)

Figure 6.10*b* shows the test and adjustment points. First perform the preliminary adjustment as described in Sec. 6.9.2. Adjust the FM generator for 97.9 MHz, modulated by 1 kHz with 75-kHz deviation (100 percent modulation, mono). Adjust the generator output amplitude for 65.2 dBf. The term *dBf* (found in some FM tuner specifications) is the power level measured in dB, referenced to 1 *femtowatt* (10^{-15}). If you are fortunate, the service literature will spell out a voltage reading (usually in the microvolt range).

Adjust the tuner to 97.9 MHz (as indicated on the front-panel fluorescent display), and adjust T_{201} for 0 V ±50 mV on the null meter (connected between TP_4 and TP_5).

Adjust T_{202} for *minimum distortion* on the distortion meter (connected to the left-channel audio-output jack).

Work between T_{201} and T_{202} until you get minimum distortion and (it is hoped) 0 V ±50 mV on the null meter.

6.9.4 FM Muting Adjustment

Figure 6.10*c* shows the test and adjustment points. This adjustment sets the signal threshold for audio muting in the FM auto mode (where the audio should be muted in between stations and on weak stations).

Adjust the FM generator for 97.9 MHz, modulated by 1 kHz with 75-kHz deviation. Adjust the generator output amplitude for 33 dBf (or the equivalent output voltage).

Adjust the tuner to 97.9 MHz (as indicated on the front-panel fluorescent display). Place the tuner in FM auto mode (press auto mode and check that the auto mode indicator turns on).

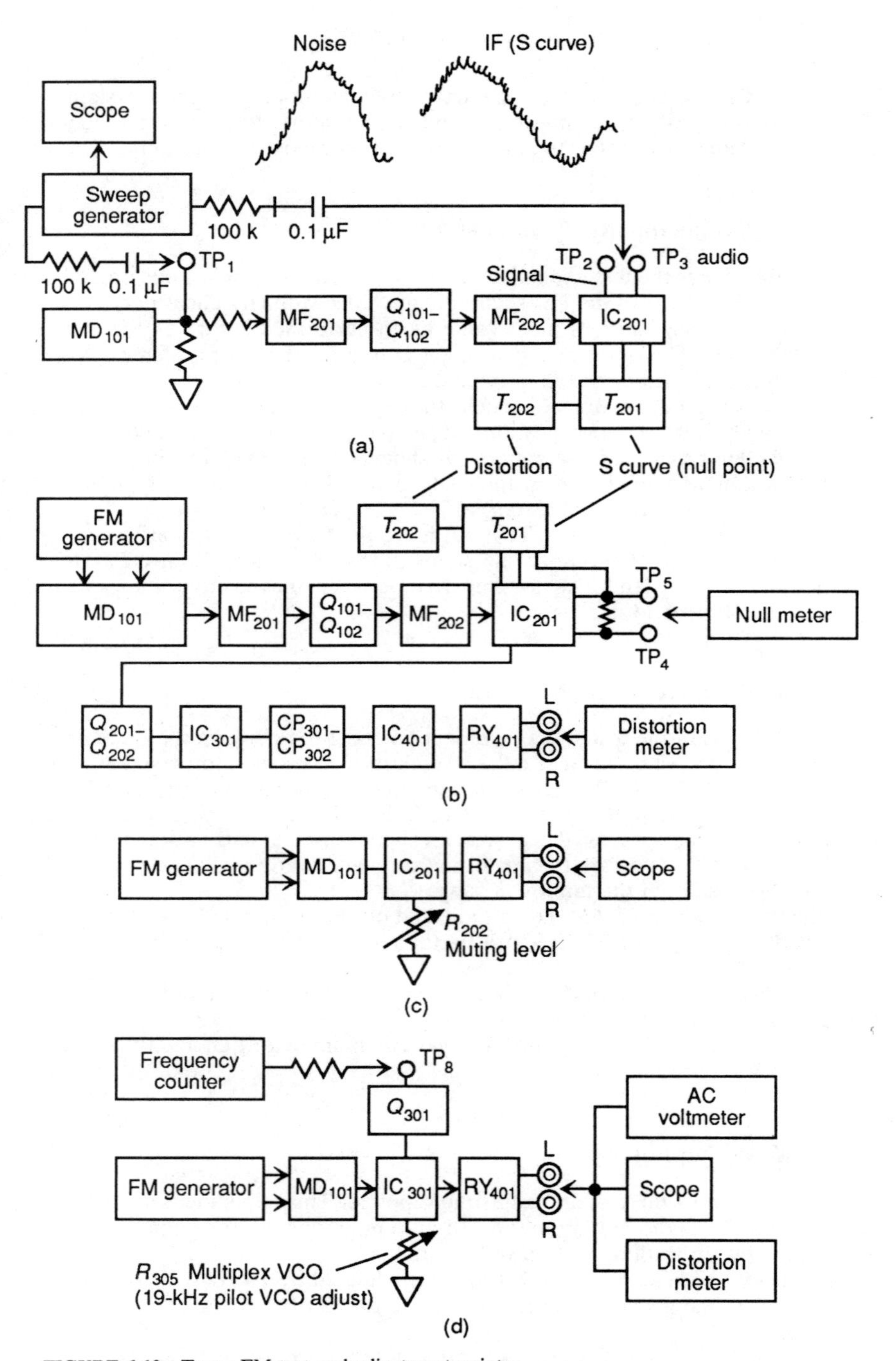

FIGURE 6.10 Tuner FM test and adjustment points.

Adjust R_{202} until the audio is muted. Then slowly readjust R_{202} until the audio is *just unmuted*.

6.9.5 FM Multiplex VCO Adjustment and Distortion Test

Figure 6.10*d* shows the test and adjustment points. This adjustment sets the 19-kHz pilot VCO in IC_{301}; check the resultant distortion.

Adjust the FM generator for 97.9 MHz with no modulation. Adjust the generator output amplitude for 65.2 dBf (or the equivalent output voltage).

Adjust the tuner to 97.9 MHz (as indicated on the front-panel fluorescent display). Place the tuner in FM auto mode (press auto mode and check that the auto mode indicator turns on).

Adjust R_{305} for 19 kHz ±50 Hz as indicated on the frequency counter connected to TP_8.

Leave the tuner and FM generator set at 97.9 MHz. Apply 1-kHz modulation to the FM generator left channel (with a pilot-carrier deviation of 6 kHz and a total deviation of 75 kHz).

Adjust the IFT of MD_{101} (using a nonmetallic alignment tool) for minimum distortion (as indicated by the distortion meter connected to the left-channel audio output jack). Do not adjust the IFT of MD_{101} more than one-quarter turn from the setting established in Sec. 6.9.2.

Remove the left-channel modulation from the FM generator, and apply the same modulation to the right channel. Monitor the right-channel audio output with the distortion meter, and check that the distortion is about the same for both channels.

6.9.6 AM IF Adjustment

Figure 6.11*a* shows the test and adjustment points. The purpose of this test is to set the IF circuit of the AM section. Adjust the sweep-generator frequency to 450 kHz. Increase the generator output until a waveform appears on the scope. Do not overdrive the IF section. Adjust T_{151} until the waveform is as shown in Fig. 6.11*a* (maximum at 450 kHz).

6.9.7 AM Local-Oscillator Confirmation

Figure 6.11*b* shows the test and adjustment points. This test is to check that the tuning voltage (error voltage from PLL IC_{503}, Sec. 6.4.2) applied to the AM section is correct. Normally, it is not necessary to adjust the local oscillator, but the tuning voltage should be checked after any service in the AM section of the tuner.

Adjust the tuner to 1630 kHz (as indicated by the front-panel display). Check that the DVM (connected to TP_9) reads less than 23 V.

Adjust the tuner to 530 kHz, and confirm that the TP_9 voltage is 1.8 V ±0.3 V. If the reading at 530 kHz is out of tolerance, adjust L_{152} for an error voltage of 1.8 V ±0.1 V. Then check that the error voltage is less than 23 V with the tuner at 1630 kHz.

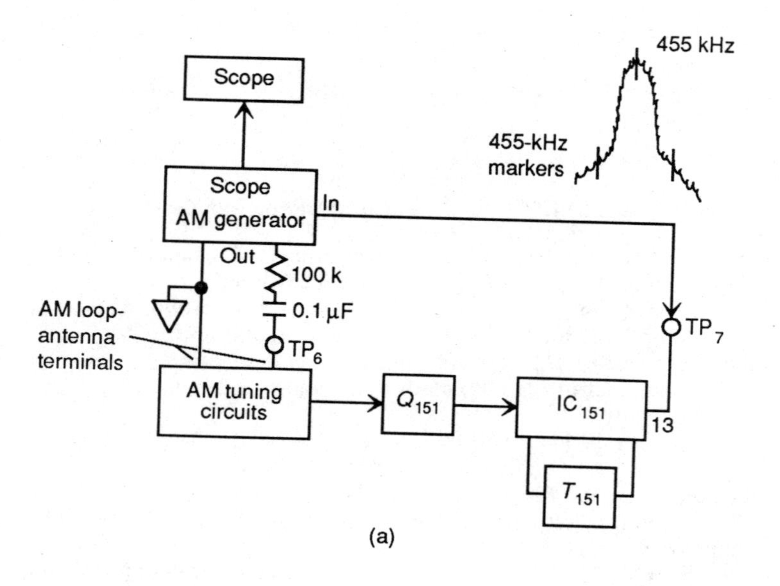

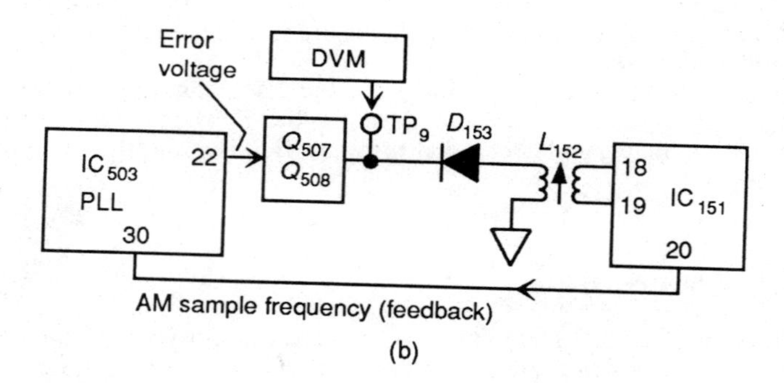

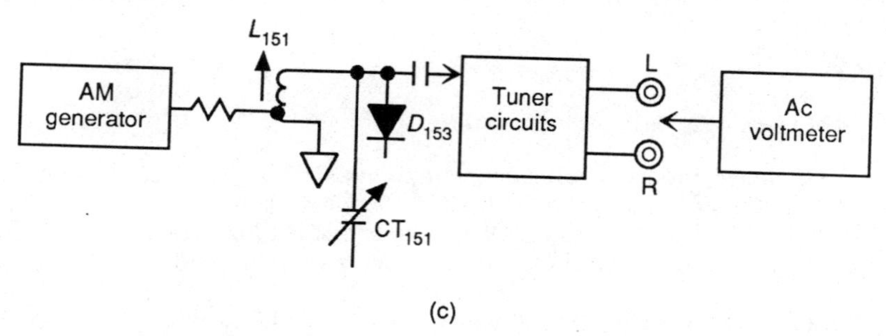

FIGURE 6.11 Tuner AM test and adjustment points.

6.9.8 AM Tracking Adjustment

Figure 6.11*c* shows the test and adjustment points. This procedure adjusts the RF input circuit of the AM section for proper tracking across the AM broadcast band.

Adjust the generator for 600 kHz, modulated with 400 Hz at 30 percent. Set the generator output level as necessary for a reading on the ac voltmeter (connected to the left- or right-channel audio output). Use the minimum output from the generator that produces a satisfactory reading.

Adjust the tuner to 600 kHz (as indicated on the front-panel display). Adjust L_{151} for maximum output on the ac voltmeter. Adjust the tuner to 1400 kHz, and adjust CT_{151} for maximum output. If necessary, work between L_{151} and CT_{151} for maximum output at 600 and 1400 kHz. These two adjustments usually interact.

6.10 PRELIMINARY TROUBLESHOOTING

Here are the checks to be made before going into the tuner circuits.

If the tuner does not turn on (either manual or remote), check that the power cord is properly connected to a known-good power source.

If the tuner operates manually but not by remote, check the remote-unit batteries and any control cables to the tuner. Then try resetting the power circuits by pressing the front-panel power button off and on.

If there is no tuner signal (as indicated by the front-panel signal-strength display), check for the following. Lead-in wires may be shorted together at the antenna terminal strip. The antenna transformer (balun) or TV/FM splitter may be defective. The coaxal lead-in may be broken or disconnected from the antenna or tuner. A broken or disconnected flat lead-in can produce no signal or a very weak signal.

If there is no tuner audio but a signal is present (with audio from other components good), check for the following. The audio output cables may be disconnected from the tuner or amplifier. The amplifier may be set for another audio source. (Make sure that the amplifier can play all other audio sources.) The station may be transmitting a signal without sound (dead air). Try another station.

If the tuner skips a desired FM station in auto mode, try tuning in the station in mono. If the station can be tuned in mono but not auto mode, the station signal may be too weak.

If FM reception is poor, make sure that the FM antenna is securely fastened to the correct terminals on the tuner. If an indoor dipole is being used for FM, reposition the dipole. Check the outside FM antenna and lead-in. The indoor dipole antenna must be disconnected when an outside antenna is used for FM. If a combination TV/FM outdoor antenna is used, check the TV reception. If TV reception is also poor, suspect the antenna and lead-in. If TV reception is good, suspect the TV/FM splitter and any other components that are only in the FM signal path.

If preset-tuning or memory-scan operation is abnormal, disconnect the power cord for 5 s or more, then reconnect the cord. If the problem persists, reprogram the preset memory. Note that the memory may be lost if power is disconnected for several days.

If AM reception is poor, the AM loop antenna must always be connected, even when an external antenna is used.

If AM reception is noisy, AM radio interference may be entering the tuner with the electrical power. Try a plug-in noise suppressor at the wall outlet. Also, TV sets, fluorescent lights, and many electric appliances emit small amounts of radio signals in the frequency band of AM broadcasts. Try moving the loop antenna (if practical) or move the tuner away from the noise source. An external AM antenna may be required in extreme cases.

If there is no FM stereo reception or one speaker is dead but there is a good front-panel stereo indication, check for loose cables in the signal path between the tuner and amplifier. Temporarily reverse the left and right cable connections *at the tuner outputs*. If the same speaker remains dead, the problem is not in the tuner (and is probably in the amplifier or speakers).

CHAPTER 7
TAPE CASSETTE AUDIO

This chapter describes the overall function, user controls, operating procedure, installation, circuit theory, typical test and adjustment procedures, and step-by-step troubleshooting for state-of-the-art audio cassette decks found in audio systems.

7.1 OVERALL DESCRIPTION

Figure 7.1 shows the front and back of a composite audio cassette deck. This deck is a four-track, two-channel stereo model that requires an amplifier-loudspeaker combination (Chap. 5) to play and record audio cassettes. The deck uses standard-size audio cassettes and can be used as part of a system (home entertainment, stereo, etc.). In this book, we are concerned with operation of the deck in the stand-alone condition (with amplifier and loudspeakers, of course).

7.1.1 Features

Types of tape: Most decks can use three types of tape: *standard or normal, chromium oxide* (CrO_2), and *metal*. Actually, CrO_2 is chromium dioxide, although it is usually called chromium oxide, chrome, chromium, or possibly CRO.

Automatic tape compatibility: Circuits within the deck adjust equalization bias for the three types of tape, as discussed in Sec. 7.3.

Dolby noise reduction: Most present-day decks include some form of Dolby noise reduction, or NR, at least the B and C systems.

Microprocessor control: Our deck is under control of microprocessors that monitor and control all functions via buttons on the front panel, as well as through a system bus.

Auto-reverse tape transport: Both sides of the tape can be played or recorded without turning over the cassette tape. The auto-reverse system automatically reverses tape direction (to play or record the opposite side) when the first side is finished. This feature is not available on all present-day decks.

Scan and play system: This system makes it possible to scan a tape and play only the first 10 s of each program, permitting the listener to sample tape seg-

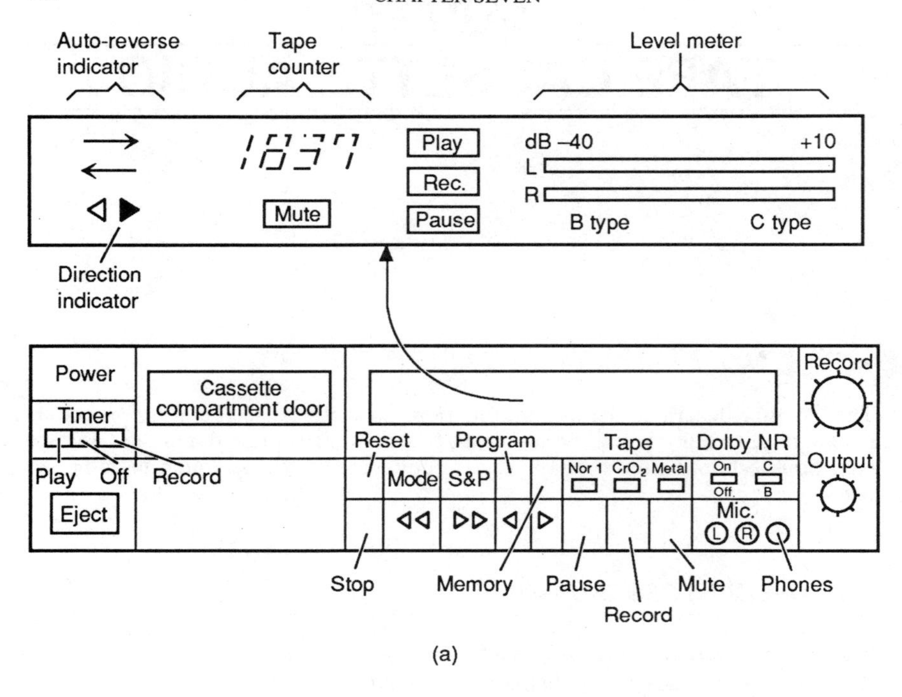

(a)

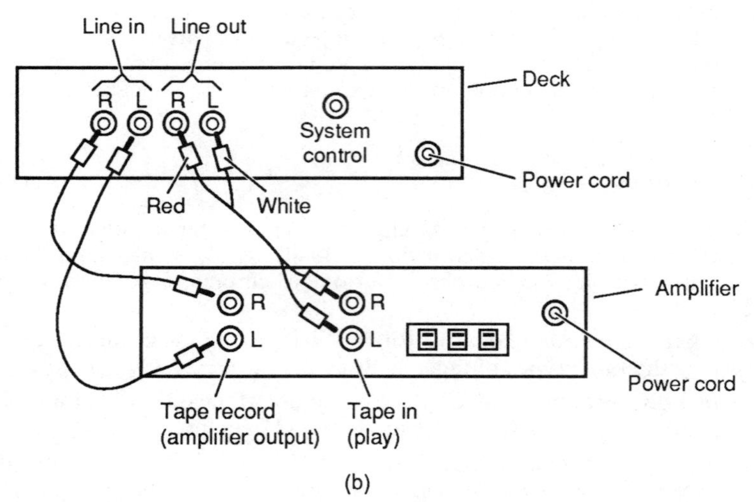

(b)

FIGURE 7.1 Operating controls, indicators, and connections for composite audio cassette deck.

ments as an aid in program selection. On our deck, up to 15 of the scanned programs can be selected for playback (in the order the programs are recorded) by entering an assigned number in memory.

Random-access memory: This feature holds up to 15 programmed selections for playback in any order.

Record protection: The deck includes circuits that prevent the deck from entering the record mode when a prerecorded cassette is installed (unless it is desired to record new information on the tape).

Automatic tape cueing: This feature permits the deck tape-transport mechanism to cue the beginning of the tape (just past the tape leader) automatically, making it possible to access blank tape quickly. For partially recorded tape, the deck "finds" the end of the previous recording and cues the starting point for the next recording.

Fluorescent audio meter: Two 20-segment light bars are used to indicate peak levels of each channel. These light bars are similar to the audio-level indicators described for the amplifier (Chap. 5).

Record mute: This feature adds a 4-s quiet interval to tapes being recorded. Note that this feature can be selected only at the deck front panel, not from system control (on our deck).

7.1.2 Operating Controls and Indicators

Figure 7.1*a* shows the operating controls and indicators. Compare the following with the controls and indicators of the deck you are servicing:

Pressing power turns the deck on. Pressing power again turns the deck off. None of the other controls function when power is off. There is a 4-s delay from turn-on until the deck starts playing.

The eject button opens the cassette door for inserting and removing cassettes.

The timer button permits operation of the deck with an external timer.

When stop is pressed, tape motion stops, but power stays on. Cassettes may be removed or inserted after stop is pressed, without turning off the power. Stop is also used while programming the memory to erase entries. The last entry is erased if stop is pressed momentarily. The entire memory is erased if stop is pressed for more than 1 s.

The button with *double arrows pointing left* moves the tape rapidly to the left (on side 1 this is rewind; on side 2 the function is fast forward).

The button with *double arrows pointing right* moves the tape rapidly to the right (on side 1 this is fast forward; on side 2 the function is rewind).

The button with a *single arrow pointing left* plays side 2 (tape moves from right to left) and is also used in conjunction with record to start recording on side 2.

The button with a *single arrow pointing right* plays side 1 (tape moves from left to right) and is also used in conjunction with record to start recording on side 1.

Pause is used to interrupt recording or playback. The tape stops moving and remains stopped until the pause is ended by pressing play. The pause indicator turns on during pause, and the rec (record) indicator stays on if there is a pause during record.

If pause is pressed while the deck is searching for a program, the search continues until the program is found and the tape pauses at the beginning of the program.

Record switches the tape deck into the recording mode but does not start the tape moving. Record is used in conjunction with one of the play buttons (single arrow) to begin recording.

Mute is used only during record. When mute is pressed momentarily, a 4-s quiet interval is recorded on tape for automatic cueing. The interval may be extended by holding the mute button. The deck automatically switches to the recording-pause mode at the end of the muted interval. Press play or pause to continue recording.

The recording input is automatically switched from the input jacks to the mic (microphone) jack when a microphone is plugged into the L or R (or both) jacks. Make certain to unplug the microphones to record from any other source. (This can sometimes cure a "no audio" trouble symptom.)

Use of the phones (headphones) jack does not affect any other function of the deck. So headphones can be plugged in or out during both record and play. (This not true for all decks.)

The output control sets the level of audio at both the phones jack and the rear-panel output jacks (to the speakers through the amplifier) but *does not affect record level*. Keep this in mind when troubleshooting an "I turned the volume all the way up, but it still is weak on record" symptom.

The two record controls (one behind the other) set the level of audio during record.

Dolby B/C selects either Dolby B or Dolby C for both recording and playback. Most deck manufacturers recommend that Dolby C be used for recording because of improved noise reduction. However, make certain to play prerecorded tapes with the Dolby type *used during record*.

Dolby on-off should be set to on during play only for tapes recorded with some type of Dolby noise reduction. During record, use Dolby as desired, but make certain to play back any Dolby-recorded tape with the correct Dolby type. Commercial tapes recorded with Dolby noise reduction are supposed to be so designated on the label.

The three tape type switches are mechanically interlocked so that pressing one of the switches releases the other two. Make certain to press the button corresponding to the type of tape in use. (This can cure many "poor sound quality" trouble symptoms.)

Memory enters the program selections into memory. Program is used to select the program number (which increases one increment each time program is pressed).

When S&P (scan and play) is pressed, the tape rewinds and begins playing the first 10 s of each recording on one side of the tape. The tape counter on the display indicates the numerical order of the programs on the tape as they are played (A-01, A-02, etc.).

The reversing mode changes each time mode is pressed. The mode is indicated in the upper-left corner of the display. Three modes are available: one-side mode, both-sides mode, and continuous mode.

With one-side mode, you may record or play in one direction or the other, and the tape stops at the end. You must manually select the reverse mode to record or play the other side of the tape.

With the both-sides mode, recording or playing begins with side 1 (tape moving from left to right). At the end of side 1, the tape reverses and plays (or records) side 2 (right to left) and stops. The deck does not reverse at the end of side 2.

With continuous mode, the tape automatically reverses at the end of each side and continues playing until the tape is stopped (up to a maximum of 16 "round trips"). If you record in the continuous mode, the tape stops after both sides are recorded.

The reset control resets the tape counter on the display to 0000. The counter resets automatically when the tape is rewound to the beginning.

When the tape is rewound, or when reset is pressed, the display counter reads 0000. The counter advances as the tape plays and reads end when the tape reaches the end. Note that if a tape is run part way in either direction and then removed before the counter reaches 0000 or end, you will get a "can't play or record a whole tape" trouble symptom when the tape is started again.

During programming, when the program button is pressed the first time, the counter reads 01. Each time program is pressed, the number increases (up to 15).

During scan and play, the selection number being played (for audition) is displayed. A-01 designates the first selection on the tape, A-02 is the next selection, and so on.

As the deck starts playing each selection from memory, the display shows which program is being played, according to the order in memory. As an example, P-03 indicates that the third selection entered into memory is playing. The number is preceded by P to indicate programmed playback.

The play indicator turns on to indicate that tape is playing or recording. The rec indicator turns on to indicate that the deck is recording (or is in the pause mode) during record. The pause indicator turns on to indicate that the deck is in the pause mode. The mute indicator turns on to indicate that the tape deck is recording with the input muted.

The L and R level meters (two 20-segment light bars) show the instantaneous level of the signal during record or playback. Note that the range for this deck is −40 dB to +10 dB.

One of the two direction-indicator arrows on the display turns on whenever power is on. The arrow pointing to the right indicates that the deck is set to record or play side 1 (left to right). The arrow pointing left indicates that the deck is to play or record side 2 (tape travel right to left).

The B- or C-type indicators turn on to indicate the corresponding Dolby noise-reduction system selected.

7.1.3 Connections

Figure 7.1*b* shows connections between the deck and amplifier. Compare the following notes to the connections for the deck being serviced. Typically, two pin cords are supplied with the deck (or with the amplifier) for connection between the L and R stereo inputs and outputs and the corresponding inputs and outputs on the amplifier. The deck outputs are connected to the amplifier inputs for playback, while the deck inputs are connected to the amplifier outputs for recording. Although the connections are very simple, certain precautions must be observed for all decks.

Look for any color coding on the pin cords. Typically, red is used for the right channel and white is for the left (but do not count on it). Also look for any ground terminals or leads that are part of the pin cord. (Ground terminals are quite common for phono turntables but usually not for cassette decks.)

Do not connect the deck to the CD, aux, or phono inputs of an amplifier. Instead, always use the tape play input (or whatever the deck input is called on the amplifier). Generally, there is no damage done if you connect the deck to the wrong input on the amplifier. However, the deck output does not match the other audio components. For example, the typical 500-mV output of a cassette deck is too high for the typical 5-mV output of a turntable and too low for the typical 2-V output of a CD player. Generally, the tape record output of an amplifier is applied through a buffer, as discussed in Chap. 5.

Do not connect the output of a cassette deck to the output of the amplifier or the input of the deck to the input of the amplifier. Damage may result, and the connections will not work.

7.2 RELATIONSHIP OF DECK CIRCUITS

Figure 7.2 shows the relationship of the deck circuits. The record/playback heads serve a dual purpose: recovery of the signal from the tape during playback and recording of the audio signal on tape during record mode. The record/playback switching circuits (Sec. 7.3) are responsible for determining whether the heads are in playback or record.

The tape-transport or mechanism processor IC_{500} provides the necessary switching signals to the record/playback switching circuits and also outputs control signals to the main signal-processing circuits (Sec. 7.3). The primary purpose of IC_{500} is to monitor and control the tape-transport mechanism (Sec. 7.6).

The mic input is applied directly to the signal-processing circuits, as is the line in-out jack input. (The line in-out is the audio line when the deck is used in an audio system.) A separate line-out is provided for stand-alone operation. The line-out is also used to supply signals to the music-detect circuits (Sec. 7.4) and to the meter-detect circuits (Sec. 7.8).

During scan and play, the music-detect circuits detect the music (or other program material) recorded on tape and cause the tape to go from search to play for 10 s and then back to search. *During random-access memory play*, the music-detect circuits cause the deck to go to the play mode and remain there until the end of the recorded material. The deck then goes to search until the next programmed selection.

The meter-detect circuits rectify the audio signal from the signal-processing circuits and generate a dc control voltage applied to counter microprocessor IC_{200}. This dc voltage represents the audio level and is used to generate the L and R bar graph (20-segment light bars) on the front-panel display.

The output from the tape-end detector circuit (Sec. 7.7) is applied to IC_{500}. Tape-transport microprocessor IC_{500} is also responsible for driving the reel-motor circuits (Sec. 7.6) and operating the three tape-transport solenoids (Sec. 7.6).

IC_{500} issues status signals to system-control microprocessor IC_{501} and receives command signals from IC_{501}. The status signals inform IC_{501} (and other

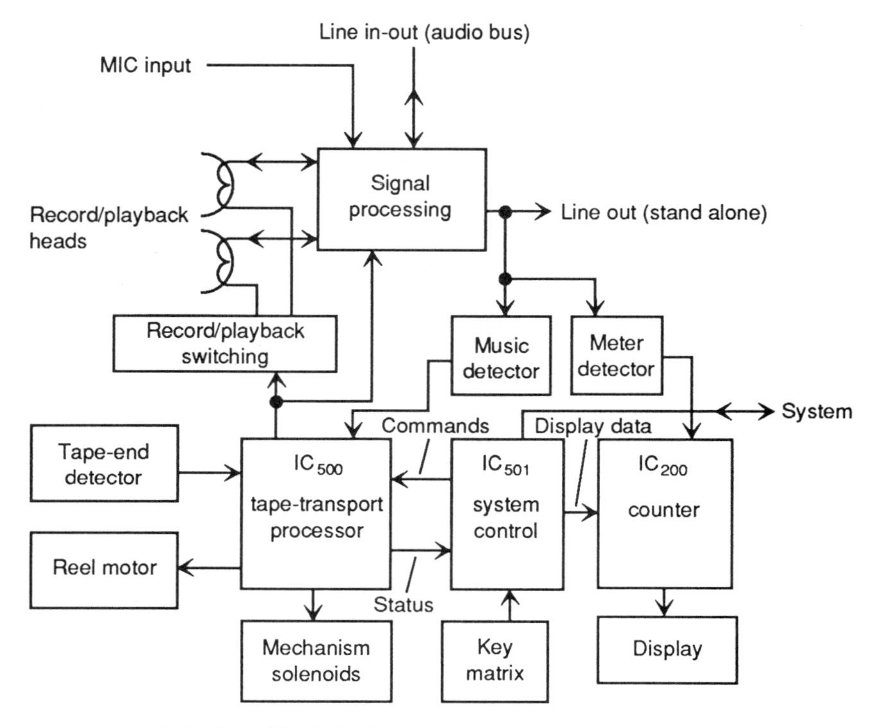

FIGURE 7.2 Relationship of deck circuits.

system components, if any) as to the mechanical and functional status of the deck. Commands from IC_{501} inform IC_{500} what function is selected (by the front panel or system). IC_{501} also issues display-data signals to IC_{200}, causing IC_{200} to generate front-panel indications on the display (Sec. 7.8).

7.3 RECORD/PLAYBACK FUNCTIONS

Figures 7.3 through 7.7 show the basic record/playback circuits.

7.3.1 Overall Record/Playback Circuits

Figure 7.3 shows the overall record/playback circuits (for one channel) in block form. The record/playback heads are placed in the playback mode, or record mode, by the record/playback switching circuit. This switching circuit is controlled by an output from pin 24 of IC_{500}.

During the playback mode (IC_{500} pin 24 low), the head recovers the recorded audio signal from tape and applies the signal to a playback amplifier. *During the record mode* (IC_{500} pin 24 high), the switching circuits place the record/playback heads in the record mode so that both line and microphone audio can be recorded on tape.

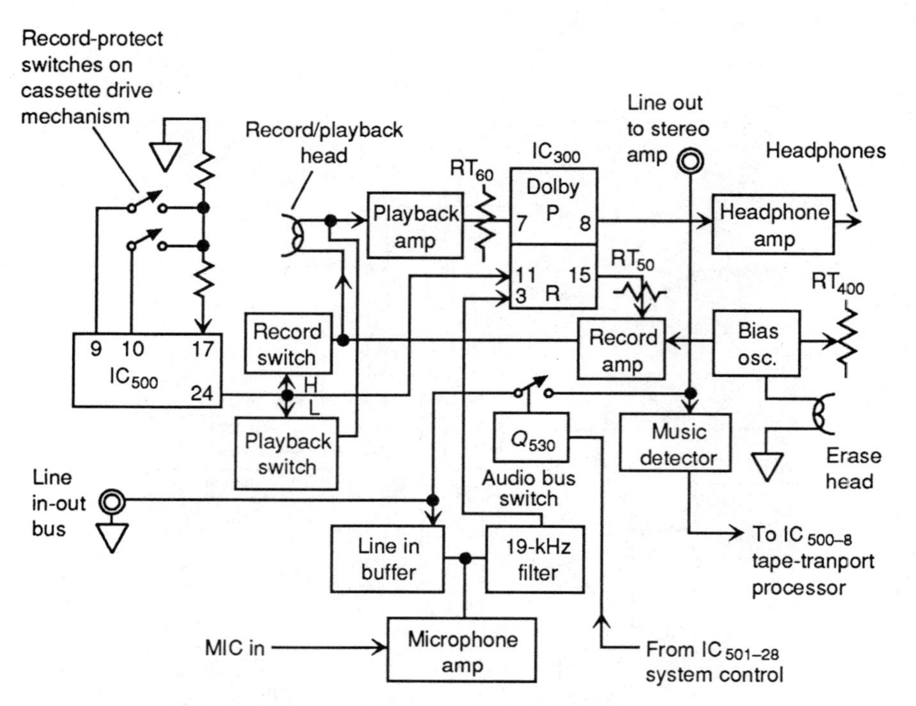

FIGURE 7.3 Overall record and playback circuits (for one channel).

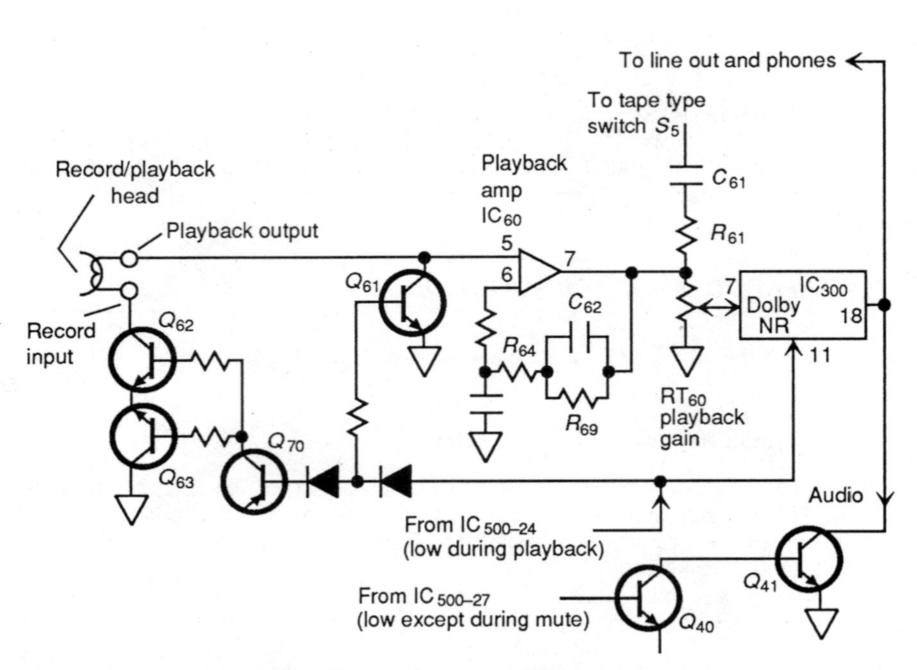

FIGURE 7.4 Record and playback switching and amplifier (playback).

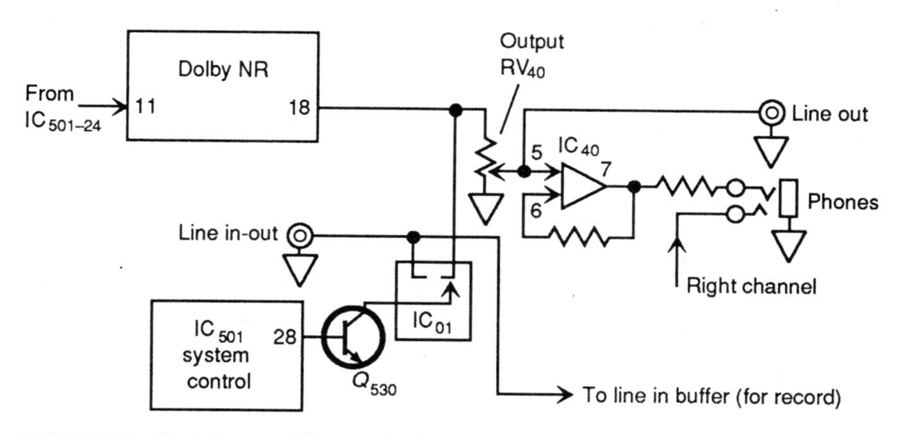

FIGURE 7.5 Headphone and line-out circuits.

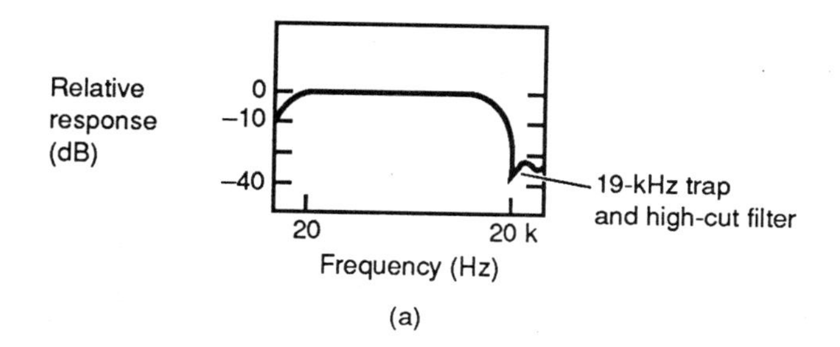

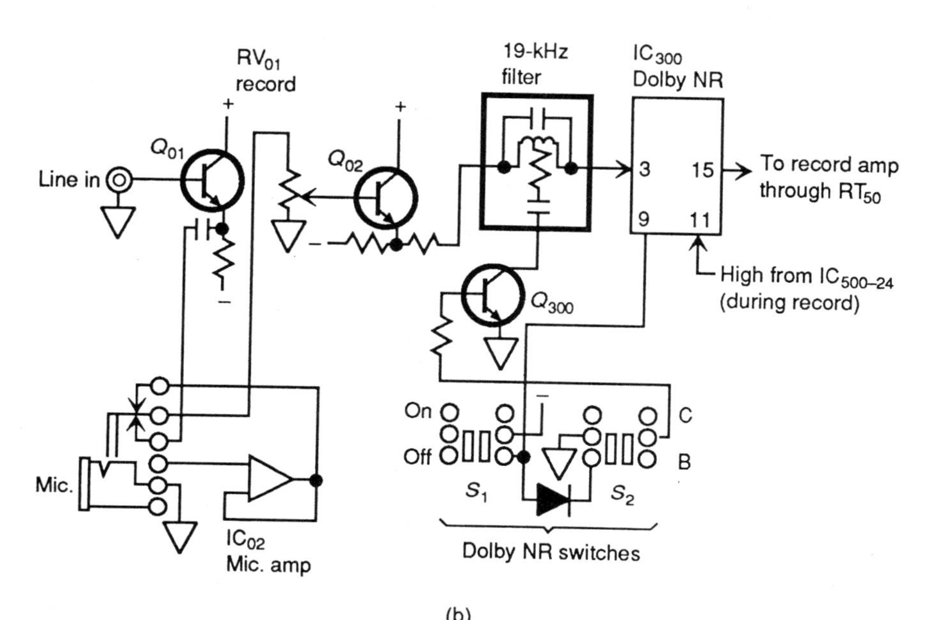

FIGURE 7.6 Record and playback switching and filter circuits.

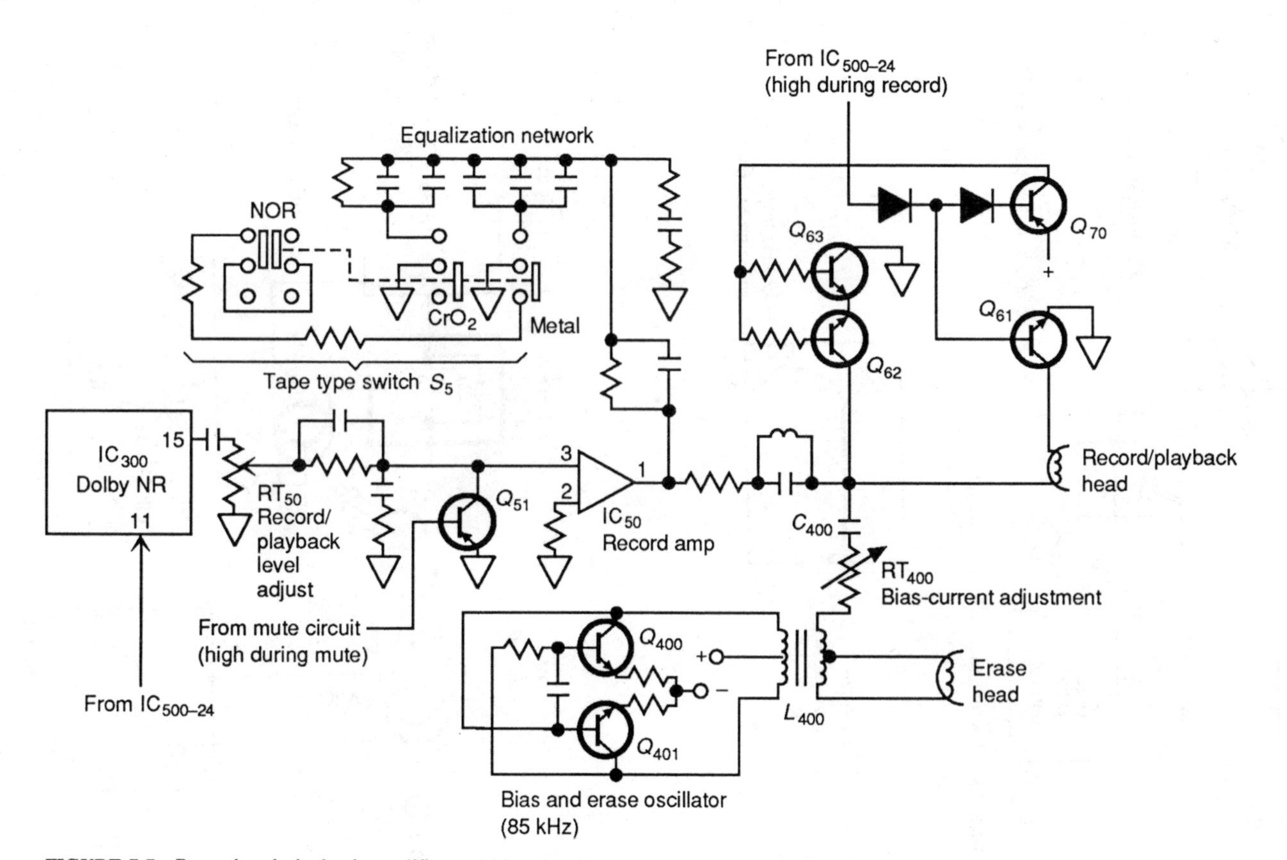

FIGURE 7.7 Record and playback amplifier and bias circuits.

7.3.2 Record/Playback Switching During Playback

Figure 7.4 shows the record/playback switching, as well as the playback amplifier used during playback. With pin 24 of IC_{500} low, both Q_{62} and Q_{63} are turned on (grounding the record input to the head), while Q_{61} is turned off (permitting the recovered audio to pass to IC_{60}). Note that the equalization network connected to IC_{60} is the standard equalization circuit for most audio cassette tapes, with 3190-μs (R_{69} and C_{62}), 120-μs (R_{64} and C_{62}), and 70-μs (R_{61} and C_{61}) time constants. The voltage gain of IC_{60} is about 49 dB at 400 Hz.

The output of IC_{60} is applied to the playback portion of the Dolby processing circuits in IC_{300} through playback-gain control RT_{60}. After processing by the Dolby circuits (Sec. 7.5), the audio is applied to the headphone amplifier and to the music-detect circuits (Sec. 7.4).

7.3.3 Headphone and Line-Out

Figure 7.5 shows the headphone and line-out circuits. Note that output level control RV_{40} controls audio to both the headphones and line (stereo amplifier). The output of the headphone amplifier is typically 80 mV into 8-Ω impedance, when the circuit is tested with a Dolby calibration tape.

Audio-bus switch Q_{530} permits the line in-out jack to be used as a bidirectional audio bus connection during system operation. Audio-bus switch IC_{01} is controlled by Q_{530} which, in turn, is controlled by the signal at pin 28 of IC_{501} (high for audio bus).

During stand-alone operation (audio-bus switch IC_{01} open), the line in-out jack acts as a line-in function only. During system operation, the line in-out jack may be used as both an in and output connection (in, switch open; out, switch closed, during playback).

7.3.4 Record/Playback Switching During Record

Figures 7.6 and 7.7 show the record/playback switching, as well as the buffer, amplifier, and filter circuits used during record. With pin 24 of IC_{500} high, Q_{70} is turned off, as are Q_{62} and Q_{63} (removing the ground from the record input to the head). Q_{61} is turned on, completing the ground connection for the record input of the head.

Record Input. As shown in Fig. 7.6, the audio at the line in-out jack (for system use) is applied to the line-in buffer circuits. The buffer keeps the line input at an impedance of about 200 Ω. The audio exits the buffer and is applied to a 19-kHz filter that removes an 19-kHz pilot signals from an FM multiplex broadcast (Chap. 6) and/or ac bias leakage. Either of these signals can trigger the Dolby NR circuits, thus upsetting the proper response characteristics of the Dolby NR function. Audio from the filter is applied to the record portion of IC_{300} (Sec. 7.5).

Audio from the microphone input can also be recorded. Operation of these circuits is identical to line-in except that audio from the microphone is applied to microphone amplifier IC_{02} as shown in Fig. 7.6.

Record Output. As shown in Fig. 7.7, audio from IC_{300} is applied to the record amplifier IC_{50} through record/playback level adjustment RT_{50}, together with signals from the bias oscillator. After processing by IC_{50}, the audio is applied to the record input of the record/playback head and recorded on tape.

Note that a fixed amount of bias current is applied to the erase head. However, the record head receives higher or lower bias current, depending on the position of the tape-type switch S_5. Bias current is also adjusted by RT_{400}.

Tape Equalization. The record amplifier consists of IC_{50}, with the associated components to compensate for record-current requirements. These components boost both the high and low ends of the frequency range. (This is commonly called *equalization* or *tape equalization*.) Tape-type switch S_5 cuts in the components as necessary to provide the correct compensation for the three basic types of tape (normal, CrO_2, and metal).

Record-Protect Functions. As shown in Fig. 7.3, record-protect switches on the cassette drive mechanism prevent IC_{500} from placing the deck in the record mode (prevent pin 24 from going high), no matter what commands are applied to IC_{500}.

When a cassette with the tabs intact is installed, the record-protect switches are closed. Scan signals from pins 9 and 10 of IC_{500} are applied to pin 17, permitting normal control by IC_{500} (that is, IC_{500} can go high if so instructed by the front-panel controls or by a remote command).

When a cassette with the tabs removed is installed, the record-protect switches are open. This prevents scan signals from being applied to pin 17, thus preventing record operation (pin 24 of IC_{500} cannot go high), no matter what commands are applied to IC_{500}.

Note that there is a separate record-protect switch for forward and reverse (on this particular deck). Remember this when troubleshooting a "no record in one direction" trouble symptom. Check that both tabs are in place on the cassette. Try another cassette or put heavy tape over the missing tab area. Refer to Sec. 7.12.

7.3.5 Record/Playback Troubleshooting

If audio is absent or abnormal during playback but the tape transport appears to be normal (tape runs properly in both directions), first check the playback gain adjustment as described in the service literature. Playback gain is adjusted by RT_{60} as shown in Fig. 7.4. Also, turn up the front-panel output control RV_{40} (Fig. 7.5). (This just might cure the problem.) If not, continue playing a known-good tape and monitor the audio at pin 7 of IC_{60} (Fig. 7.4). The output should be about 100 mV.

If there is no audio output from IC_{60}, check the circuits from the heads to IC_{60}, including Q_{61} (off) and Q_{62} and Q_{63} (on).

If there is audio from IC_{60}, check for audio at pin 18 of IC_{300} (Dolby). If there is no audio from IC_{300}, check the audio path from IC_{60} to pin 7 of IC_{300}.

Make certain that pin 24 of IC_{500} and pin 11 of IC_{300} are low (about -7 V). If they are not, suspect IC_{500}. If pin 24 of IC_{500} goes high, IC_{300} goes into the record mode rather than the playback mode.

If there is audio from pin 18 of IC_{300}, check for audio at RV_{40} (Fig. 7.5). If it is absent, make sure that Q_{40} and Q_{41} (Fig. 7.4) have not been turned on (mute condition) by a signal at pin 27 of IC_{500}. If Q_{40} and Q_{41} are not on, check the audio path from IC_{300} to RV_{40}. Then check audio from RV_{40} to the line-out jack and phones jack. If audio is available at RV_{40} but not at the jacks, suspect IC_{40}.

If playback audio is present but there is background noise that increases with time or there is a decrease in playback audio at high frequencies with time, the heads may want degaussing. Refer to Sec. 7.11.

If audio is absent or abnormal during record but the tape transport appears to be normal (tape runs properly in both directions), first check the playback circuits as just described. Always clear any playback problems before you check record. This applies to virtually all cassette decks. If there is no audio (or poor audio) during playback, the problem can be common to both record and playback. On the other hand, if the problem is only in record, you have quickly isolated the trouble to a few circuits (bias oscillator, record amp, etc.).

If it is not possible to record any audio (with good playback), start by checking for audio at pin 15 of IC_{300} (Figs. 7.6 and 7.7). Also try to cure the problem by turning up the front-panel record control RV_{01} and by adjusting the record-level RT_{50} and bias-current RT_{400}, as described in Sec. 7.10.

If audio is present at pin 15 of IC_{300}, check the audio path from IC_{300} to the heads. Also make certain that pin 24 of IC_{500} and pin 11 of IC_{300} are high (about +6 V) to place IC_{300} in record mode. If there is no audio at the input to IC_{50}, make sure that Q_{50} and Q_{51} have not been turned on (mute condition) by a signal at pin 27 of IC_{500}.

If there is audio at the heads but the deck does not record, check the bias oscillator (Fig. 7.7). As a general rule, if the tape can be erased, the oscillator and erase head are good. However, the bias signal may not be reaching the record/playback heads. Check for an 85-kHz bias signal on both sides of C_{400} and RT_{400}, as well as adjustment of RT_{400}.

Remember that each type of tape requires a different amount of bias current. For example, the bias signal measured at C_{400} on our deck is about 3.4 V for normal tape, 5.2 V for chrome, and over 10 V for metal tape. This is determined by the tape-type switch S_5 setting.

If there is audio to the heads and the bias voltage is correct, suspect the heads. Before pulling the heads (a tedious job) make sure that Q_{61} is turned on and Q_{62} and Q_{63} are off, placing the heads in a condition to record. If not, suspect Q_{61}, Q_{62}, Q_{63}, or Q_{70}.

If there is audio from pin 15 of IC_{300}, trace the audio path from the line-in and mic jacks (Fig. 7.6) through Q_{01}, Q_{02}, Q_{300}, and IC_{01}. Obviously, if you can record from line-in but not from mic, suspect IC_{02}.

Also remember that Dolby NR switch S_1 controls Q_{300}, which in turn controls the Dolby filter. However, even if the filter circuit or Q_{300} and S_1 fail, you will probably be able to record, even though the recording will be poor.

7.4 MUSIC-DETECT FUNCTIONS

Figure 7.8 shows the music-detect circuits. These circuits locate music passages on the tape. This function is necessary for scan-and-play capabilities as well as for random-access operation. Although the function is called music-detect, the circuits operate on signals of any kind (voice, tone, etc.) as long as the signals are substantially above the background level.

The left and right audio signals are summed together through R_{30L} and R_{30R}. The summed signals are applied to the music-detect circuits within IC_{500} through

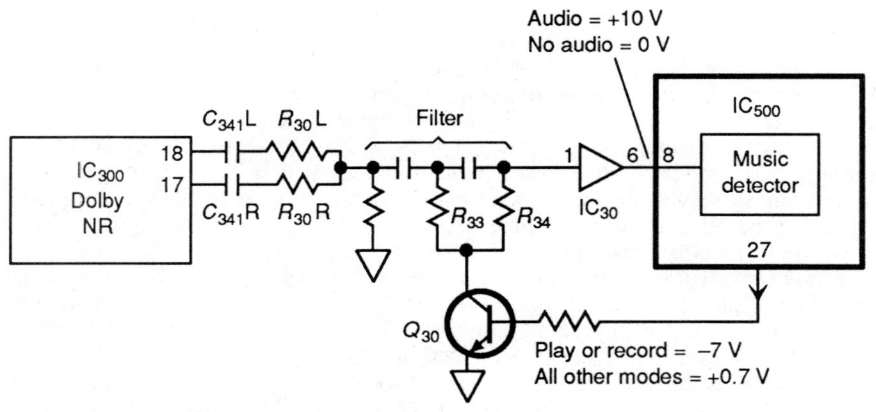

FIGURE 7.8 Music-detect circuits.

a filter R_{33} and R_{34} and amplifier IC_{30}. The frequency-response characteristics of the filter are determined by the status of Q_{30}.

The output of IC_{30} is either about +10 V when there is music (or other recorded material) or 0 V when there is no music (or other signal substantially above the background level). This output is applied to pin 8 of IC_{500} and causes IC_{500} to take the appropriate action (supply signals to the tape-transport mechanism to produce scan-and-play and/or random-access operation).

Note that *during normal play*, the line-mute output at pin 27 of IC_{500} goes low, turning Q_{30} off. This keeps the filter attenuation characteristics normal. *During high-speed search*, Q_{30} is turned on by a high at pin 27 of IC_{500}. This grounds R_{33} and R_{34} and increases attenuation provided by the filter. More attenuation is necessary since signal amplitude is increased when tape is moved across the head at search speed. Such an increase in signal amplitude can be mistaken for the presence of music (or other signals).

7.4.1 Music-Detect Troubleshooting

If there is no scan-and-play operation (tape does not stop at programmed selections), the music-detect circuits are logical suspects. However, before you go directly to music-detect, make sure that there is audio at pins 17 and 18 of IC_{300}. If not, troubleshoot the playback circuits as described in Sec. 7.3.5.

If there is audio from IC_{300} at R_{30L} and R_{30R}, check at pin 6 of IC_{30} (or pin 8 of IC_{500}) for about +10 V when music is present and 0 V when there is no music. If practical, play a tape known to have both recorded and blank (no audio) passages.

If the voltage at pin 6 of IC_{30} switches between +10 V and 0 V but there is no music-detect, suspect IC_{500}.

If there is no change at pin 6 of IC_{30}, check for a change in audio at pin 1 of IC_{30}. The signal is very low at this input, but there should be a change between audio and no audio. If there is a change at pin 1 but not at pin 6, suspect IC_{30}.

If the voltage at pins 1 and 6 of IC_{30} does not switch, select scan-and-play (S&P) and check for about +0.7 V at the base of Q_{30}. If it is absent, suspect IC_{500}

(check for a mute signal of about +0.7 from pin 27 of IC_{500}). If the base of Q_{30} is at +0.7 V but Q_{30} is not turned on, suspect Q_{30}. Remember that the signal at pin 1 of IC_{30} is attenuated drastically when Q_{30} is turned on.

7.5 DOLBY PROCESSING

Figure 7.9 shows the Dolby processing circuits. Note that Dolby processing functions are combined in a single module (IC_{300}), commonly called an MD_{300} Dolby noise reduction (NR) module. The use of a single module for Dolby processing is quite common for decks used in present-day home-entertainment systems.

Figure 7.9*a* shows the functions performed within the Dolby module. Figure 7.9*b* shows the basic functions of a tape deck with Dolby. In any tape deck, undesirable "hiss" noise (made up predominantly high frequency) is introduced into the audio path at various points (typically at the tape record/playback heads, playback amplifier, and bias circuit). This annoying noise can be reduced by passing the signals through processing circuits as is done in many non-Dolby decks. However, the objectionable noise does not have the same effect on signals of various amplitudes and frequencies. For example, the noise is most objectionable and noticeable in signals of *low amplitude* and *high frequency*.

With Dolby NR, the low-level and high-frequency signals are boosted by the same amount during playback that the signals were suppressed during record. This makes the output signal an exact reproduction of the input signal, but with the noise component suppressed.

As shown in Fig. 7.9*a*, the Dolby module IC_{300} is controlled by mode-select signals at pin 9 and record/playback signals at pin 11. *During record*, pin 11 of IC_{300} is set to about 6 V by IC_{500}. This causes the Dolby record circuits to turn on and the playback circuits to turn off. Audio signals to be recorded are then applied through amplifiers and the 19-kHz filter to pins 3 and 4 of IC_{300} (Sec. 7.3). *During playback*, pin 11 of IC_{300} is set to about −7 V by IC_{500}, causing the Dolby playback circuits to turn on and the record circuits to turn off.

The audio signals can be processed for either Dolby B or Dolby C or can be recorded without any Dolby processing, depending on the voltage at pin 9 of IC_{300}. In turn, the voltage at pin 9 is set by Dolby NR switches S_1 (on-off) and S_2 (B/C), as shown in Fig. 7.6.

Audio signals taken from tape by the head during playback are amplified and applied to pins 7 and 8 of IC_{300}, as discussed in Sec. 7.3. These signals can be processed for either Dolby B or Dolby C or can be played back without any Dolby processing, depending on the voltage at pin 9 of IC_{300} (as set by Dolby NR switches S_1 and S_2).

7.5.1 Dolby Circuit Troubleshooting

Before you condemn any Dolby circuit, make certain that the tape is being played back in the *same mode as during record*. If Dolby is not used during record, do not use Dolby during playback (in a hopeless attempt to improve quality). If Dolby C is used during record, play back in Dolby C, not Dolby B, and so on. If there is no substantial difference in sound quality, with and without Dolby (of the correct type, B, or C), make the following checks.

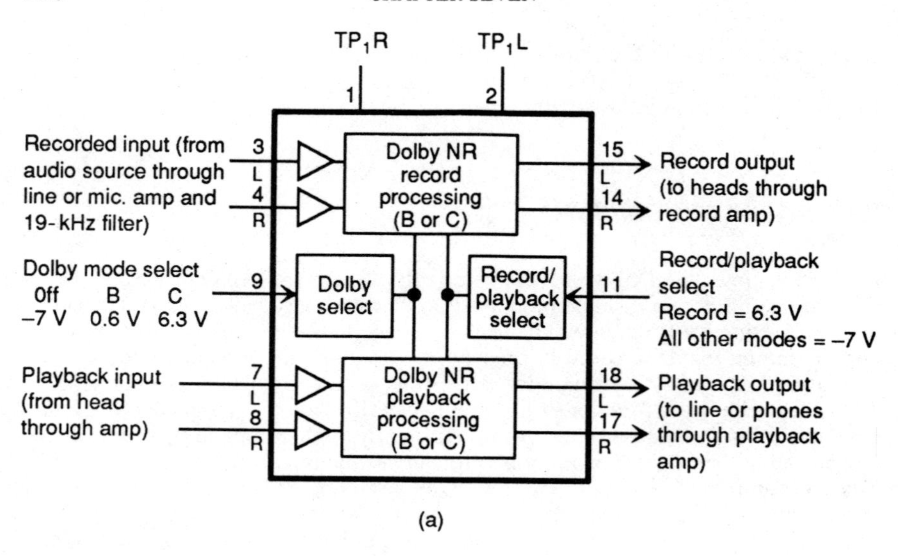

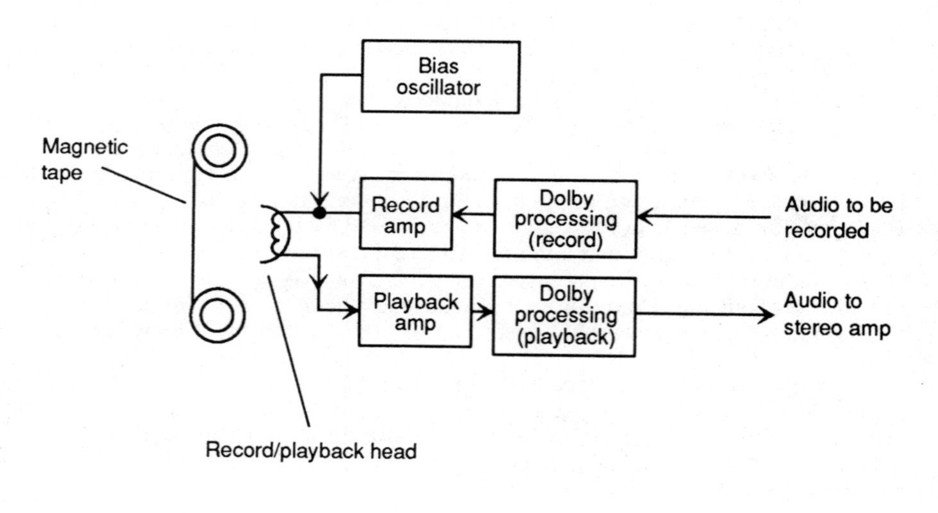

FIGURE 7.9 Dolby processing circuits.

During playback, check for correct voltage at pin 9 of IC_{300} in each mode (−7 V for Dolby off, 0.6 V for Dolby B, and 6.3 V for Dolby C). If voltages are not correct, suspect S_1, S_2, and Q_{300}. If voltages are correct at all modes but there is no substantial difference in sound quality between Dolby and non-Dolby, suspect IC_{300}.

7.6 TAPE-TRANSPORT FUNCTIONS

Figure 7.10 shows the tape-transport circuits, and Figure 7.11 shows the location of the major tape-transport components. We do not show full details of the tape-transport here since the transport is well illustrated in the service literature for our deck (and in the literature of most decks). Instead, we concentrate on the components involved in troubleshooting.

The *forward* and *reverse solenoids* operate the pinch *rollers* that hold the tape against the corresponding capstans, while the *pause solenoid* stops tape drive (even though the *capstan motor* continues to rotate when the tape transport is turned on).

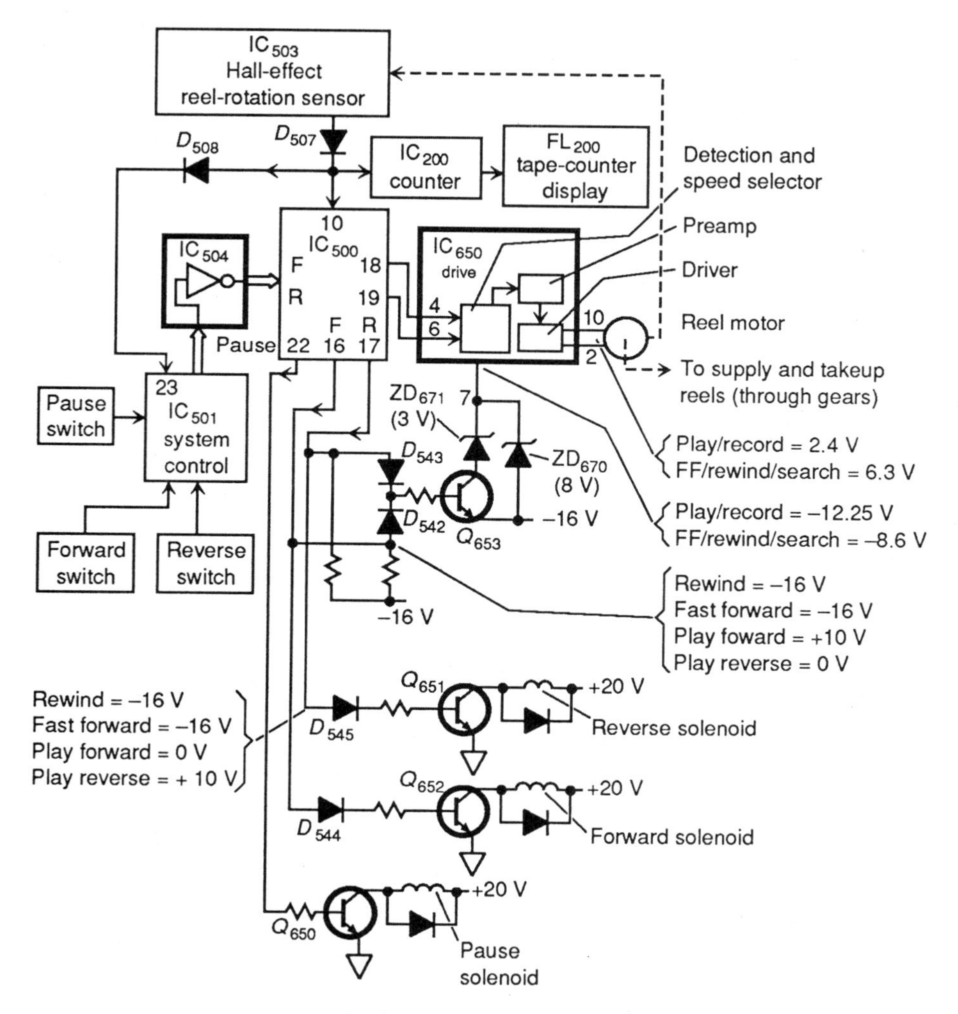

FIGURE 7.10 Tape-transport circuits.

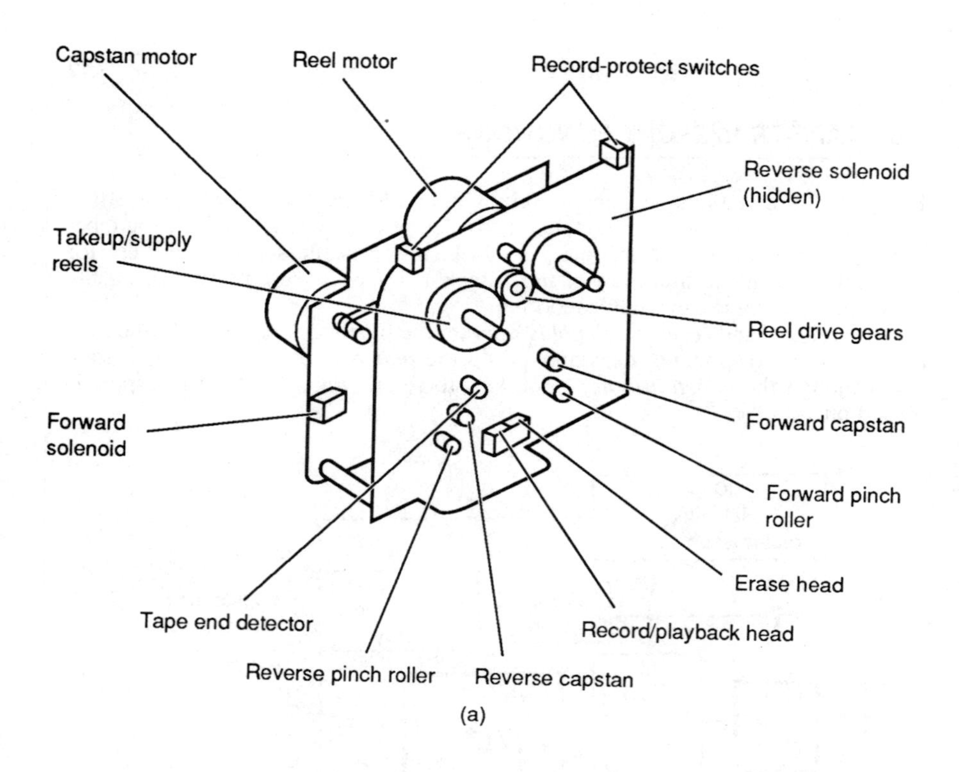

(a)

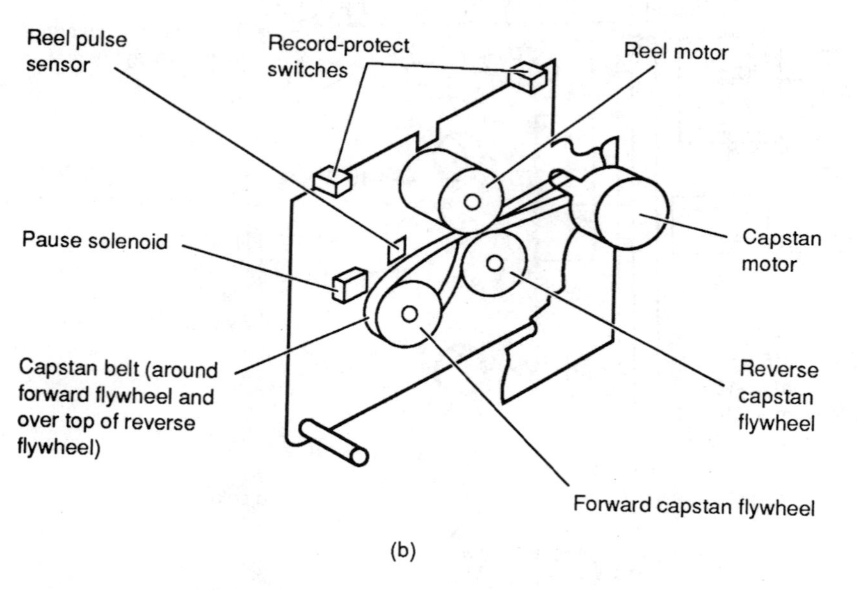

(b)

FIGURE 7.11 Tape-transport components.

The three solenoid-control circuits are essentially the same. When pins 16, 17, or 22 of IC_{500} are high, the corresponding transistors Q_{652}, Q_{651}, and Q_{650} are turned on, the solenoids are energized, and the functions (forward, reverse, or pause) are activated.

The two *reels* (*supply* and *takeup*) are geared to a single motor. Each reel assumes the supply or takeup function, depending on the operating mode, forward or reverse. The reel motor can be driven in either direction and at two different speeds through operation of the reel-motor drive IC_{650}. In turn, IC_{650} receives control signals and voltages from IC_{500}. The direction of reel rotation (forward or reverse), as well as the reel speed (fast or normal) depends on the signals applied to IC_{650} from IC_{500}.

The direction of reel-motor rotation is determined by the logic at pins 18 and 19 of IC_{500}. With pin 18 high, the reel motor rotates forward. With pin 19 high, the motor reverses direction. These signals are applied to the reel motor through circuits within IC_{650}. With both pins 18 and 19 of IC_{500} low, the reel-motor voltage at pins 2 and 10 of IC_{650} drops to zero, and the reel motor stops.

Reel-motor speed is determined by the logic at pins 16 and 17 of IC_{500} (which is also applied to the forward and reverse solenoids). The logic is tri-state. That is, pins 16 and 17 can be high (+10 V), low (0 V or ground), or "floating" (where the output of IC_{500} can be pulled in either direction). For example, when pins 16 and 17 are both floating, both pins go to −16 V. When one pin is floating, that pin goes to −16 V, but the other pin goes to 0 V or +10 V, as determined by the command within IC_{500}.

During fast forward, where high speed is required, both pins 16 and 17 float and thus go to −16 V. This turns Q_{653} off, removing the 3-V zener ZD_{671} from the circuits. With ZD_{671} out, −8.6-V zener ZD_{670} applies about −8.6 V to pin 7 of IC_{650}. This causes the reel-motor drive voltage at pins 2 and 10 of IC_{650} to go to about 6.3 V, driving the reel motor at fast speed (about 3 times normal speed).

During forward play, where normal speed is required, pin 17 goes to 0 V. Since this is well above the Q_{653} emitter bias of −16 V, Q_{653} turns on, producing about −12.25 V at pin 7 of IC_{650}. This causes the reel-motor drive voltage at pins 2 and 10 of IC_{650} to go to about 2.4 V, driving the reel motor at normal speed.

As the reel rotates in either direction, Hall-effect generator IC_{503} produces pulses. These pulses are applied to the front-panel fluorescent tape-counter display FL_{200} through IC_{200}, producing the four-digit tape count (Sec. 7.8).

The same IC_{503} pulses are applied to IC_{500} through D_{507}, and to IC_{501} through D_{508}. If the reel stops because of some mechanical problem (jammed reel, stuck tape, etc.), the pulses stop. This causes IC_{500} to remove the drive to the reel motor and IC_{501} to shut the system down.

7.6.1 Tape-Transport Troubleshooting

From a troubleshooting standpoint, the tape transport consist essentially of the reel motor, capstan motor, and solenoids (pinch rollers and pause), all of which function to drive the tape across the heads in both forward and reverse directions. So let us examine a possible failure in any of these components.

If the deck stops shortly after start in all modes, check the reel-rotation pulses. The fact that the transport is shut down when reel-rotation pulses from IC_{503} are stopped creates a troubleshooting problem (on many decks). For example, if IC_{503} is defective or if the pulses do not reach IC_{500} and IC_{501}, the transport stops, even though there is no mechanical problem. Once the reel stops, there are no further pulses, so the system stops anyway.

The simplest way to check this condition is to monitor the pulses at pin 10 of IC_{500} and/or pin 23 of IC_{501}, when the reel *first starts to rotate* after a stopped condition. If the pulses are present (even briefly) but the transport starts and then shuts down, IC_{500} and IC_{501} are the most likely suspects. (If the reel motor stops first, suspect IC_{500}.) However, if there are no pulses when the reel starts, suspect IC_{503} and/or D_{507} and D_{508}.

If the tape-transport operation appears abnormal (sluggish, erratic, tape slippage, etc.), the mechanism may require cleaning and/or lubrication, as described in Sec. 7.10.

If the capstan motor does not rotate, check both the motor and power wiring. On our deck, the capstan motor should start rotating when power is first applied and should continue regardless of the operating mode. If it does not, check the power wiring to the capstan motor. If power is available but the capstan motor and belt are not turning, suspect the motor.

If the tape fails to move in either direction, with the capstan motor running, check the solenoids and reel motor. On most decks, the forward and reverse solenoids actuate and pull the pinch roller against the tape and capstan when forward or reverse is selected (in both playback and record). Similarly, the reel motor should drive the supply and takeup reels through gears when forward or reverse is commanded.

If the reel motor does not operate when forward operation is selected, check for +10 V at pins 16 and 18 of IC_{500}. If the voltage is absent at either pin, check that the forward command is applied to IC_{500} from IC_{501} through IC_{504}. If it is not, suspect the forward-play switch, IC_{501}, or IC_{504}.

Next, check that the voltage at pin 7 of IC_{650} is −12.25 V. If it is not, suspect Q_{653}, ZD_{671}, or ZD_{670}. If the voltage is correct at pin 7 (and pin 4) of IC_{650} but not at pins 2 and 10, suspect IC_{650}. If the voltage is correct at pins 2 and 10 but the reel motor does not turn, suspect the reel motor (or possibly jammed reel drive gears).

If the reel motor does not operate when reverse operation is selected, check for +10 V at pins 17 and 19 of IC_{500}. Then check Q_{653}, ZD_{671}, ZD_{670}, IC_{650}, IC_{501}, IC_{504}, the reverse switch, and the reel motor, as described for forward operation.

If the forward solenoid does not actuate when forward operation is selected, check for +10 V at pin 16 of IC_{500}. If the voltage is present, suspect Q_{652}, D_{544}, or the forward solenoid. If the voltage is absent at pin 16 of IC_{500}, check that the forward-play command is applied to IC_{500} from IC_{501} through IC_{504} when the forward command is given. If it is not, suspect the forward play switch, IC_{501}, or IC_{504}.

If the reverse solenoid does not actuate when reverse operation is selected, check for +10 V at pin 17 of IC_{500}. Then check Q_{651}, D_{545}, the reverse solenoid, and for a reverse command through IC_{501} and IC_{504}, as described for forward operation.

7.7 TAPE STOP AND REVERSAL CIRCUITS

Figure 7.12 shows the tape end-detect circuits. These circuits provide for both automatic stop and reversal of the tape.

The tape end-detect circuits are used primarily to sense the leader at both ends

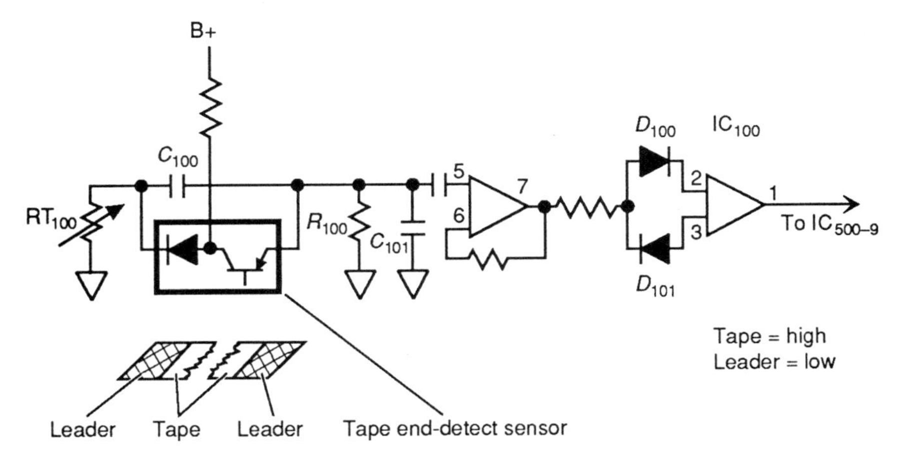

FIGURE 7.12 Tape end-detect circuits.

of the cassette tape. Most decks have some similar circuit to stop the tape automatically when the tape reaches either end. In our deck, the function also permits the tape to reverse directions and keeps the gap in program material to a minimum (during reversal).

Input to the circuit is from the tape end-detect sensor (shown in Fig. 7.11). The sensor is a combination LED and phototransistor. The LED produces a light source that is applied to the tape surface, while the phototransistor receives light reflected from the tape. The output of the sensor is applied to pin 9 of IC_{500} through two amplifiers within IC_{100}. The amount of current through the LED (and thus the amount of light generated) can be adjusted by RT_{100}.

The reflected light remains constant as long as oxide-coated tape passes by the sensor. When the leader is reached (at either end), there is a drastic change in reflected light. This changes the output to pin 9 of IC_{500}, indicating that the tape is near the end. IC_{500} then causes the tape transport to reverse direction, or to stop, depending on the operating mode.

One benefit of the reflective system is that it works equally well on any type of leader (metal, clear, or translucent). Also, the automatic-reverse feature occurs even if no leader is present on the tape. When the tape moves to the end of the reel, the reel hub stops rotating. This is sensed by IC_{500}, which reverses or stops the tape.

During normal operation, a fixed amount of light is reflected onto the phototransistor, producing a fixed voltage across R_{100}. This allows C_{100} and C_{101} to charge to the same voltage. The voltage remains constant as long as the reflected light is constant (oxide coating). When the leader moves under the phototransistor, there is a drastic change in reflected light, changing the voltage across R_{100}.

If the leader is metal, the reflectivity goes up, causing the drop across R_{100} to increase rapidly. In turn, this causes C_{100} and C_{101} to charge to a higher level, producing a positive "pulse" at pin 5 of IC_{100}. If the leader is clear or translucent, the reflectivity goes down, causing the drop across R_{100} to decrease, producing a negative "pulse" at pin 5 of IC_{100}.

No matter what type of leader is involved, a pulse appears at pin 7 of IC_{100}

when the leader is reached. The positive or negative pulse is applied through D_{100} or D_{101} to pins 2 or 3 of IC_{100}. In turn, this produces a pulse at pin 1 of IC_{100}. The pulse (applied to pin 9 of IC_{500}) is high when the oxide-coated tape is passing the sensor and goes low when the leader is reached.

7.7.1 Tape Stop and Reversal Troubleshooting

If the tape does not stop (or reverse) when the end is reached, look for a low at pin 9 of IC_{500} when the leader is over the sensor. If pin 9 is low but the tape does not stop (or reverse), suspect IC_{500}. Remember that IC_{500} causes the reel motor to stop (or reverse direction). However, the capstan motor always rotates (even when the tape is stopped) and always in the same direction. Tape reversal is accomplished when the solenoids (Sec. 7.6) release the tape from one capstan and hold the tape against the other capstan. Tape is stopped when the solenoids release the tape from both solenoids (and the reel motor stops). If any of these functions do not occur, with pin 9 of IC_{500} low, suspect IC_{500}. Refer to Sec. 7.6.

If there is no low at pin 9 of IC_{500} when the tape is over the leader, try correcting the problem by adjusting RT_{100} as described in Sec. 7.10. If the problem is not corrected by adjustment, suspect the tape end-detect sensor, IC_{100}, D_{100}, D_{101}, or the associated circuit. Note that there should be about 5.5 V across C_{101} without a cassette in the deck and with the play button held down.

7.8 FRONT-PANEL DISPLAY

Figure 7.13 shows the front-panel display drive circuits. Note that counter microprocessor IC_{200} is responsible for generating all grid- and segment-drive signals applied to the front-panel fluorescent display FL_{200}. Also note that operation of FL_{200} is very similar to the operation of the display on our amplifier (Chap. 5). Therefore we do not duplicate these descriptions here. Instead, we concentrate on drive signals to FL_{200}.

There are essentially three drive sources or inputs to IC_{200}: one from the counter circuits, one from the audio-level circuits, and one from the tape-transport control.

The counter circuit pulses or inputs originate at reel sensor IC_{503} (Sec. 7.6). Also, a signal from system control IC_{501} is applied to IC_{200} through Q_{519} as a reset pulse for the counter.

The audio-level circuits receive both left- and right-channel audio (samples from either playback or record circuits) through RT_{200L} and RT_{200R}. These controls adjust the left and right audio channels to the same level. The audio-level circuits generate left and right dc control signals applied to pins 6 and 7 of IC_{200}. The circuits also generate a reference voltage applied to pin 5.

The control and reference signals are applied to audio-level circuits in IC_{200}. The output of the audio-level circuits is applied to the decoder and display-drive logic. The drive logic produces the necessary signals to both the grids and segments of FL_{200}. These signals activate the 20-segment light bars (bar graph) to indicate the audio level of both channels.

The tape-transport microprocessor IC_{500} transmits status signals to pin 13 of IC_{200}. These status signals are a combination of data and sync information from

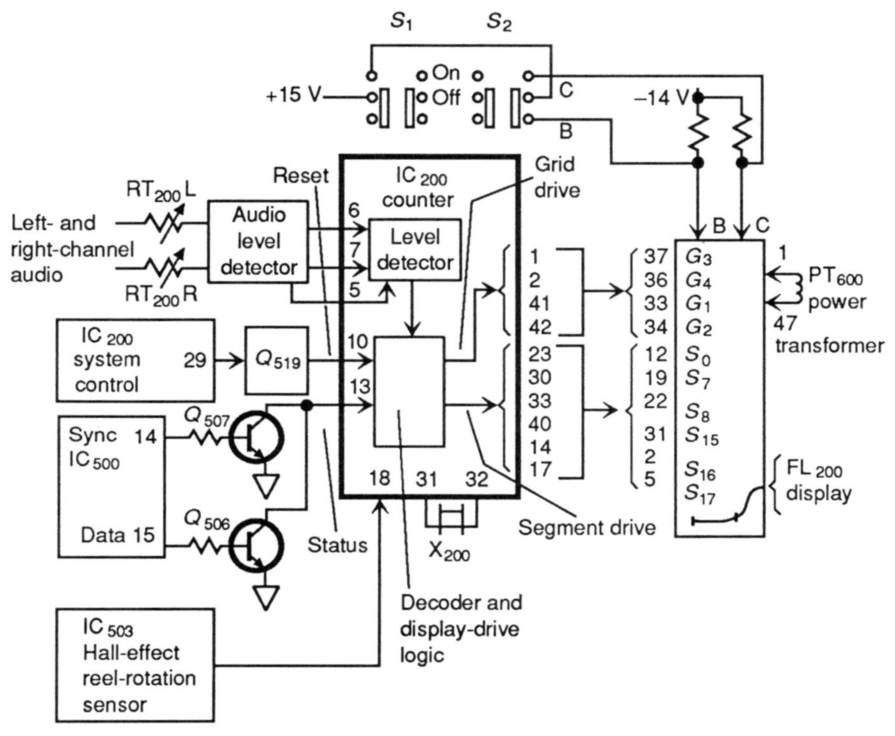

FIGURE 7.13 Front-panel display drive circuits.

pins 15 and 14 of IC_{500}, respectively. The input to pin 13 of IC_{500} is applied to the decoder and display drive logic. In turn, the logic generates four grid-drive signals and 18 segment-drive signals. These signals activate various function indicators on FL_{200} (such as play, record, pause, mute, tape counter, etc.).

7.8.1 Front-Panel Display Troubleshooting

If the front-panel display is inoperative, suspect both the fluorescent display FL_{200} and the display drive IC_{200} since IC_{200} controls FL_{200}. Of course, if a particular input is absent or abnormal, the display cannot operate properly. So always check the input before you condemn the display.

If the fluorescent display is totally dark, check for ac power to the filaments at pins 1 and 47. Then check for grid drive or voltage at pins 33, 34, 36, and 37. (As a general tip, if any part of the display is on, it is reasonable to assume that FL_{200} is probably good and getting power, but do not count on it.)

If the fluorescent display is on but a particular display in inoperative, check the input signal to FL_{200} through IC_{200}. If the input to IC_{200} is correct but the display is not, check the segment signals from IC_{200} to FL_{200} (if you can find which segments are involved from the service literature; this is not always easy). For example, assume that there is no counter display, but other display functions are normal. Since the counter pulses originate at reel sensor IC_{503}, suspect IC_{503} if the pulses do not appear at pin 18 of IC_{200}. Also, suspect IC_{501} or Q_{519} if counter reset pulses are absent at pin 10 of IC_{200}.

Note that on the fluorescent display of our deck, the settings of the Dolby switches S_1 and S_2 are fed directly to FL_{200}, bypassing IC_{200}. Always look for such unusual connections.

7.9 COMMAND CIRCUITS

Figure 7.14 shows the command circuits. Note that the commands can come from the remote unit (through the system) or from the front-panel buttons. Also note that these circuits are similar (but not identical) to those for the amplifier (Chap. 5).

Most modern decks have some similar circuits for converting front-panel commands (pushing a key, button, or switch) or system commands (possibly on a control bus) into electrical signals that operate the tape transport and corresponding front-panel display. In our deck, key-scan pulses are produced in sequence from pins 9 through 12 of IC_{501} and applied to the key-scan matrix (front-panel operating switches). These four scan signals (S_0 through S_3) are coupled back into IC_{501} through four input ports (K_0 through K_3) at pins 13 through 16 of IC_{501} when the appropriate button is pressed.

For example, if the stop button is pressed, the S_1 output is applied to the K_0 input. This causes IC_{500} to stop the tape transport in both playback and record.

Note that all of the function-command outputs from IC_{501} are single control signals, except for program, memory and S&P. For example, a mute command requires a single signal at pin 35. All of these single control signals are inverted (by IC_{504} or Q_{517} and Q_{518}) before application to IC_{500} (the mute command high at pin 35 of IC_{501} appears as a low at pin 36 of IC_{500}).

The program, memory, and S&P functions use a separate scan-out/scan-in matrix. This is necessary to synchronize IC_{501} with IC_{500} during such complex functions as memory and scan-and-play. For example, when S&P is pressed, the high at pin 30 of IC_{501} is applied to a NAND gate in IC_{502}. The other input of the NAND gate receives continuous scan pulses from pin 32 of IC_{500}. These scan pulses are produced by IC_{500} and are synchronized with the IC_{500} clock. The combination of two highs at the NAND-gate input produces a low at the base of Q_{515} and causes pin 33 of IC_{500} to go high (each time pin 32 of IC_{500} goes high). With these two signals synchronized, IC_{500} causes the tape-transport (and play) circuits to perform the scan-and-play function.

7.9.1 Command Troubleshooting

If there are no front-panel or remote commands, try selecting a front-panel command and check for a high at the corresponding output pin of IC_{501}. If the appropriate pin is not high, suspect IC_{501}. Next, check for a low at the corresponding pin on IC_{500} when a front-panel command is selected. If the appropriate pin is not low, suspect IC_{504}, Q_{517}, or Q_{518} (depending on the mode selection). If inputs to IC_{501} are low when the mode is selected but the command is not performed, suspect IC_{500}.

In the case of commands that require synchronization, check both the com-

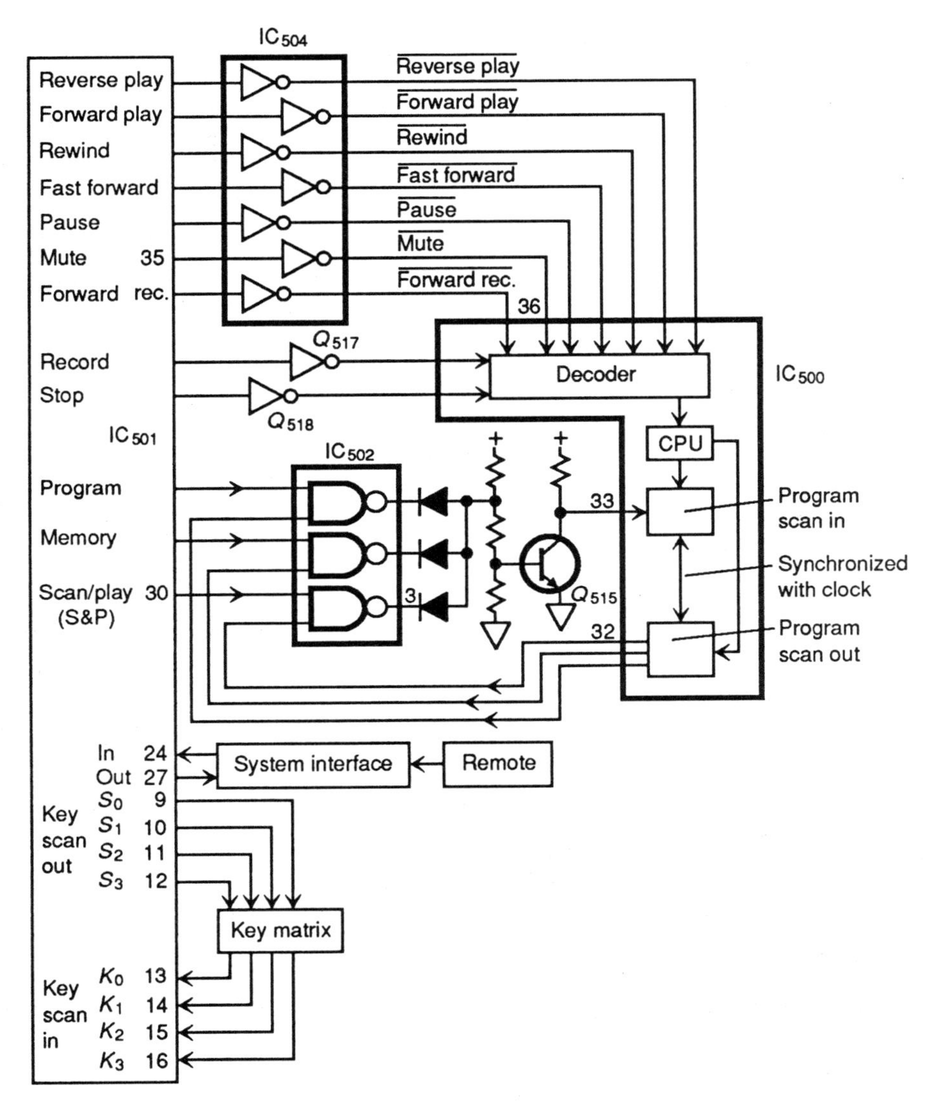

FIGURE 7.14 Command circuits.

mand and sync signals at IC_{500}. For example, press S&P, and check for a high at pin 30 of IC_{501}. If it is absent, suspect IC_{501}. Next, check for a scan pulse at pin 32 of IC_{500}. If it is absent, suspect IC_{500}. If it is present, check for a low at pin 3 of IC_{502}. If it is absent, suspect IC_{502}. Next, check for a high at pin 33 of IC_{500}. If it is not high, suspect Q_{515} or the coupling diode. If pin 33 of IC_{500} is high but the deck does not enter scan-and-play, suspect IC_{500}.

If remote commands are absent or abnormal, with good front-panel operation, check for signals at pins 24 and 27 of IC_{501}.

7.10 TYPICAL TESTING AND ADJUSTMENTS

This section describes typical electrical testing and adjustments for our deck. Most decks require one or more of the test and adjustment procedures described here, even though the adjustment points may be different. On the other hand, there are virtually no universal or typical mechanical adjustments, so always consult the service literature. Generally, the mechanical test and adjustment procedures found in service literature are well written and illustrated. This is not always true for electrical adjustments (which often omit the purpose of the test or adjustment).

7.10.1 Test Tapes and Cassettes

You need test tapes to properly test and adjust any deck. Our deck requires four special test tapes or cassettes, in addition to three blank cassettes (standard or normal, chrome, and metal). They are a mirror-tape for tape-travel checks, an 8000-Hz tape for azimuth adjustment, a 3000-Hz tape for motor-speed adjustment, and a Dolby tape for Dolby NR checks.

7.10.2 Preliminary Setup

Make the following checks and control settings before and tests and/or adjustments:

1. Clean the heads, pinch rollers, tape guide, and capstan using alcohol, as discussed in Sec. 7.11.
2. Set the record and output level controls to maximum.
3. Set the Dolby NR switch to off. (Always make all tests and adjustment with Dolby off, unless you are checking a Dolby function.)
4. Set the tape-type switch to NOR 1, CrO_2, or metal as required by the type of tape installed.

7.10.3 Tape Speed Testing and Adjustment

The purpose of this test and adjustment is to make certain that the tape motor is rotating at the correct speed. Some decks have adjustment controls for drive circuits to the motor. Our deck has a semifixed variable resistor located inside the motor. Generally, adjustment is not required unless the motor is replaced. However, the speed should be checked at every service period.

Connect a frequency counter to the line output terminals (or at any convenient point where playback audio can be measured).

Load a 3000-Hz test tape, and play the tape in the normal forward mode (do not record).

After a warm-up of about 20 min, check the frequency reading the frequency counter. The ideal frequency reading is 3000 Hz, which indicates that the motor speed is correct. A typical speed (frequency) tolerance is ± 1 percent (30 Hz), so any reading between 2970 and 3030 Hz is satisfactory (but check this against the service literature).

If necessary, adjust the motor speed resistance until the motor speed is within tolerance (preferably with a frequency reading of exactly 3000 Hz). Always allow sufficient time for the motor speed to stabilize as you make this adjustment.

Make the initial test and/or adjustment at some point near the middle of the tape. Then check to see if there is any difference in frequency reading at points near the beginning and end of the tape. Any differences in frequency reading indicate tape speed fluctuations.

If you notice any tape-speed fluctuations (almost all decks have some speed fluctuations), find the percentage of fluctuation using the following equation:

$$\text{Tape speed fluctuation} = \frac{F_1 - F_2}{3000} \times 100\%$$

where F_1 = maximum reading, and F_2 = minimum reading.

A typical speed fluctuation tolerance is ± 1 percent. This means that the difference between the maximum and minimum frequency readings should be 30 Hz or less.

If the motor speed cannot be brought within tolerance or if the speed fluctuations are greater than 1 percent this usually means that the motor must be replaced.

7.10.4 Azimuth Test and Adjustment

The purpose of this test is to make certain that the heads are properly adjusted in relation to tape travel. Ideally, the heads should be centered so that the output (amplitude and phase) is the same from each head, in both forward and reverse modes. Typically, if the heads are not centered, one head produces more output when the tape moves in one direction and less output when the tape reverses direction. Remember this when troubleshooting such a symptom. Of course, if there is severe unbalance, the problem is most likely a defective head rather than azimuth adjustment.

Most decks have some form of azimuth adjustment. Figure 7.15*a* shows the head assembly (and adjustment screws) for our deck. Note that there are two azimuth screws and two W-nuts. The W-nuts are for adjusting the tape guides so that tape travel is straight and true in *relation to the assembly*. The azimuth screws are for adjusting the tape *in relation* to the tape path.

A word of caution before going on. *Never make any mechanical adjustments to the heads or head assembly* until you are certain that such adjustment is required. The heads do not often "go out of alignment" with normal use, particularly if the head adjustment screws are sealed with "screw lock" at the factory or at last service. However, the heads should be tested for proper azimuth at each service.

Connect a meter or scope to the line output (or any point where playback can be measured). A dual-channel scope is convenient since you can monitor both channels simultaneously. You can also use a dual-channel meter (Fig. 4.1).

Load an 8000-Hz test tape (or a tape of the frequency specified in service literature). First play the tape forward and note the reading on both channels. Then play the tape in reverse and note both channel readings.

In our deck, the maximum deviation in amplitude between the L and R channels and/or between the forward and reverse operation is 2 dB (which is much easier to read on a meter than a scope). If the difference is 2 dB or less, *leave the heads alone*.

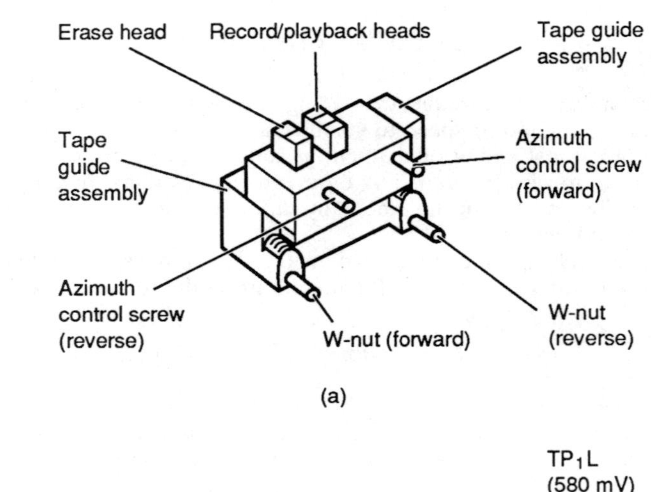

(a)

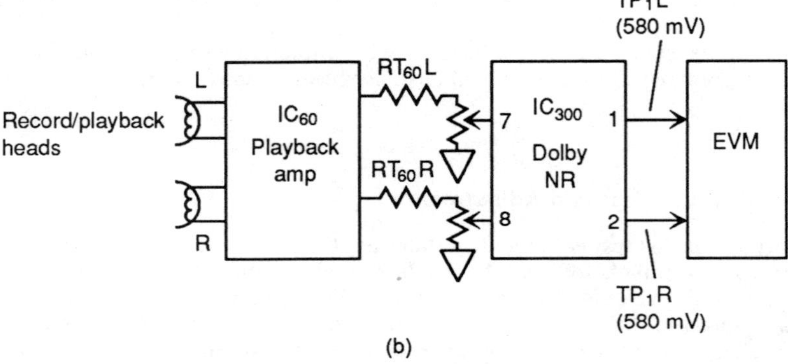

(b)

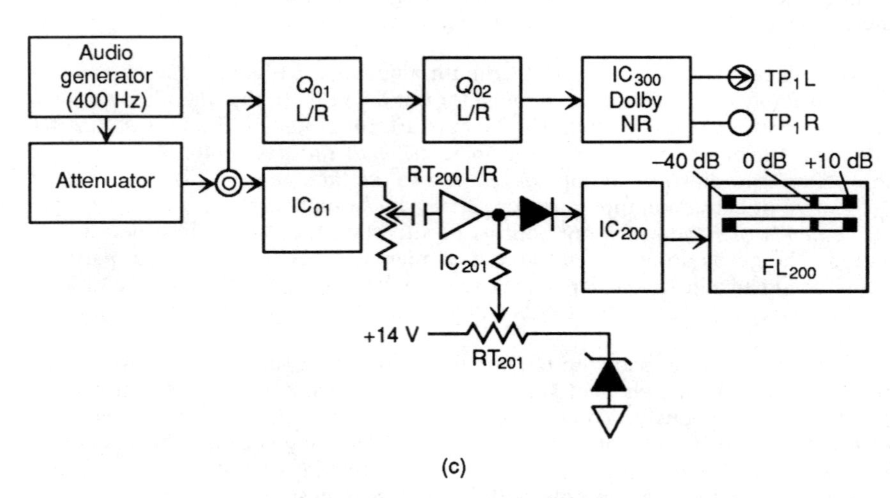

(c)

FIGURE 7.15 Typical head assembly, playback gain adjustment, and front-panel meter adjustment.

If you must adjust the azimuth, here are some tips to make it easier. Use one channel as a reference point for both voltage amplitude measurement and phasing comparisons. Adjust the azimuth screws (Fig. 7.15*a*) for maximum voltage output. In our deck, the right-side azimuth screw is adjusted in the forward mode, while the left-side screw is for reverse mode. Remember that there is interaction between these adjustments (another good reason for not tampering with head adjustments).

Once both heads are adjusted for maximum output, compare the phase of the two output signals on the dual-channel scope. If the phase is within 90° (between the left and right channels), quit while you are ahead. If not, make a *slight* adjustment of one channel (but not both) to bring the phase difference within 90°.

Remember that if there is a severe unbalance condition in the audio circuits, do not try to correct the problem by head adjustment. Find the problem.

After the azimuth adjustments are complete, remove the 8000-Hz test tape. Load a mirror tape and play it in the forward direction. Check tape travel across the heads. The tape travel should be straight and true. There should be no "tape curl." If necessary, adjust the tape guides (with the W-nuts, in our deck) so that the mirror tape reflects straight and true tape travel.

Always recheck the azimuth adjustments after making the tape path adjustments (or any other adjustments your particular head assembly requires). As a general rule, all head adjustments interact.

On our particular deck, where the heads are physically rotated in forward and reverse, check that the azimuth screws are not loosened when the heads are rotated. When you are certain that the heads are properly adjusted, apply screw lock to all adjustment screws.

7.10.5 Playback Gain Adjustment

Figure 7.15*b* is the adjustment diagram. The purpose of this test is to set playback gain for both channels. Connect a dual-channel meter (or two meters) to TP_{1L} and TP_{1R} of Dolby noise-reduction IC_{300}. Load a Dolby calibration tape. Play the tape in the forward direction, and adjust RT_{60L} and/or RT_{60R} to get 580 mV at each test point. Note that the output of most Dolby units is typically in the ½-V range, but always check the service literature.

7.10.6 Output-Level Readout Adjustments

Figure 7.15*c* is the adjustment diagram. The purpose of this test is to set the front-panel light bars to produce the correct indication for a given audio level. This adjustment should not be performed until the playback gain is set. The first step involves setting the readout for 0 dB.

Connect an audio generator to the line-in terminals and inject a 400-Hz signal. Adjust the output level of the generator to produce the same reading at TP_{1L} and TP_{1R} of Dolby IC_{300} as is produced by the Dolby calibration tape during playback gain adjustment (580 mV, Sec. 7.10.5). Then adjust RT_{200L} and/or RT_{200R} until the 0-dB segment of the readout *just begins to turn on*.

The second step involves setting the readout of −40 dB. Leave the generator at 400 Hz but connect an atteunator between the generator and line-in terminals (or use the generator attenuators) to attenuate the generator output by −37 dB. Adjust RT_{201} until the −40-dB lamp *just turns off*.

7.10.7 Bias-Current and Record/Playback Output-Level Adjustments

Figure 7.16*a* is the adjustment diagram. Both the bias-current and record/playback output level should be adjusted together (on our deck). The two functions are usually interactive on most decks.

On some decks, the actual head current (bias or erase) is measured and adjusted. This is done by inserting a resistor (typically 1 k so that the readings are in millivolts) at a test point in the head circuit and measuring the voltage drop across the resistor. The head current is then adjusted to a given value.

On our deck, the audio is recorded and played back at a low frequency and then at a high frequency. Head current is adjusted until the playback level is the same (within a given tolerance) at *both low and high frequencies*. Even on decks where the service literature specifies a given head current, it is a good idea to check head current over the full audio range. Drastic differences in head current at high- and low-frequency limits often indicate problems.

For bias-current adjustment, connect an audio generator to the line-in terminals and adjust the generator for 1.2 kHz with an output level of 0 dB. Adjust the attenuator for −23 dB. Place the deck in record. Using normal (NOR) tape, record at 1.2 kHz for about a minute. Readjust the generator frequency to 12 kHz, and continue recording for another minute.

Play back the recordings. The output difference between 1.2 and 12 kHz should not exceed +2 or −1 dB. If the difference is greater, adjust RT_{400L} and/or RT_{400R} until the 12-kHz playback is no greater than +1 dB more than the 1.2-kHz playback.

For recording/playback level adjustment, leave the audio generator connected to line-in but reset the generator frequency to 400 Hz with an output of 0 dB. Then set the attenuator to −3 dB. Record the 400-Hz signal. Play back the tape and note the playback level. The output difference between record and playback should not exceed ±0.5 dB. If the difference is greater, adjust RT_{50L} and/or RT_{50R} for the correct level.

With the record/playback level properly set, recheck the bias current adjustment for each of the three different types of tape. Set the tape select switch to each position (NOR, CrO_2, and metal) as required. Typical playback levels are normal +2.5 dB, −1.5 dB; chrome +4 dB, −3 dB; metal +4 dB, −3 dB.

7.10.8 Tape Leader Adjustment

Figure 7.16*b* is the adjustment diagram. The purpose of this adjustment is to make sure that the tape-end detect circuits reverse (or stop) the tape when the leader is reached at either end.

Before starting this adjustment, make sure that the tape guides (Fig. 7.15*a*) are clean. On our deck, the top of the flat surface on the right-hand guide must be clean to provide good light reflection. The light must pass from the sensor LED through the leader to the reflective surface and then back to the phototransistor when the leader is over the flat surface.

Connect a dc voltmeter across R_{100}. Press and hold down the forward play

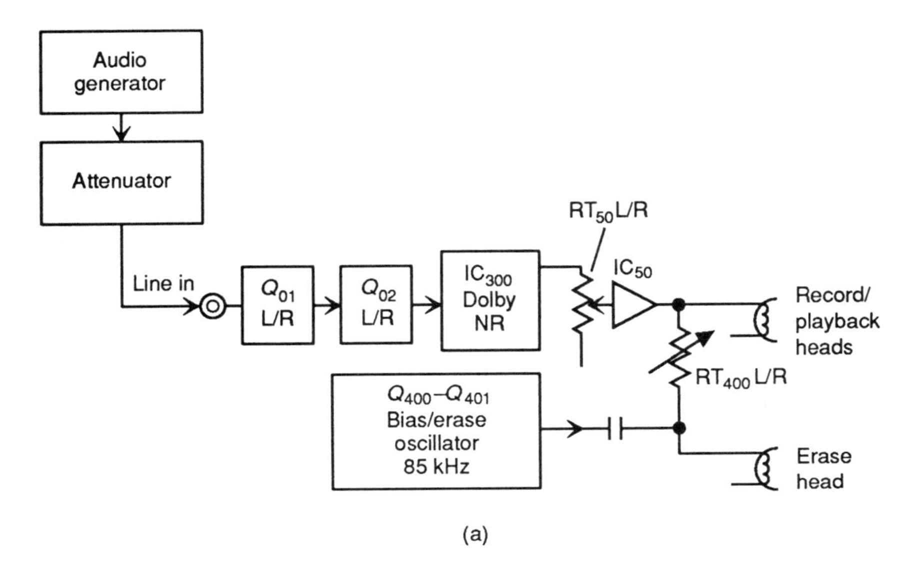

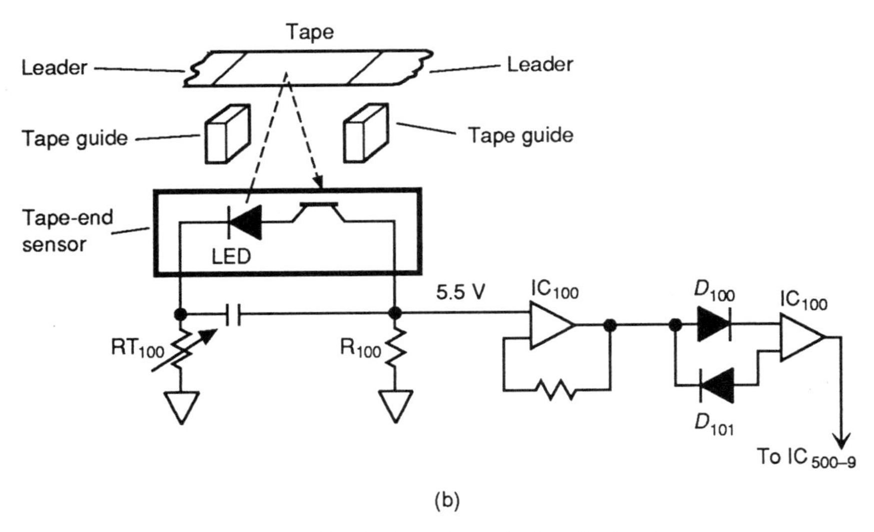

FIGURE 7.16 Bias-current, record/playback, and tape-leader adjustments.

button to prevent the tape-end detector from being activated. Adjust RT_{100} for 5.5 V as measured across R_{100}.

Load any standard tape and play forward to the tape end. When the deck automatically reverses, press stop immediately, and check that the tape reversal takes place as soon as the leader crosses the tape guide.

7.10.9 Dolby NR Operation Checks

The purpose of these tests is to check operation of the circuits within Dolby NR IC_{300}. Both Dolby B and Dolby C functions must be checked (if the deck has both functions).

Connect the audio generator to line-in. Adjust the generator to 5 kHz, attenuated to −40 dB. Make a recording on metal tape for about 1 min with Dolby NR off and for a similar time with Dolby B on. Play the tape back with Dolby NR off. The output difference between the Dolby off and Dolby B recordings should be about 10 dB.

Leave the generator connected to line-in. Adjust the generator to 1 kHz, attenuated to −40 dB. Make a recording on metal tape for about 1 min with Dolby NR off and for a similar time with Dolby C on. Play the tape with Dolby NR off. The output difference between the Dolby off and Dolby C recording should be about 16 dB.

7.11 ROUTINE MAINTANANCE

Use the following procedures in the absence of maintenance instructions provided by the service literature (or as a supplement to the instructions).

7.11.1 Lubrication

Never lubricate (or clean) any part not recommended in the service literature. Most tape decks use many sealed bearings that do not require either cleaning or lubrication. A drop or two of oil in the wrong places can cause problems, even possible damage. Clean off any excess or spilled oil using gauze soaked in alcohol (Sec. 7.11.2).

In the absence of specific recommendations, use a light machine oil (such as sewing-machine oil) and a medium grease or lubricating paste.

Lubricate the *rotary portion* of the cassette mechanism (tape transport) with *one or two drops of oil*. Apply *grease to the sliding portions* of the mechanism. Be very careful to keep oil or grease from getting on the belts, flywheels, or pinch rollers. Oil or grease on these surfaces can cause slippage (resulting in erratic tape drive and poor audio).

Perform lubrication once a year or after each 1000 hours of operation (sure you will).

7.11.2 Cleaning

After about 10 h of operation (actually, when the customer brings in the deck for service), clean the heads, capstans, and pinch rollers with a head-cleaning bar or wand and alcohol.

Although there are spray cans of head cleaner or solvent, most manufacturers recommend alcohol and cleaning sticks (or Q-tips, cotton swabs, etc.) for all cleaning. Methyl alcohol does the best cleaning job but can be a health hazard (especially if you drink it). Isopropyl alcohol is usually satisfactory for most cleaning (don't drink that either).

7.11.3 Head Degaussing

Some technicians never degauss the heads, while others degauss at each service (using a *commercial head eraser*). In between these extremes, some technicians degauss only when certain symptoms occur. The two most common symptoms are *background noise* that increases with time and *decrease in output* at high frequencies (treble) with time.

The author has no recommendations on head degaussing, except: Never degauss the heads with a cassette in place (particularly a customer's favorite tape or an expensive shop test tape).

7.12 PRELIMINARY TROUBLESHOOTING

Here are the checks to be made before going into the deck circuits.

If the deck operates manually but not by remote, check the remote batteries and the control cable. Then try resetting the power circuit by pressing the front-panel power button off and on.

If the left or right speaker is dead when the deck is used in a system, try playing another component connected to the audio line (tuner or CD player). If operation is normal with the other component, check for a loose cable between the deck and other components. Then temporarily reverse the left and right speaker leads. If the same speaker remains dead, the speaker is at fault.

If the left or right speaker is dead when the deck is used in a nonsystem configuration, temporarily reverse the left and right cable connectors at the amplifier input from the deck (deck output). If the same speaker remains dead, the amplifier (or speaker) is probably at fault. If the other speaker goes dead, suspect the deck or the deck output cable.

If there is no sound from either speaker but operation is good with another component (tuner or CD), suspect the deck or the cable between deck output and amplifier input.

If the sound is distorted, the deck output level may be too high, the tape may be recorded with Dolby and played back without Dolby (or recorded in one type of Dolby and played back in another type), the tape selector may be set for the wrong type of tape (normal, chrome, or metal), or the deck output may be accidentally connected to the phono input of the amplifier.

If there is hum or noise (only when the deck is used), check for the following. The shield of the audio cable from the deck may be broken, or the connector may not be firmly seated in the jacks. The deck may be too close to the amplifier. The magnetic fields produced by the amplifier may induce hum into the deck circuits.

If there is hum or noise when recording from a tuner, the noise may be produced by the deck erase head. In many cases, this can be eliminated by separating the deck and tuner or by repositioning the tuner loop antenna. In extreme cases, it may be necessary to use an external AM antenna. (Note that FM signals should not be affected by noise produced at the deck erase head.)

If the customer is having difficulty with audio cassette tapes, explain the fol-

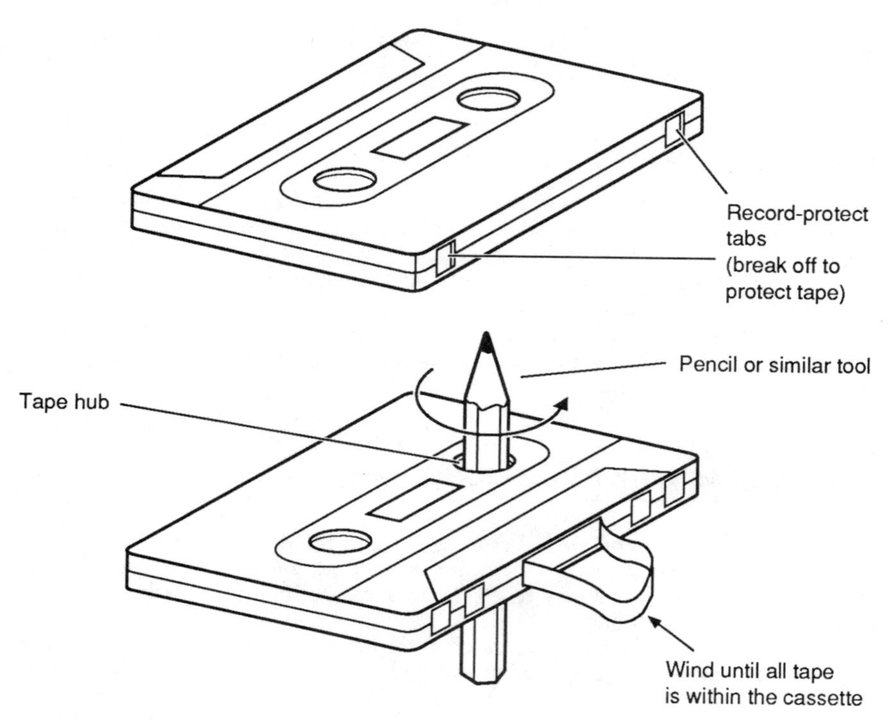

FIGURE 7.17 Record-protect tabs and cassette winding.

lowing. The deck erases whatever is on a tape when the deck records the new material. To protect a tape from accidental erasure by recording on the tape a second time, break off one or both plastic tabs on the back of the cassette, as shown in Fig. 7.17. There is one tab for each side of the tape. To record on a tape protected in this way, seal over the hole with adhesive tape (heavy Scotch tape).

Remind the customer that audio cassettes should not be exposed to direct sunlight or stored in hot places. High temperatures may warp the cassette or damage the tape itself. Keep the cassettes in boxes when not in the deck. This keeps dust from entering the cassettes.

Tapes can be erased by magnetic fields produced by permanent magnets, electric motors, electrical transformers, and other devices. The amplifier may produce a field strong enough to partially erase tapes or affect the fidelity of the tapes. Do not lay cassettes on top of the amplifier, even for a short time, and keep cassettes away from all magnets and magnetic fields.

Make a habit of winding the tape to one end before removing the cassette from the deck. In this way, only the tape leader is exposed. Avoid touching the tape or leader at all times. Even the natural oils from fingers can damage a tape and can be deposited on the tape-transport mechanism, eventually contaminating other tapes.

If a tape is accidentally pulled from a cassette, insert the tapered end of a large pencil or similar tool into the center of the tape hub, and carefully wind the tape back into the cassette, as shown in Fig. 7.17. Be careful not to twist or wrinkle the tape.

CHAPTER 8
CD AUDIO

This chapter describes the overall function, user controls, operating procedures, installation, circuit theory, typical test and adjustment procedures, and step-by-step troubleshooting for state-of-the-art CD players found in audio systems.

8.1 OVERALL DESCRIPTION

Figure 8.1 shows the front and back of a composite CD player. This player is designed to operate either as part of a system or as a stand-alone player that can be used with stereo amplifiers and speakers. In this book, we are concerned with operation of the player in the stand-alone condition.

8.1.1 Features

Random access: This feature allows preprogramming of up to 15 different CD selections for playback, in any sequence desired, and allows listeners to tailor the playback sequence of any CD to their own tastes.

Adjacent band access: This feature allows the listener to skip forward or backward to the next band and to begin playing at the beginning of the selection; it also allows convenient sampling of the CD selections at the listener's convenience.

High-speed scan: With this feature, the listener may quickly scan over the CD in either a forward or reverse direction while the front-panel digital counter displays the exact position of the pickup. This allows quick and convenient access to any point on the CD.

Index search: This feature allows cueing in either direction to a desired location within an index-coded band and provides virtually instantaneous access to any desired section on the CD.

Programmed repeat: The CD player can be programmed to repeat an entire CD, a single CD selection, or any desired portion of a CD, allowing continuous playback of favorite selections.

Pause control: This feature immediately halts playback until instructed to continue, permitting temporary interruptions of the playback. The feature also allows instantaneous return to listening when pause is released.

Digital readout: The digital readout on the front panel indicates the current

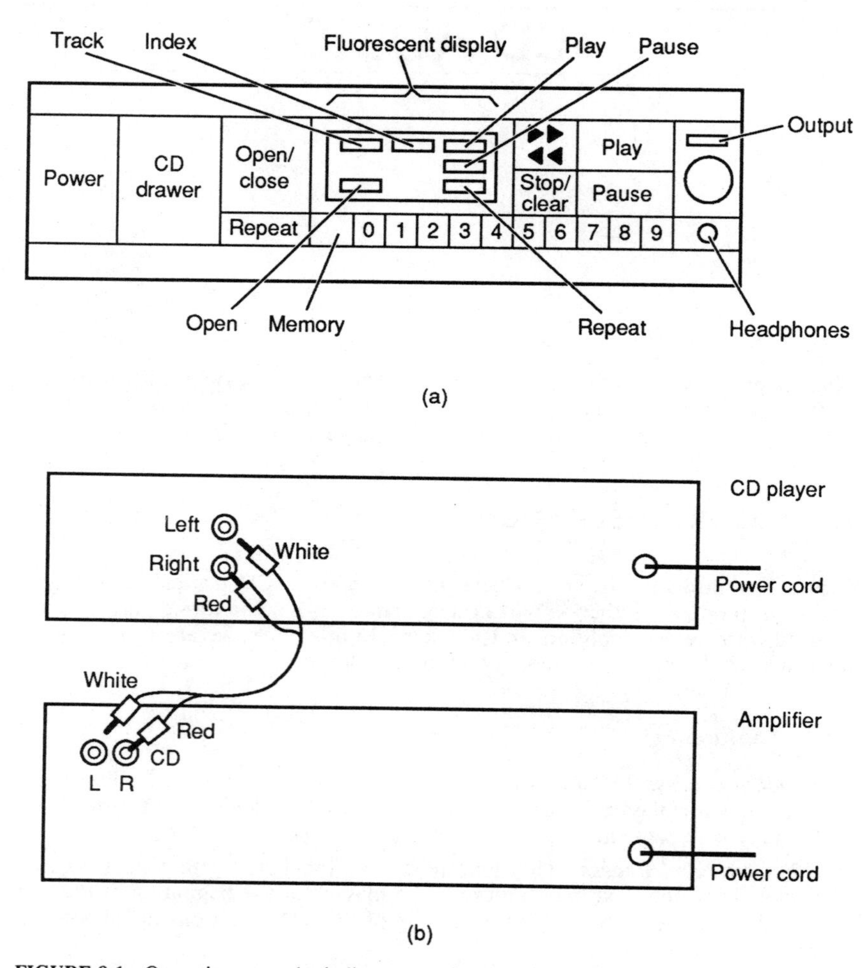

FIGURE 8.1 Operating controls, indicators, and connections for composite CD player.

band number, index number (if available), elapsed playing time, and status of the repeat operation. The display also alerts the listener to the status of all player operations.

Note that all of these features are under the control of a system-control microprocessor (IC_{301}) and an interface microprocessor (IC_{901}). The features can be selected by means of the front-panel controls or by a remote unit (on most players).

8.1.2 Special Precautions

In addition to the usual operating and service precautions for any electronic equipment, there are special precautions that apply to CD players.

Transit Screw. Most CD players have a transit or transport screw to hold the laser optical pickup in place when the player is moved or shipped. Without such a screw, the pickup can move back and forth, causing possible damage to the delicate optics. The transit screw is accessible from the bottom on most CD players.

Make certain to remove or loosen the transit screw before using the player. (Generally, the screw is of the captive type that is not removed, but not always. Therefore do not lose the screw if it is of the free type.)

Most important, be sure to install the transit screw when transporting or shipping the player. On our player (a horizontal front-load type) the transit screw can be installed only when the optical pickup is in one position (the at-rest or secured position). This usually means turning on the power, shutting the CD tray, and then tightening the screw.

Obviously, the transit screw can cause problems for the service technician. When the customer brings in the player, the customer will probably forget to tighten the screw, possibly damaging the optics. The opposite occurs if the service technician tightens the screw upon returning the player to the customer. The customer will get the player home and will probably call with a complaint that the player "no longer works after you fixed it." You must (patiently) explain these facts to the customer.

Laser Diodes. Direct exposure to a laser beam of any type can cause *permanent eye injury or skin burns*, as can any very intense light source. To make this problem more of a hazard, the laser-diode light beam is invisible (in contrast to the red light beam produced by the laser tubes used in some older videodisc players).

Since you cannot see the actual laser-diode light beam, you are never really sure when the beam is present. The operator (or customer) of a CD player is never exposed to the laser beam, even when the CD compartment is open. (There are interlocks that cut power to the laser when the compartment is not fully closed.) This is not necessarily true for the servicer. If you go inside to get at the laser diode (by overriding interlocks, jamming switches, etc.), the beam may get you. Don't panic, but please keep on your toes.

Remember that the laser diode is an electrostatically sensitive (ES) device, similar to any MOS/CMOS part, and should be so treated during service. This includes placing a conductive sheet on the workbench, using wrist straps, and so on. Often, manufacturers ship replacement laser pickups in a bag made of conductive material that can be used on the bench to prevent static breakdown (which must be avoided with any ES part). Note that the laser diode is usually part of the optical system or pickup. Most CD player manufacturers recommend replacement of the complete optical assembly as a package and do not supply replacement parts.

Laser Warning Labels. Always start CD player troubleshooting by looking for any laser warning labels on the player. A typical warning label is a triangle containing a bright star, possibly with a beam radiating from the star. This is combined with a note such as *Danger, invisible laser radiation when open and interlock failed or defeated. Avoid direct exposure to the beam*. Then check the service literature for any additional precautions concerning the laser diodes. Whatever precautions you take for yourself, also make sure that all shields and covers are in place and that interlocks are working before you turn a CD player over to the customer.

Electromagnetic Radiation. Besides a potentially dangerous beam, a laser diode produces strong electromagnetic radiation. This will not hurt people but can be dangerous to magnetic tape, some wristwatches, and anything affected by magnetic fields.

Object Lens. The laser beam from the diodes is focused onto the CD through a lens that is part of the optical system or pickup assembly. The lens is usually called the *object* or *objective lens*. No matter what it is called, the lens must be clean and free of moisture. Try not to touch the lens surface, and keep the CD compartment closed except when inserting and removing the CD. Sound quality can be degraded if too much dirt or dust accumulates on the lens. Dirt and moisture can be wiped away with a soft cloth. Dust can be removed with an air blower (such as used on a camera lens).

Moisture. If the player is brought directly from a cold to a warm place or is placed in a very damp room, moisture may condense on the optical pickup lenses. The player will not operate properly (if at all) should the lenses become fogged. You can wipe off the objective lens but not the other lenses in the optical system. Try removing the CD and leave the player turned on for about an hour to evaporate the moisture.

Test and Adjustment Precautions. There are some very specific precautions you must observe when adjusting and testing laser diodes. These precautions are covered in Sec. 8.10.

8.1.3 Operating Controls and Indicators

Figure 8.1*a* shows the operating controls and indicators for our CD player. Compare the following with the controls and indicators of the player you are servicing (which, of course, will be different). Press power to turn the player on or off. Press open/close to alternately open or close the CD drawer. When the drawer opens, a red open indicator turns on. Pressing open/close while a CD is playing stops the mechanism and opens the drawer, the repeat function is canceled (if on), and program memory is erased.

Display track number: When the CD drawer is closed with the open/close button, when the CD has played to the end and stopped, or when the stop/clear button is pressed, the total number of tracks on the CD is displayed. At all other times, the track number being played is displayed.

Display index number: If index coding exists on the CD, the index number is displayed while the CD is played.

Display time indicator: When the number indicating the total number of tracks is displayed, the total playing time is also shown. At other times, the elapsed playing time is displayed for the track that is being scanned.

Display play indicator: Play turns on during play.

Display repeat indicator: Repeat turns on when the repeat function is active.

Display pause indicator: Pause turns on when the pause function is active.

Display memory indicator: Memory turns on when a program sequence is in memory.

Display open indicator: Open turns on when the CD drawer is opening.

The button with double arrows pointing to the right: scans the CD in a forward direction. When first pressed, the audio material is heard at about 3 times normal speed. If the button is held, the player begins sampling short segments. Release the button to resume normal play. The button may also be pressed with the play button to skip to the beginning of the next track.

Pressing *the button with the double arrows pointing to the left* scans the CD in a reverse direction. The button may also be pressed with the play button to restart the selection (being played) at the beginning.

The stop/clear button stops playback and clears any program from memory.

The play button initiates play. The play button may also be used to close the CD drawer and begin play.

Press pause to interrupt playback. Press play to resume playback. The pause button may be used to close the CD drawer. The *directory of selections* (also called the CD directory or disk directory) is read, and the player pauses at track 1.

The output control adjusts the output level to the headphone jacks and to the rear-panel output jacks. There is a special problem for those not familiar with CD players that involves the setting of the output control (or volume control as it is called on some players). This is because the background noise on a CD is practically nil. If you crank up the volume, either on the player or the amplifier while listening to a portion of the CD where no audio signals are recorded (or to a very low-level audio), the speakers may be damaged when that portion of the CD with peak signals is played (and you can imagine what happens to your eardrums when listening with headphones).

The 0 through 9 buttons are used to select a track or index number.

After a track number has been selected, the display flashes that number. Push memory to enter the track into memory. Repeat for each track to be programmed.

Push repeat to repeat the entire CD or the selections that have been programmed into memory.

Note that most of the front-panel controls have remote-control counterparts. Most operating sequences can be initiated or ended using the front-panel controls, the remote unit, or a combination of both (on most players).

8.1.4 Connections

Figure 8.1*b* shows connections between the CD player and amplifier. Compare the following notes to the connections for the player being serviced.

External connections between the CD player and amplifier are made from the back of the player. In most cases, a pin cord is supplied with the player (or amplifier) for connection between the player L and R stereo outputs and the corresponding inputs on the amplifier. Although the connections are very simple, certain precautions must be observed (for all players).

Never connect a CD player to the phono input of an amplifier. Instead, use the CD, aux, or possibly the tape play inputs. The CD player output is about 2 V, which can damage the amplifier and/or speakers and will overdrive the amplifier.

Even on those CD players that have adjustable output, the player impedance (about 50 kΩ) is best matched to the CD, aux, and tape inputs. As is the case with any audio component, always switch off the power before making or breaking connections between the player and amplifier.

Before connecting the player to a power source, check that the operating voltage of the player is identical to the voltage of the local power supply. Unlike most home electronic components (TV, VCR, etc.), many CD players are designed for worldwide use and can be operated at 120, 220, and 240 V. The operating voltage is selected by special connections and/or switch settings, so check the service or operating literature.

8.1.5 Preliminary Checks

Always make a few preliminary checks before launching into a full troubleshooting routine for any player. Start by making the preliminary check in Sec. 8.11. Then go on with the following:

Make certain that the transit screw is removed or loosened *before* operating the player. Tighten the transit screw only when the CD player is to be moved.

If practical, check that the customer's stereo system is operating normally before you do any extensive service on the CD player.

Cleaning the objective lens should be a routine part of servicing. A dirty objective lens can cause a variety of symptoms (intermittent or poor focus, skipping across the CD, erratic play, excessive dropouts, to name a few). These same symptoms can also be caused by a defective CD. Try a known-good CD first.

Do not replace the pickup assembly or make any adjustments on the pickup before checking for mechanical problems that can affect the pickup. For example, look for binding at any point on the pickup travel, which indicates that the rails or guides are adjusted too tightly (Sec. 8.3). At the other extreme, if you hear a mechanical "racheting" or "chattering" when the pickup is moved, the rails may be too loose. (Note that player literature generally gives all mechanical adjustment procedures in great detail, so we do not repeat the procedures here.)

8.2 RELATIONSHIP OF CD PLAYER CIRCUIT

Figure 8.2 shows the relationship of player circuits. The laser beam is generated by laser diodes located on the pickup assembly. The beam is passed through a lens assembly and focused on the CD. The reflected beam from the CD is then applied to phototransistor sensors through the lens assembly.

The low-amplitude output of the sensors is applied through preamplifiers, filters, and waveshaping circuits IC_{601}, through IC_{603} back to the laser drive IC_{604}, as described in Sec. 8.4. The output is also applied to data-strobe, signal-generator, and error-correction circuits IC_{401} through IC_{403}.

The purpose of the data-strobe circuit (Sec. 8.5) is to identify the signals correctly as to digital 0 or 1 status and to clock the sync, control, and audio information. Once this is done, the circuits perform signal processing (error correction and decoding) to produce four major outputs.

Two of the outputs are left and right audio signals as described in Sec. 8.6. A third output is the necessary control information for the CD motor (turntable) circuits (Sec. 8.9). The fourth output represents information about where the laser beam is located on the CD and is applied to the system-control microprocessor IC_{301}. In turn, IC_{301} generates signals applied to servo control IC_{101} and IC_{102}.

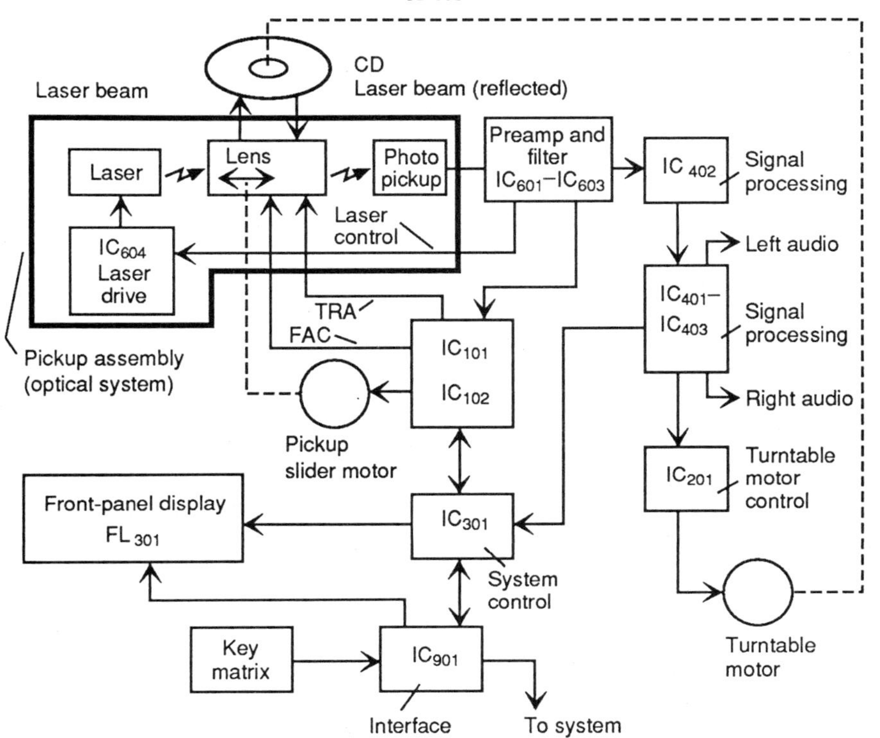

FIGURE 8.2 Relationship of CD player circuits.

The servo circuits IC_{101} and IC_{102} are responsible for moving the laser pickup assembly across the CD, keeping the beam in focus, and maintaining the focus properly centered on the CD tracks, as described in Secs. 8.7 and 8.8.

IC_{301} and IC_{901} provide for interfacing among the player circuits. IC_{901} is responsible for communications between the CD player and the audio system (if any). IC_{901} also receives and decodes all inputs from the front-panel keyboard matrix, transferring such commands to both FL_{301} (for front-panel display) and IC_{301} (for system control).

8.3 MECHANICAL FUNCTIONS

Figure 8.3 shows the major mechanical components of our CD player. Figure 8.4 shows the associated circuits. The mechanical components of a CD player perform two major functions: loading and unloading the CD and driving the optical pickup (and laser beam) across the CD.

8.3.1 Loading and Unloading

In our player, the tray is opened by a loading motor, or LDM (in response to pushing the open/close button), a CD is inserted within the tray, the tray and CD

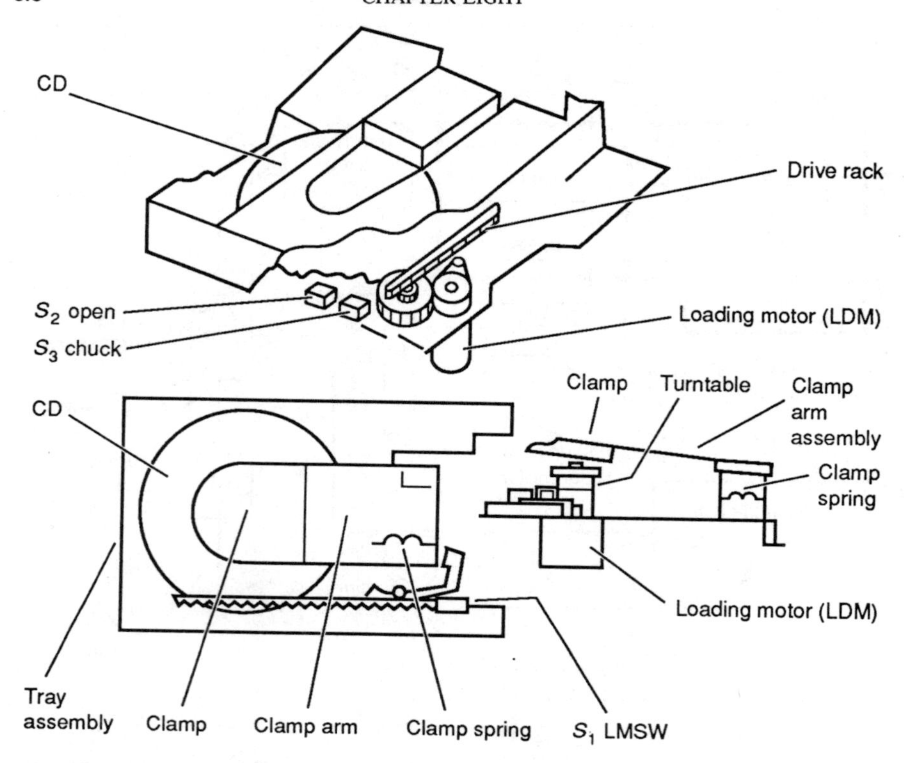

FIGURE 8.3 Major mechanical components of a CD player.

are returned within the player (by the LDM), and then the CD is installed on the turntable (by the same drive mechanism and LDM).

Operation of the LDM is controlled by limit switches S_2 and S_3 and by IC_{301}. So when you are troubleshooting any mechanical functions, you must study both the mechanical drawings and circuits. The following paragraphs describe both mechanical and electrical functions of our player. Again, remember that these descriptions must be compared with the descriptions found in (or omitted from) the service literature.

Note that most of the mechanical components are part of a mechanism secured to the left side of the player mainframe by two rails. The tray is moved out of the player front panel on rollers by the LDM. This action also raises the spring-loaded clamp. A CD is installed manually on the tray, and the tray is pulled within the player by the LDM. This action also lowers the clamp so that the CD is pressed against the turntable motor.

The LDM receives open/close drive signals from IC_{301} through IC_{102}. In turn, IC_{301} receives indicator signals from the tray-open switch (open) S_2 and the tray-closed (chuck) switch S_3. (In the literature, the tray-open switch S_2 is called the open switch for obvious reasons. The tray-closed switch S_3 is called the chuck switch since the switch is actuated only when the CD is clamped or "chucked" onto the turntable.)

Note that the open/close drive signals from IC_{301} are also applied to IC_{901}, at

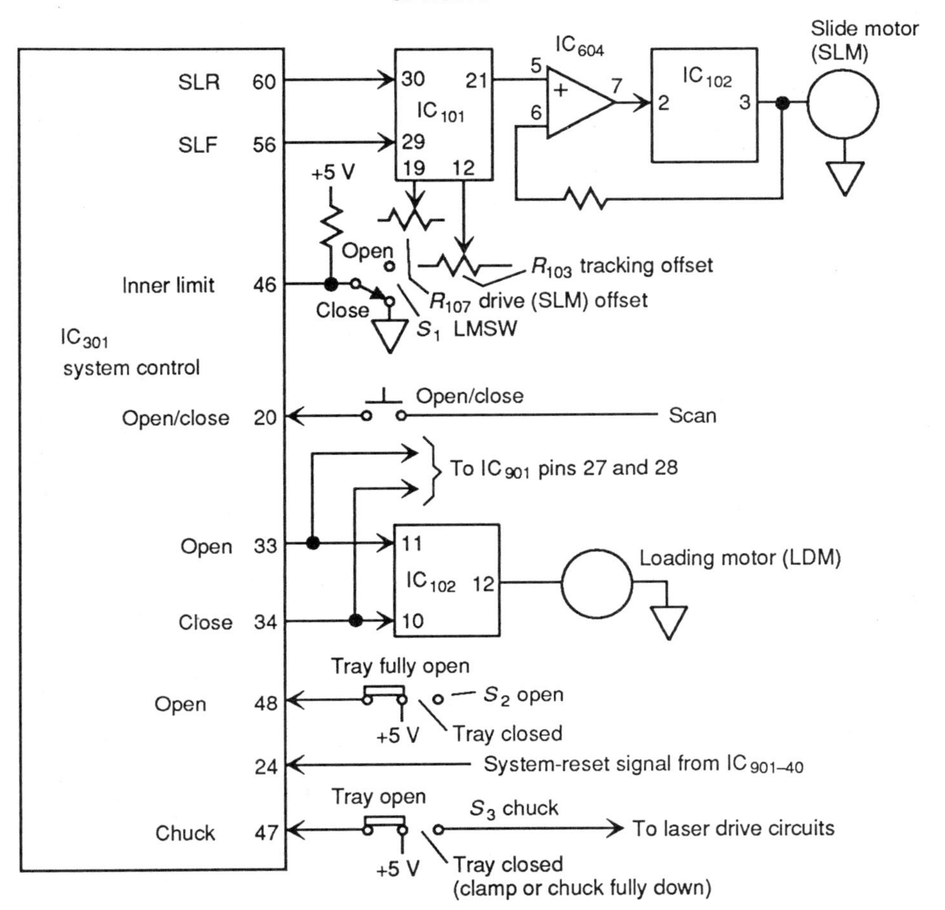

FIGURE 8.4 Circuits associated with mechanical components.

pins 27 and 28. This informs IC_{901}, and the system (if any), that the tray is moving in or out or has stopped.

When the tray is open and the front-panel open/close button is pushed, IC_{301} produces a close signal at pin 34. This signal is applied to the LDM through IC_{102} and causes the LDM to close the tray. Until the tray is in and the clamp is fully down, S_3 remains in the tray-open position. This applies +5 V to pin 47 of IC_{301}, informing IC_{301} to continue applying a close signal.

When the tray reaches the fully in position and the clamp is in the fully down position, S_3 actuates (moves to the tray fully closed position). This removes the signal from pin 47 of IC_{301}, the close signal is removed, and LDM stops driving the tray and clamp. Note that S_3 also applies +5 V to the laser drive circuits as described in Sec. 8.4.

When the tray is closed and the front-panel open/close button is pressed, IC_{301} produces an open signal at pin 33. This signal is applied to the LDM through IC_{102} and causes the LDM to open the tray.

Until the tray is fully out, S_2 remains in the tray-closed position. This leaves pin 48 of IC_{301} open, telling IC_{301} to continue applying an open signal. When the tray reaches the fully open position, S_2 actuates (moves to the tray fully open position). This applies +5 V to pin 48 of IC_{301}, and the LDM stops driving the tray.

8.3.2 Optical Pickup (Beam) Drive

A motor is required to keep the beam moving across the CD at a constant rate, even though the CD speed changes (as described in Sec. 8.9). In our player, the optical pickup (also called the *slide* or *sled* in some literature) is moved across the CD by the slide motor (SLM). In turn, the SLM is operated by slide forward (SLF) and slide reverse (SLR) signals from pins 56 and 60 of IC_{301} through IC_{101}, IC_{604}, and IC_{102}.

Note that R_{107} provides an offset adjustment for the drive signals applied to the SLM. The adjustment sets the point where the pickup accesses the beginning of the CD (the directory). If the adjustment is not correct, the program information may not be read properly. R_{103} provides an offset adjustment for the tracking signals applied to the SLM and is adjusted to produce the best response (minimum audio dropout), as discussed in Sec. 8.10.

When a CD is first installed (with the tray fully in and clamp fully down), IC_{301} produces a temporary SLR signal at pin 60. (The temporary SLR is generated by a system-reset signal from pin 40 of IC_{901}.) The SLR is applied to the SLM and causes the SLM to move to the inner limit of the CD (the CD start position). As you may know, CDs start playing from the center and play to the outside, opposite to the play of an LP record.

Until the pickup and beam are at the inner limit, S_1 remains in the open position. This applies +5 V to pin 46 of IC_{301}, informing IC_{301} to continue applying the temporary SLR signal to the SLM. When the pickup reaches the inner limit, S_1 actuates (moves the to closed position). This grounds the signal at pin 46 of IC_{301}, the SLR signal is removed, and the SLM stops.

When the pickup is resting at the inner-limit position and the front-panel play button is pressed, IC_{301} produces an SLF signal at pin 56. The SLF is applied to the SLM through IC_{101}, IC_{604}, and IC_{102} and causes the SLM to drive the pickup and beam across the CD.

8.3.3 Mechanical Troubleshooting

In most players, the entire mechanism can be replaced as an assembly. Some manufacturers also recommend replacement of the motor and limit switches (and they describe the procedures for replacement and adjustment). The mechanical section is one area where most CD player service literature is very good (if only the theory and troubleshooting sections were that clear).

We do not dwell on mechanical replacement and adjustment here. However, as a practical matter, *never disassemble the player mechanism beyond that point necessary to replace or adjust a part*. Similarly, never make any adjustments unless the troubleshooting procedures lead you to believe that adjustment is required.

If the tray refuses to open or close, first check that IC_{301} is getting the proper key-scan signals when the front-panel open/close button is pressed. If not, suspect the front-panel key matrix and/or IC_{901} (Sec. 8.2).

Next, make sure that the LDM receives a signal from pin 12 of IC_{102} when open/close is pressed. If it does, but the motor does not turn, suspect the motor (or possibly a jammed tray). If there is no signal at pin 12 when open/close is pressed, you have a problem between IC_{301} and the LDM.

Check for signals at pins 10 and 11 of IC_{102} each time open/close is pressed, and make sure that the signals invert (pin 10 high and 11 low, then vice versa). Check for corresponding inverted signals at pins 33 (open) and 34 (close) of IC_{102}. If they are absent or if the signals do not invert when open/close is pressed, suspect IC_{301} (or possibly IC_{901} and the key matrix).

If the tray opens but not fully, check when the open switch S_2 actuates, as indicated by a low-to-high change at pin 48 of IC_{301}. If necessary, adjust S_2 (as described in the service literature). Before making any adjustments, check for a mechanical condition that might prevent the tray from opening fully (binding gears, jammed cross-roller, improperly adjusted rails, etc.).

If the tray opens fully but the loading motor does not stop, the problem is always an improperly adjusted S_2 (although there is an outside chance that IC_{301} is at fault).

If the tray closes but not fully and the clamp does not hold the CD in place on the turntable, check that S_3 actuates, as indicated by a high-to-low change at pin 47 of IC_{301}. If necessary, adjust S_3 as described in the service literature. Before adjustment, check for mechanical problems (something in the clamp hinges or tray, wiring that has worked its way out of place, etc.).

If the tray closes and the clamp goes fully down but the LDM does not stop, the problem is likely to be an improperly adjusted S_3, but it could be IC_{301}. Look for a high-to-low change at pin 47 of IC_{301}, which should occur when the tray is fully in and the clamp is down on the CD.

If the pickup does not move (*with the tray in*) *when power is first applied* (you may not be able to see the pickup, but you *should hear the motor*), check for SLR at pin 60 of IC_{301}. If it is absent, suspect IC_{301} or the lack of a reset signal at pin 24 of IC_{301} (from pin 40 of IC_{901}). If you get SLR but the motor does not run, suspect IC_{101}, IC_{604}, IC_{102}, and the motor itself.

If the pickup appears to move to the inner limit when power is applied (you hear the motor stop and start) but the directory is not read properly (say that the total playing time, or number of programs on the CD, is not given on the front-panel display), try correcting the problem by adjustment of SLM offset R_{107} before going into the circuits. Follow the service-literature procedures for adjustment of R_{107} (or the procedures described in Sec. 8.10).

Note that on any problem with the pickup drive (forward or reverse) you should check for SLM drive voltage at pin 3 of IC_{102}. If the motor runs but the pickup does not move or movement is erratic, look for mechanical problems (jammed gears, binding rollers, improperly adjusted rails, etc.).

If the pickup moves but does not reach the inner limit, check when S_1 actuates, as indicated by a high-to-low change at pin 46 of IC_{301}. If necessary, adjust S_1 as described in the service literature.

Before adjusting S_1, check the adjustment of the SLM offset R_{107}. If R_{107} can be adjusted so that the pickup accesses the CD properly (the directory can be read in full), S_1 is probably adjusted correctly. On most players, S_1 does not go out of adjustment except when the pickup is replaced (or when there has been tampering).

If the pickup reaches the inner limit but the motor does not stop, the problem is almost always one of an improperly adjusted S_1. A possible exception is where IC_{301} is defective and is not responding to the low at pin 46.

8.4 LASER OPTICS AND CIRCUITS

Figure 8.5 shows the laser optics and circuits. These components are responsible for recovering, preforming preliminary preamplification, and waveshaping the audio signal recovered from the CD. Before we get into the circuits, let us review the basic functions of laser optics (as applied to CD players).

8.4.1 Laser Optics

As shown in Fig. 8.5*a*, the solid-state laser used in CD players emits a beam that is first passed through a defracting lens. This lens splits the main laser beam into three separate beams, which are next passed through a prism, a coupling lens, and an objective (or focus) lens to the CD. The light reflected back from the CD is first passed back through the objective and coupling lenses to the prism. The beams are then reflected (by a mirror in the prism) to a photodetector through another pair of coupling and cylindrical lenses. The purpose of the cylindrical lens is to cause the normally round laser beam to distort in an oblong fashion. The beam is elongated if the objective lens is not properly focusing the beam on the underside of the CD (as discussed in Sec. 8.7).

The photodetector serves three primary purposes: (1) to generate the signals necessary for autofocus, (2) to generate the tracking signals that allow the beams to follow the tracks on the CD accurately, and (3) to generate the digital audio signal that is converted back to an analog audio signal.

8.4.2 Laser Control

The conduction rate of the laser diode is controlled by driver Q_{601}. Note that Q_{601} receives emitter current through R_{623} and the chuck switch S_3. As discussed in Sec. 8.3, S_3 is actuated only when the tray is fully in and the clamp is fully down. This removes 5 V from pin 47 of IC_{301} and applies 5 V to the emitter of Q_{601}, turning Q_{601} on.

With this safety circuit, the laser remains off until the CD tray is fully closed and the CD is in place to receive the laser beam. Once power is available to the laser through S_3 and Q_{601}, the laser is then turned on and off by a control signal from pin 51 of IC_{301}.

When pin 51 of IC_{301} goes high (laser off), the high is applied to pin 3 of IC_{604} through D_{601}, causing pin 1 of IC_{604} to go high and turning off Q_{601}. This removes power from the laser diode. With pin 51 of IC_{301} low (laser on), the low is applied to pin 3 of IC_{604} through D_{601}, causing pin 1 of IC_{604} to go low and turning on Q_{601}. This applies power to the laser diode.

As discussed in Sec. 8.4.1, the laser beam is reflected from the CD onto the photosensor. The beam is also applied directly to a monitor diode that regulates the amount of current through the laser diode (and thus the intensity of the laser

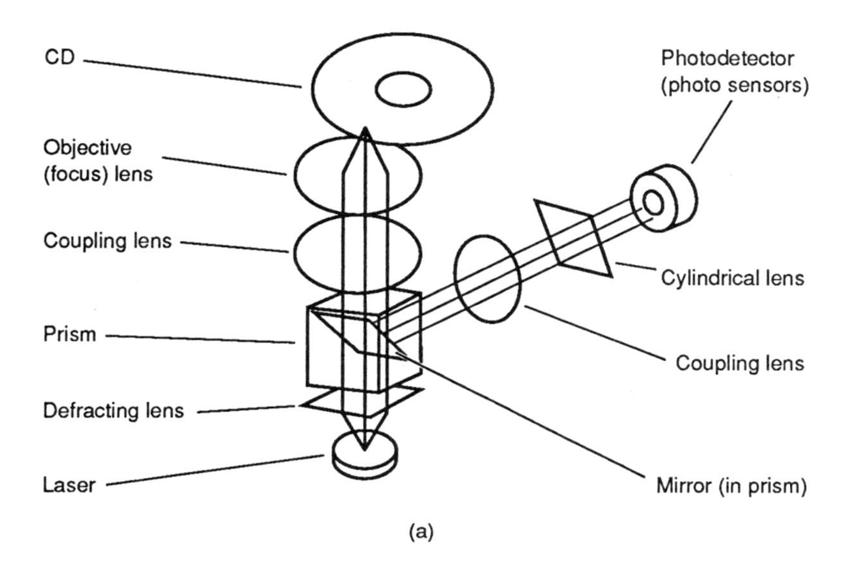

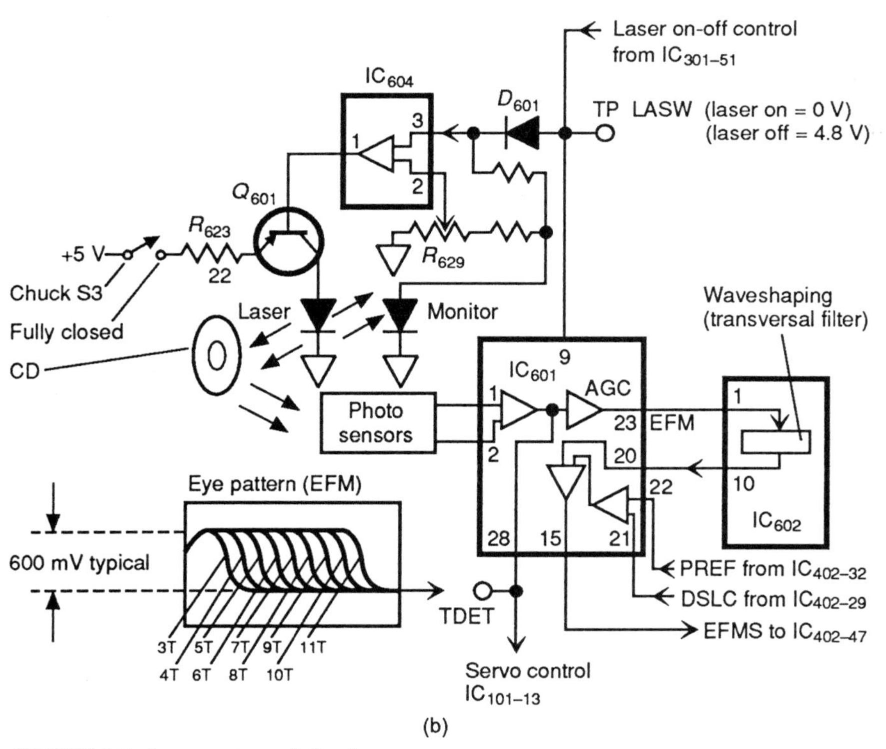

FIGURE 8.5 Laser optics and circuits.

beam). Most CD players include circuits to monitor and control the amount of light emitted by the laser. This is necessary for proper performance of the optical system. For example, a low output from the laser diode can produce tracking errors as well as audio dropout.

The output of the monitor diode is applied to the input at pin 3 of IC_{604}. The other (inverting) input at pin 2 receives an adjustable reference voltage, set by R_{629}. The output of IC_{604} is applied to the laser diode through Q_{601}. If the laser diode output goes below the desired reference level, the monitor diode output decreases and the IC_{604} output goes less positive. This increases laser drive current supplied by IC_{601}, increasing the laser diode output back to normal. The opposite occurs if the laser output increases. The laser diode output can be set to an optimum value with R_{629}.

8.4.3 Laser Adjustment

Always follow the service-literature instructions when adjusting the laser diode. Typical adjustment procedures are described in Sec. 8.10.

8.4.4 Laser Beam Processing

The laser output is reflected from the CD onto the photosensors. The outputs of the photosensors are applied to pins 1 and 2 of IC_{601}. Within IC_{601}, the outputs are applied to a differential amplifier. The output from the differential amplifier is applied to pin 13 of servo control IC_{101}, through pin 28 of IC_{601}, and to an amplifier within IC_{601}.

The signal at pin 28 of IC_{601} is known as the main eight-to-fourteen modulation (EFM) signal and represents the primary signal output from the CD. [Note that this signal is also called the *high-frequency signal* (HF) or the RF signal and produces what is known as the *eye pattern* shown in Fig. 8.5*b*.]

Although the EFM signal appears to be a series of sine waves at this point (at the TDET test point), the signals are digital. The designations 3T, 4T, and so on, refer to 3 times the period required to read a pit (Sec. 8.8) on the CD, 4 times the period, and so on. The limits of 3T to 11T are set by CD specifications.

The EFM signal also exits IC_{601} (at pin 23) through an AGC amplifier and enters waveshaping (transversal filter) circuits within IC_{602} through pin 1. (If you are interested, the waveshaping circuits assure that the 3T signal, or high frequency, is equal in amplitude to the 11T signal, or low frequency.) The waveshape output is applied to the signal processing circuit at pin 47 of IC_{402} through pin 10 of IC_{602}, pin 20 of IC_{601}, a comparator within IC_{601}, and pin 15 of IC_{601}. The main signal is then processed in IC_{402} as described in Sec. 8.5.

The comparator within IC_{601} shapes the EFM signal into square waves (known as the EFMS or EFM square-wave signal). The EFM signal is compared to a dc threshold voltage developed by circuits within IC_{402} and IC_{601}. In brief, the EFMS is applied to IC_{402} which, in turn, develops two square-wave signals (the data slice level control, or DSLC, and the preference pulse, PREF). DSLC and PREF are combined within IC_{601} and produce an error voltage that becomes the threshold voltage for the EMF comparator.

8.4.5 Laser Troubleshooting

The laser diode must produce a proper beam if the CD player is to perform all functions correctly. If the beam is absent, there is no EFM signal. If the beam is weak, EFM is weak. If the monitor diode does not monitor the laser properly, the beam can shift to an incorrect level (high or low) without being sensed by the laser-drive circuits. Any of these conditions can cause improper tracking which, in turn, can produce an even weaker EFM.

Always look at the laser circuits first when you have mysterious symptoms with no apparent cause (improper tracking that cannot be corrected by adjustment, excessive audio dropout with a known-good CD, etc.). Start with laser diode adjustment. This should show any obvious problems in the laser circuits and also tell if the EFM signal is good. (An EFM signal of proper amplitude generally means a good laser.)

If the laser appears to be completely dead (no glow at the objective lens, no EFM, or no movement of the focus when power is first applied as described in Sec. 8.7), make sure that Q_{601} is getting 5 V through S_3. If it is not, suspect S_3 (or adjustment of S_3). Note that when the tray is open, pin 47 of IC_{301} goes high, shutting down many functions of IC_{301}, including the laser switch (LASW) signal at pin 51.

If power is applied to the laser diode, look for an LASW signal at pin 51 of IC_{301}. If LASW is high (absent), suspect IC_{301}. If it is present, check for a signal at pin 3 of IC_{604} from the monitor diode. If it is abnormal, suspect the monitor diode and/or R_{629}.

If the signals are present at both pins 2 and 3 of IC_{604}, look for drive signals at pin 1 of IC_{604} and at the base of Q_{601}. If they are absent, suspect IC_{604}. If they are present but there is no laser beam output, suspect Q_{601} or the laser diode. An alternate method for checking the laser diode is described in Sec. 8.10.1.

Finally, check for EFM signals at pins 28, 23, and 20 of IC_{601} and EFMS signals at pin 15 of IC_{601}. (Note that the EFM signals are high frequency, while the EFMS signals are square waves.

If the EFM signals are absent at pins 23 and 28 of IC_{601}, suspect IC_{601} or the photosensors. If the signals are present at pins 23 and 28 but not at pin 20, suspect IC_{602}.

If the signals are present at pin 20 of IC_{601} but not at pin 15, suspect IC_{601}. However, make sure that the DSLC and PREF signals (both square waves) from IC_{402} are present at pins 21 and 22 of IC_{601}. If they are not there, suspect IC_{402}. (Unfortunately, IC_{402} does not generate PREF and DSLC signals if there is no EFMS applied to pin 47 of IC_{402}.)

8.5 LASER SIGNAL PROCESSING

Figure 8.6 shows the basic signal processing circuits. The EFM signal is applied to pin 47 of IC_{402} as described in Sec. 8.4. IC_{402} decodes the digital audio signal recovered from the CD and corrects any error that may be in the signal. The EFM signal is first applied to the data-strobe portion of IC_{402}, where audio and control information is separated. The control information is applied to circuits within IC_{402} that perform sync detection, de-interleaving, and the interpolation of audio-data signals.

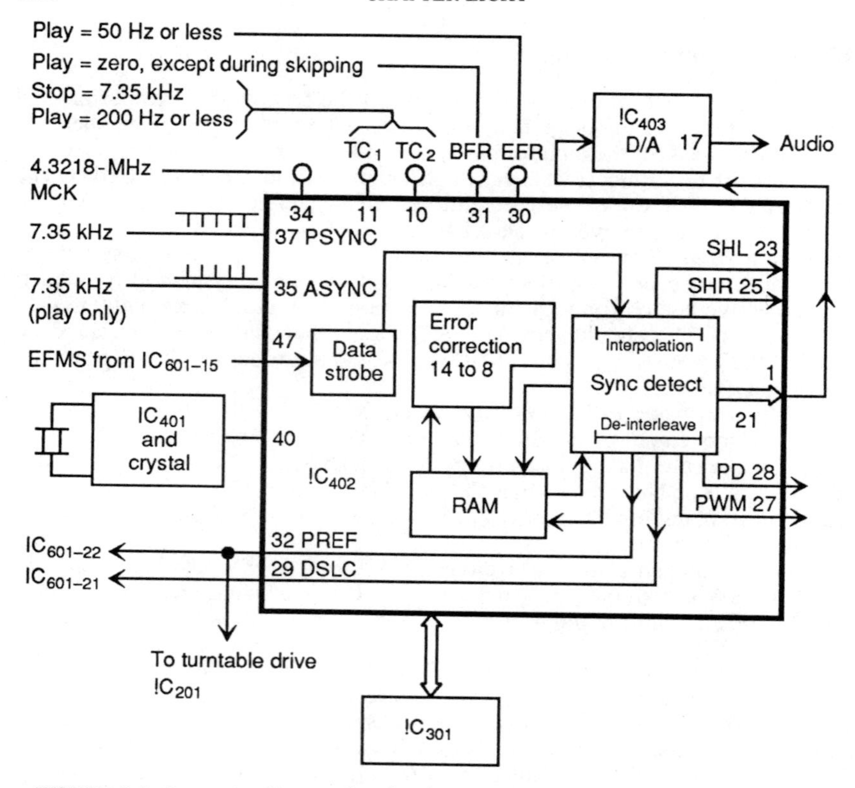

FIGURE 8.6 Laser signal processing circuits.

The signal containing the audio is applied to an error-correction circuit, which converts the 14-bit modulation back to an 8-bit modulation. (During CD manufacture, the audio to be recorded is first converted to an 8-bit digital format and then to a 14-bit format for recording on the CD.) The restoration process in the CD player is done by shifting the digital data bits into and out of RAM. The audio data bits are then shifted through the RAM circuit into the circuits where the de-interleaving and interpolation are performed.

In brief, de-interleaving is the reverse of interleaving where the audio words (in digital form) are rearranged (or scrambled) and recorded in a predetermined pattern on the CD. This interleaving is necessary to prevent adjacent audio words or bits from being altered by a single dropout (say from a spec of dust on the CD that overlaps several bits). With interleaving, the original adjacent audio bits are no longer adjacent (when recorded) so even a large dropout has less effect on the audio signal. The interleaving (or scrambling) process provides a form of error correction known as cross-interleaved Reed-Solomon code (CIRC). Actually, CIRC involves both interleaving and the addition of parity bits.

Interpolation is a process of averaging the recovered digital audio signal, where a digit or audio sample may be lost. An average value that falls between the two last-known points or values is generated by the interpolation circuits. From a practical troubleshooting standpoint, interpolation and de-interleaving, sync detection, and modulation restoration are all performed automatically by IC_{402}, as discussed in Sec. 8.5.1.

The audio is output in a 16-bit parallel digital form at pins 1 through 21 of IC_{402}. The 16-bit audio is then applied to a digital-to-analog (D/A) converter IC_{403}, where digital audio is restored to conventional analog audio.

IC_{402} also generates sample/hold left (SHL) and sample/hold right (SHR) signals. The SHL and SHR signals are very similar in function to the head-switching signals found in a VCR. The SHL and SHR signals are necessary to convert the serially recorded stereo information back to parallel stereo, as discussed in Sec. 8.6.

IC_{402} also generates a number of other signals, as shown in Fig. 8.6. For example, PD, PWM, and PREF signals are applied to the CD turntable motor drive IC_{201}, as discussed in Sec. 8.9. Similarly, the functions within IC_{402} are under control of IC_{301}. This creates certain troubleshooting problems, as we discuss next.

8.5.1 Signal Process Troubleshooting

A failure in the signal process circuits can cause a variety of failure symptoms in both audio and turntable motor circuits. Similarly, a failure in system control can appear as a failure in signal processing. From a practical standpoint, there is no sure way to tell if the problem is in signal processing, system control, turntable motor, or audio. However, there are some checks that can help you pin down the problem.

First check for audio at pin 17 of D/A converter IC_{403}. You should get both left- and right-channel audio (at a very low level). If you get no measurable audio at this point, suspect IC_{401}, IC_{402}, or IC_{403}. If there is measurable audio, the problem is likely to be in the audio circuits (Sec. 8.6).

Next, if there are excessive audio dropouts (with a known-good CD) and the front-panel indications are not normal (such as the time code changing as the CD continues to rotate), the problem is probably in the signal processing. Check all of the waveforms to and from the signal processing circuits shown in the service literature. Pay particular attention to the following (using Fig. 8.6 as a guide):

Check for a 4.3128-MHz master clock (MCK) signal at pin 34 of IC_{402}. If it is missing, check the crystal and IC_{401} connected at pin 40 of IC_{402}.

Check for 7.35-kHz PSYNC and ASYNC signals at pins 37 and 35 of IC_{402}. The ASYNC signal should be present only during play, but PSYNC should be available in both stop and play.

Make certain that PREF and DSLC are supplied to IC_{601} (Sec. 8.4) and returned to IC_{402} as square-wave EFMS signals. If EFMS is missing, check for high-frequency EFM signals at pin 20 of IC_{601}, as described in Sec. 8.4.5.

Check all signals between IC_{301} and IC_{402}. It is not practical to analyze the waveforms of these signals. However, if you can measure a data stream on each line with a scope, it is reasonable to assume that the signal is correct. If one or more of these signals are missing, suspect IC_{301}, IC_{402}, or both. Remember that a signal from IC_{301} can depend on a signal from IC_{402}, and vice versa. So you may have to replace both ICs to find the problem. Also remember that IC_{301} may not produce the signals unless other signals (such as FOK and TOK) are applied to IC_{301}, as discussed in Secs. 8.7 and 8.8.

Before you pull IC_{402}, check the TC_1 and TC_2 signals at pins 11 and 10 of IC_{402}. Both of these test points (which indicate the accuracy of the decoding process within IC_{402}) should produce a 7.35-kHz signal during stop but then drop to 200 Hz or less when play is selected. If not, suspect IC_{402}.

Finally, check the BFR and EFR signals at pins 31 and 30 of IC_{402}. Both of these test points show the accuracy of the sync and detection functions within IC_{402}. In play, BFR should always show zero, except under conditions of excessive skipping across the CD. In play, EFR may produce a signal but at a frequency below 50 Hz. The EFR and BFR signals are not defined during stop.

8.6 AUDIO CIRCUITS

Figure 8.7 shows the audio circuits. Note that the right-channel circuits are stressed. Audio output from pin 17 of IC_{403} (Sec. 8.5) is amplified by IC_{501}, and multiplexed into right and left channels by sample/hold circuits within IC_{502}, under control of IC_{402}.

Note that the audio output from pin 17 of IC_{403} is passed through front-panel output control R_{524}. Also note that the audio signal is still in a "serial" left, right, left, right format and must be converted to "parallel" stereo audio. The audio from IC_{403} also contains a certain amount of digital noise, which must be filtered out to produce a high-quality signal. This is done by the audio-output circuits.

8.6.1 Serial-to-Stereo Conversion

The SHR and SHL signals generated by IC_{402} (Sec. 8.5) are applied to pins 9 and 11 of IC_{502}; they close the proper switch at the correct time to route the left-audio information through the left-channel processing circuits and the right-audio information to the right channel. When one switch in IC_{502} is closed, the other switch is connected to ground, thus preventing any noise from passing to the processing circuits.

8.6.2 Filtering

The right-channel audio exits IC_{502} at pin 3 and is applied to IC_{503R}. The capacitor between the input (pin 2) and output (pin 6) of IC_{503R} removes much of the digital noise present at pin 3 of IC_{502}. The audio is then applied to pin 2 of IC_{504R}, an "analog" low-pass filter (LPF). The audio exists IC_{504R} at pin 6 and is applied to pin 3 of IC_{505R}. The first stage of amplification within IC_{505R} occurs between pins 1 and 3. The audio reenters IC_{505R} at pin 6, is amplified once again, and exits at pin 7.

An *RC* network is connected across pins 6 and 7 of IC_{505R} for deemphasis of the high-frequency signals. The time constant is cut in and out of the circuit by a switch in IC_{506}. The switch is controlled by the emphasis (EMP) signal at pin 41 of IC_{402}. Deemphasis is only required on CDs that have preemphasis during the

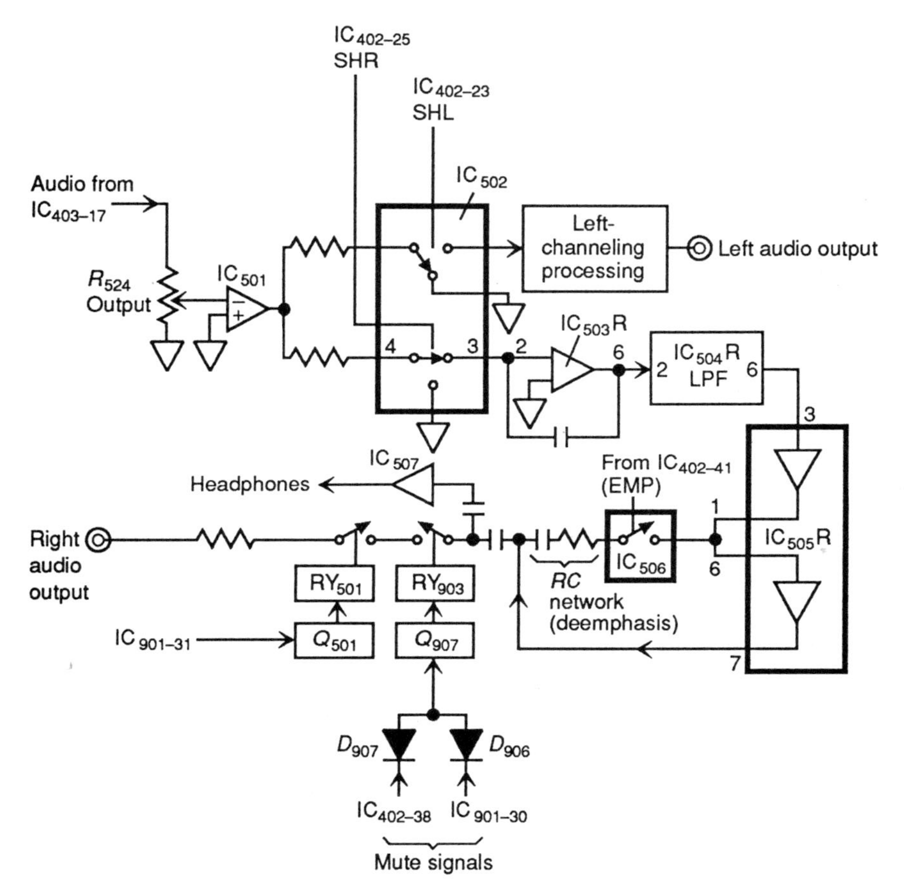

FIGURE 8.7 Audio circuits.

record process. IC_{402} recognizes a CD with preemphasis by a "flag" signal recorded on the CD. When the flag is present, IC_{402} switches in the deemphasis *RC* network.

8.6.3 Output Control

The audio at pin 7 of IC_{505} is coupled to the rear-panel output jacks through two relays. Relay RY_{903} is an internal muting relay, operated by mute signals from IC_{901} and IC_{402}. Relay RY_{501} is the audio-bus relay operated by signals from IC_{901}. The front-panel headphone jacks receive audio ahead of the relays through IC_{507}.

8.6.4 Audio Troubleshooting

If there is no audio at any output, including the headphone jack, start by checking for audio at pin 17 of IC_{403}. If it is absent, suspect the signal processing circuits (Sec. 8.5). Next, check the SHR and SHL signals from IC_{402} (pins 23 and 25). If SHR and SHL are present and there is audio at pin 17 of IC_{403}, trace the audio signal from IC_{403} to the headphones and/or rear-panel jacks. Note that the level for audio at both the rear-panel jacks and headphones is controlled by output control R_{524}.

Note that if relays RY_{501} and RY_{903} are not operating properly, the audio is cut off from the rear-panel jacks (but not the headphones). However, if the switches in IC_{506} are not responding properly to emphasis signals from pin 41 of IC_{402} or if the signal is missing, the audio passes but may appear distorted. So if you get a "the audio appears distorted when I play certain CDs" trouble symptom, start by checking the emphasis network and IC_{506} (as well as the emphasis signal from pin 41 of IC_{402}.

8.7 AUTOFOCUS

Figure 8.8 shows the principles and circuits involved for CD player autofocus. Even though CDs are manufactured to very tight tolerances, a certain amount of eccentricities may occur during the manufacturing process. To compensate for this condition, a method of automatic focusing of the laser beam on the CD surface is incorporated into all CD players.

8.7.1 Autofocus Principles

As shown in Fig. 8.8*a*, the laser beam reflected from the CD surface passes through an objective focus lens. The reflected beam then passes through coupling and cylindrical lenses. As discussed in Sec. 8.4, the cylindrical lens causes the beam to become elongated if the beam is not properly focused.

After the beam passes through the cylindrical lens, the beam is focused onto a photosensor (also called an *optical detector*, phototransistor, etc.). The detector is divided into four zones. If the lens is focused correctly on the CD surface, the reflected beam is perfectly round, and it generates equal signals from zones A, B, C, and D. When these equal signals are applied to a differential amplifier within IC_{601} (Fig. 8.8*b*), a zero autofocus signal is generated.

Now assume that the CD deviates in a downward, or negative, direction. This causes the beam striking the detector to become slightly elongated, because of the cylindrical lens. With an elongated beam, more light is applied to zones B and D, causing the noninverting input of IC_{601} to go high. As a result, the output of IC_{601} goes positive, repositioning the objective lens as necessary to refocus the beam and get an equal output from all zones of the detector.

If the CD deviates in an upward or positive direction, zones A and C produce a higher voltage than zones B and D. This raises the input voltage to the inverting input of IC_{601}. With the inverting input high, the autofocus voltage from IC_{601} goes in a negative direction, refocusing the objective lens to get equal outputs from all four zones of the detector.

8.7.2 Autofocus Circuit

Note that the autofocus sensors or detectors shown in Fig. 8.8 are also used to recover the main audio signal (EFM), as discussed in Sec. 8.4. The four zones of

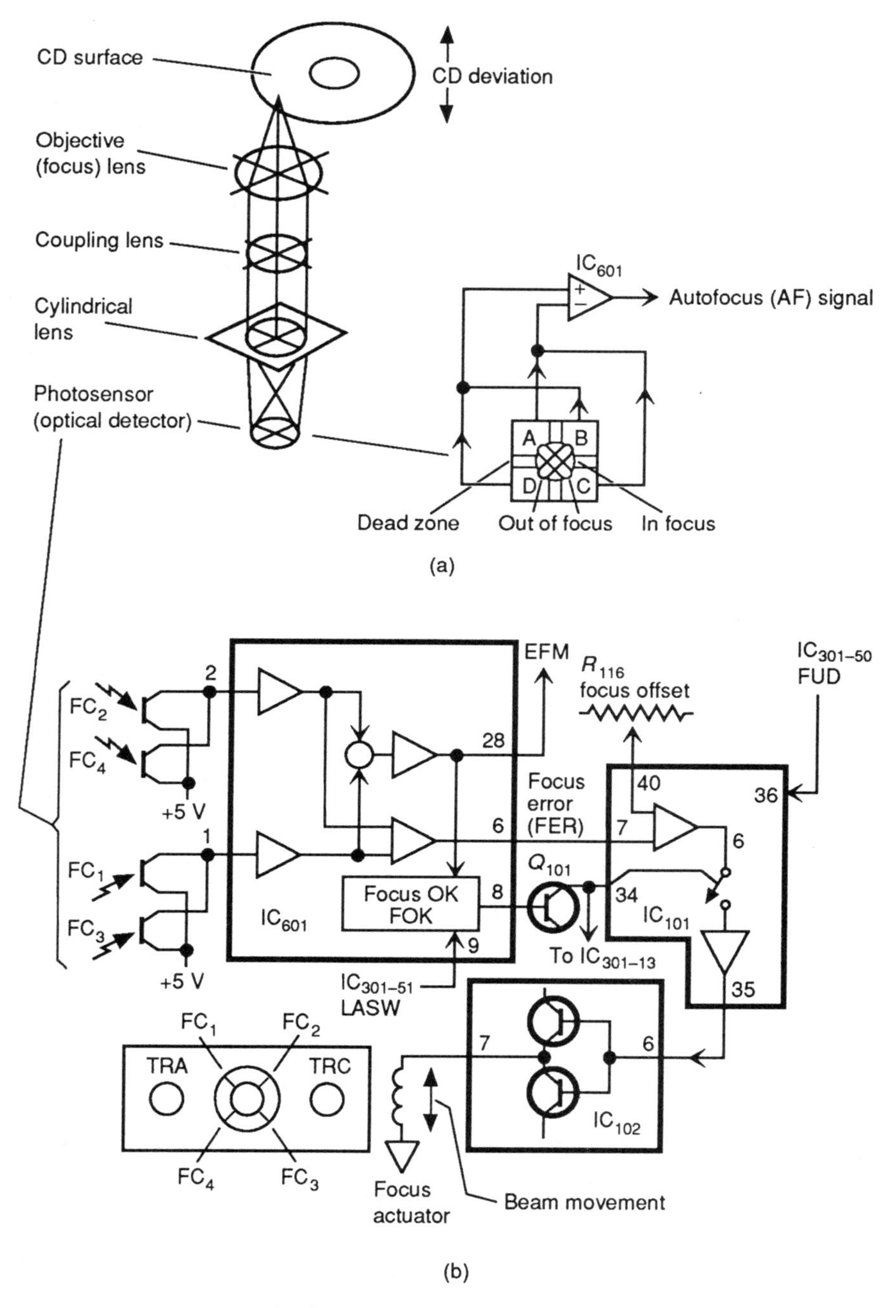

FIGURE 8.8 Autofocus circuits.

Fig. 8.8*a* are represented by four separate phototransistors FC_1 through FC_4 in Fig. 8.8*b*. The sensors are transistors that turn on when light strikes the base. The amount of current through the transistors is determined by the amount of light. Sensors FC_2 and FC_4 are placed 180° apart, with the electrical outputs applied to pin 2 of IC_{601}. The outputs of FC_1 and FC_3 are applied to pin 1 of IC_{601}.

When the laser beam is in focus, the outputs from FC_1 through FC_4 are all equal. As a result, the focus error (FER) output from pin 6 of IC_{601} is zero. However, there is always an EFM output from pin 28 of IC_{601}, with the beam in or out of focus. (The EFM signal is removed only when there is no beam or there is no CD and the beam is not reflected back from the CD onto FC_1 through FC_4.)

When the laser beam is out of focus, the outputs from FC_1 through FC_4 are not equal. As a result, there is FER output from pin 6 of IC_{601}. This FER output is applied to the focus actuator through IC_{101} and IC_{102}. The focus actuator moves the focus lens up or down to achieve focus. R_{116} provides an offset adjustment for the FER signal.

When play first begins, the focus actuator receives a focus up/down (FUD) signal from pin 50 of IC_{301} through IC_{101} and IC_{102}. The FUD signals move the focus actuator up and down two or three times as necessary to focus the beam on the CD. Once focus is obtained, a focus-ok (FOK) signal is generated by IC_{601} and applied to both IC_{301} and IC_{101} through Q_{101}. Note that IC_{601} does not produce an FOK signal unless there is an LASW signal at pin 9, as well as EFM signals at pin 28.

If an FOK signal is not received after two or three tries, IC_{301} shuts the system down and play stops (the turntable stops and the pickup moves to the inner limit). Note that if there is no CD in place, there is no EFM signal and thus no FOK signal. In this way, the FOK function also serves as a CD detector.

8.7.3 Autofocus Troubleshooting

If you suspect problems in the autofocus system, put in a CD, press play, and check that the pickup moves up and down two or three times and then settles down. If it does not, check that the laser is on, as discussed in Sec. 8.4. If the laser is on but there is no focus, try adjusting the focus servo as described in the service literature and/or Sec. 8.10.

Next, check the focus actuator by measuring the resistance of the focus and tracking coils. Typically, the focus coil is about 20 Ω, while the tracking coil is 4 Ω. The actual resistance depends on the particular assembly. However, if you get an open or short indication or a resistance that is drastically different from these values, the actuator is suspect. In some players, you can see a slight movement of the actuator when the ohmmeter is connected to the coils. This usually indicates that the actuator is good.

If the actuator coils appear to be good and the problems cannot be corrected by adjustment, check the autofocus circuits as follows:

If the focus actuator does not move up and down, check for FUD pulses just after play is selected. If the FUD pulses are absent at pin 50 of IC_{301} and/or pin 36 of IC_{101}, suspect IC_{301}. Next, check for pulses at pin 35 of IC_{101}, pins 6 and 7 of IC_{102}, and at the focus coil.

If the focus actuator moves but focus is not obtained, check for FOK signals at pin 34 of IC_{101}, pin 13 of IC_{301}, and pin 8 of IC_{601}. (If the FOK signals are not present, IC_{301} should shut the system down.)

If the FOK signals are absent, suspect IC_{601} or possibly the four pickup sensors FC_1 through FC_4. If FOK signals are absent at $IC_{101\text{-}34}$, suspect Q_{101}. Also note that the FOK signal is not generated unless there is an LASW signal applied to pin 9 of IC_{601} from IC_{301}.

Next check for FER signals at $IC_{601\text{-}6}$ and $IC_{101\text{-}6}$. If FER signals are absent at $IC_{601\text{-}6}$, suspect IC_{601} or possibly the sensors FC_1 through FC_4.

If you suspect the sensors FC_1 through FC_4, monitor the EFM signal at pin 28 of IC_{601}. If the EFM signal is good, it is reasonable to assume that all four sensors are good.

8.8 LASER TRACKING

Figure 8.9 shows the principles and circuits involved for CD player tracking of the pits (audio information) on a CD. In brief, tracking is done by passing the laser beam through a defracting lens, which separates the main beam into three separate beams that are slightly staggered. The TRA and TRC beam spots are used to generate the tracking error signal, or TER (similarly to FER). The main beam spot is used to provide both the focus signal and main audio signal.

8.8.1 Tracking Principles

As shown in Fig. 8.9*a*, TRA and TRC are shifted slightly off-center of the main spot in opposite directions. The TRA and TRC beams are reflected back from the CD track into the prism mirror and onto the tracking photosensors (on the same block as the focus and audio sensors but located on either side of focus and audio).

The outputs of the tracking sensors are applied to a differential amplifier, which generates the correction signal. For example, if the main spot shifts to the left of the desired track, TRC moves completely off the pits, thus reflecting very little light. At the same time TRA moves over the pits, thus reflecting more light. This causes more laser light to fall on TRA, generating a higher TRA output and causing the inverting input of the differential amplifier to go high. This allows the output of the differential amplifier to generate a positive voltage, causing the tracking actuator (and all three beams) to move to the right (and thus restore proper tracking).

Now assume that the main spot shifts to the right of the pits. The TRA output goes low while the TRC output goes high. Since TRC is directly over the pits, more of the laser beam is reflected onto the TRC sensor. This TRC output is coupled into the noninverting input of the differential amplifier, causing the amplifier to generate a negative output, moving the tracking actuator (and beams) to the left (to again restore proper tracking).

8.8.2 Tracking Circuits

As shown in Fig. 8.9*b*, the outputs from TRA and TRC apply signals to pins 2 and 6 of IC_{603}. The TRA and TRC signals are coupled to pins 1 and 7, respectively, through amplifiers within IC_{603}. When the laser beams track the CD correctly, the outputs at pins 1 and 7 are equal.

The TRA output is coupled through tracking offset R_{603} to pin 4 of IC_{601}. R_{603}

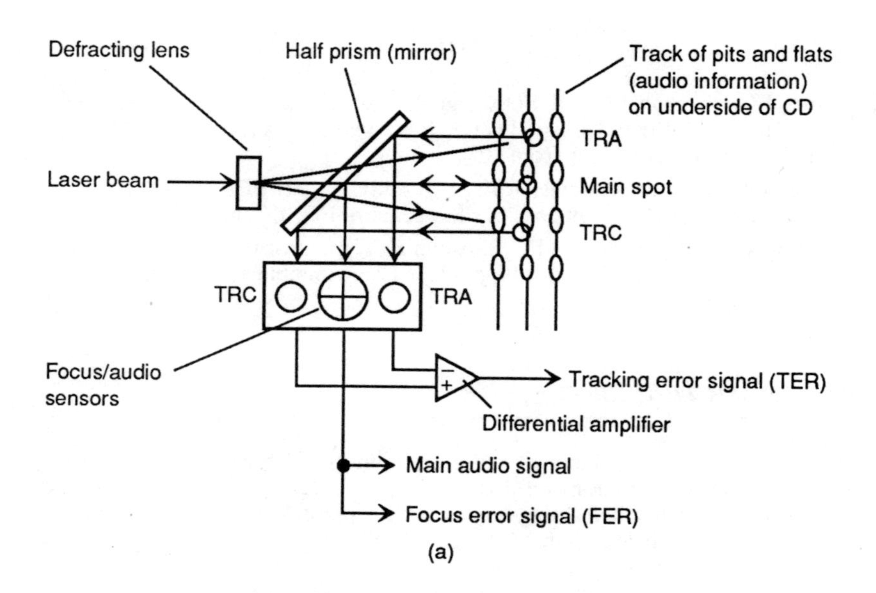

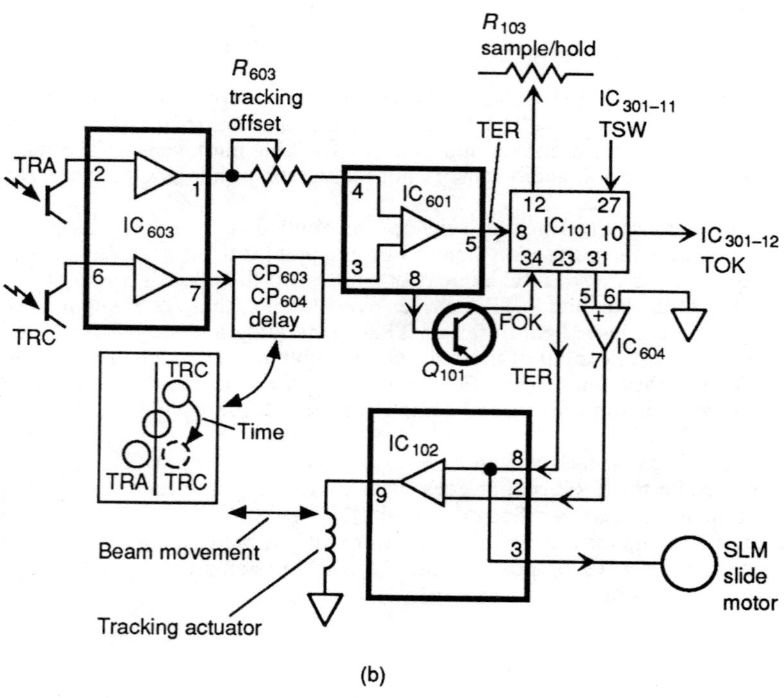

FIGURE 8.9 Laser tracking circuits.

provides a fine adjustment so that the output at pin 5 of IC_{601} is zero when the CD is tracking properly.

The TRC signal from pin 7 of IC_{603} is passed through delay line CP_{603} and CP_{604} before entering IC_{601} at pin 3. CP_{603} and CP_{604} serve to delay the TRC signal as necessary so that TRC arrives at IC_{601} simultaneously with TRA. The time delay is necessary because the tracking circuits require that the TRA and TRC beams analyze the same point on the CD, even though the beams are separated by the optical system.

The tracking error signal, or TER, (if any) at pin 5 of IC_{601} is applied through IC_{101} and IC_{102} to the tracking actuator coil and causes the actuator to move the beams right and/or left as necessary to produce proper tracking. Note that IC_{101} requires various signals (FOK, TSW, etc.) to process the TER signals properly. Also note that sample/hold R_{103}, discussed in Sec. 8.3.2, used by the pickup or slide motor (SLM) has an effect on the tracking circuits. We discuss both of these factors next.

8.8.3 Tracking Circuit Troubleshooting

It is often very difficult to separate tracking and focus servo problems. For example, unless there is an FOK signal applied to $IC_{101\text{-}34}$ through Q_{101}, the TER signal does not pass to the tracking actuator. Both the focus and tracking servos use the laser beam as a source of error signal, although different sensors are used. To complicate the problem further, the TER signal is also used by the SLM as a fine speed control. (This is done in IC_{101}.) If TER is lost, both the tracking actuator and SLM have no control signals. Either condition can produce symptoms of improper tracking.

The TER signal is passed through variable-gain and error-detection circuits in IC_{101}. These circuits can interrupt the TER signal when errors are detected (in the CD or because of improper tracking). Failure in any of these circuits can cut off, or alter, the TER signal, making it appear that either tracking or SLM servos are at fault (when actually the servos are good).

First try correcting any tracking problems with adjustment of R_{603} and R_{103}, as described in the service literature or in Sec. 8.10. Then make a quick check of the tracking actuator coil, as described for the focus actuator coil in Sec. 8.7.3. Finally, see if the SLM moves to the inner limit when power is first applied, as described in Sec. 8.7.3 (possibly adjust R_{107} and S_1). If the SLM moves to the inner limit, this confirms that the SLM, reset circuit, and basic servo circuits of IC_{101} and IC_{102} are good.

If the SLM is operating and the tracking actuator coil appears to be good but tracking problems cannot be restored by adjustment, check the tracking and pickup (SLM) motor.

Check the TER signal from the source to the tracking actuator coil (and drive signal to the SLM). Pin 5 of IC_{601} is a good place to start. Then check the tracking actuator coil.

If TER is present at pin 5 of IC_{601} but not at the coil, suspect IC_{101} and IC_{102}. Look for a drive signal at pin 23 of IC_{101}. If it is present, the problem is probably in IC_{102}. If it is absent, the problem is localized to IC_{101}.

Check for drive signals at the SLM. If TER is present at the tracking coil but there is no drive at the SLM motor, suspect IC_{102} and IC_{604} (Sec. 8.3). Also check for drive signals at pin 21 of IC_{101}.

Before you pull IC_{101}, remember that IC_{101} must receive a number of signals before TER signals can pass. For example, IC_{101} must be turned on by FOK and TSW signals, and IC_{301} must receive TOK signals from IC_{101} before returning the TSW signals. Also, the TER signals are analyzed by error-detection circuits within IC_{101}. If any of these signals or voltages are absent or abnormal, IC_{101} remains cut off, and the TER signals do not pass. So always check the signals and voltages at the pins of IC_{101} (using service-literature values) before you decide that IC_{101} is defective.

If TER is absent at $IC_{601\text{-}5}$, suspect IC_{601}, CP_{603}, CP_{604}, IC_{603}, and the TRA and TRC sensors.

8.9 TURNTABLE MOTOR CIRCUITS

Figure 8.10 shows the CD turntable motor circuits. In CD players, the turntable is rotated at a variable speed. This keeps the rate at which the track moves (in relation to the pickup) constant. (The track is the series of pits on the CD that represent audio information.) The speed variations are necessary since there is less information on the tracks near the inside of the CD (start) than near the outside (stop or end).

Most CD players use some form of unitorque motor with Hall-effect elements to get the variable speed. This is similar to the speed-control circuits for LP turntables (Chap. 9). Of course, with LP turntables, you want constant speed or constant angular velocity (CAV), instead of the constant linear velocity (CLV) required for CD players. Typically, the CD speed varies from about 480 rpm (inside) to 210 rpm (outside) so as to maintain a CLV of about 1.25 to 1.3 m/s.

In the circuit of Fig. 8.10, the Hall-effect outputs are fed through IC_{201} to the motor drive windings (A and B) and thus maintain the desired speed. The Hall-effect elements are also fed currents through IC_{201} (from controller IC_{402} under the direction of IC_{301}) to vary the speed at the desired rate. CLV circuits within IC_{402} monitor the EFM signal to determine the rate at which information is passing and then produce the necessary signals to maintain the desired speed. R_{201} sets the gain of the Hall-effect signals in IC_{201} and thus sets motor speed.

Operation of the motor control can be divided into two phases: when power is first applied (sometimes called the start-servo phase) and when the motor reaches the desired speed (the regular-servo phase).

When power is first applied, the motor runs free. DMSW and CLVH signals (applied to IC_{402} and IC_{201} from IC_{301}) are low, and the ROT signal (from IC_{301} to IC_{402}) is high. Under these conditions, the motor begins to accelerate and turn at a constant velocity. IC_{402} then produces essentially similar outputs at PWM, PREF, and PD.

After a free-run period (set by IC_{301}), ROT goes low and the motor starts to accelerate. The EFM signal is read by IC_{402} and compared to a reference. The difference between the reference and the EFM is the PWM output from IC_{402}.

During the acceleration portion of this startup period, the PWM duty cycle (which varies above and below 50 percent, as determined by motor speed) is compared with the PREF signal (which is a fixed duty cycle). The result of this comparison is applied to the motor circuits to control speed.

When the motor reaches the desired speed (the regular-servo phase), the PWM signal has a 50 percent duty cycle and the pickup reads the CD data at a constant linear velocity.

This CLV condition is maintained within a ±1 percent accuracy by means of

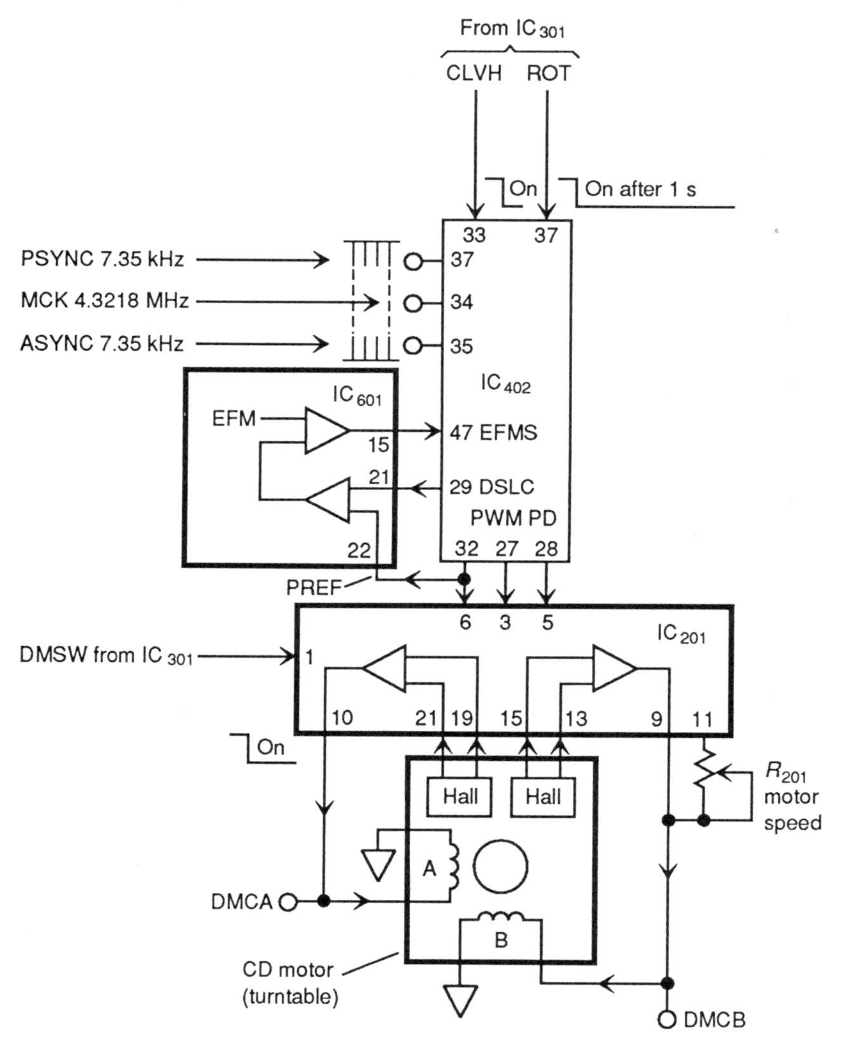

FIGURE 8.10 CD turntable motor circuits.

the PD pulse from IC_{402}. The duty cycle of the PD pulse is set by comparison of the EFM signal to a reference within IC_{402}. In turn, the PD signal is compared with the output from the PWM and PREF comparator. The result of this comparison is applied to the motor and maintains the 1 percent accuracy.

8.9.1 CD Turntable Troubleshooting

It is not difficult to tell if the turntable motor fails to rotate. Similarly, the cause of such a total failure is generally simple to locate. For example, you can check

DMCA and DMCB at pins 9 and 10 of IC_{201} for drive signals to the motor windings. If the drive signals are present but the motor does not turn, suspect the motor. If either of the drive signals is not present, trace from the motor to IC_{201} and from IC_{402} to IC_{201}.

Before you decide there is a problem in the turntable motor circuits, remember that DMSW, CLVH, and ROT signals must come from IC_{301}. Also, if IC_{301} does not receive an FOK signal from the autofocus circuits (Sec. 8.7), the DMSW, CLVH, and ROT signals are set to prevent IC_{402} and IC_{201} from passing the PREF, PWM, and PD signals to the motor circuits.

Both DMSW and CLVH are made low to turn on the motor when play is selected. ROT goes low for about 1 s after play is selected. If DMSW, CLVH, and ROT all remain high after play is selected, check for FOK signals to IC_{301} (at pin 13) and TOK signals to $IC_{301\text{-}12}$. Of course, if only one of the three signals remain high, IC_{301} is most likely at fault.

The problem is not quite that simple if the motor rotates, but you are not sure of the correct speed. This is especially true when you consider that the motor speed is constantly changing. You must rely on waveform measurements and adjustments. So the first logical step in motor troubleshooting is to perform adjustment of R_{201} as described in the service literature or Sec. 8.10.

If you get the DMCA and DMCB drive signals and the motor is turning (indicating that the DMSW, ROT, and CLVH from IC_{301} are good) but you are unable to set the output levels as described, check all the waveforms associated with the motor control circuits as follows. Check the PWM, PREF, and PD signals from IC_{402}. If any of these are absent or abnormal, suspect IC_{402}. Next, trace signals between IC_{201} and the motor (Hall-effect signals and drive signals). If any of these signals are absent or abnormal, suspect either IC_{201} or the motor.

Note that PREF from IC_{402} is also applied to IC_{601}, along with the DSLC signal from IC_{402}, to form the EFMS signal that is returned to IC_{402} (Sec. 8.4). If the EFMS signal is absent, IC_{402} does not produce PREF, PWM, and PD signals. Of course, if the EFMS signal is not applied to IC_{402}, several other problems occur (no audio, etc.).

One way to check to see if the EFMS signal is being processed properly is to compare the PSYNC and ASYNC signals at pins 37 and 35 of IC_{402}, using a dual-trace scope. As shown in Fig. 8.10, both signals should be synchronized as to time. If they are not or if either signal is missing, suspect IC_{402} (or possibly IC_{601}).

As you can see, the turntable motor circuits are closely interrelated with the signal processing IC_{402} circuits (Sec. 8.5). A failure in signal processing can also cause the motor circuits to appear defective. So if you are unable to locate a problem in the motor circuits, try checking the signal processing circuits.

8.10 TYPICAL TESTING AND ADJUSTMENTS

This section describes the test and adjustment procedures for a typical CD player. Compare these procedures to those found in service literature. We start with laser diode adjustment, which is always a good point to start on any type of CD player.

8.10.1 Laser Diode Test and Adjustment Procedures

Normally, the laser diode need not be adjusted or tested unless (1) the pickup has been replaced or (2) troubleshooting indicates a laser problem. So, before you suspect the laser, consider the following points.

Even though the laser beam is invisible, the diffused laser beam is often visible at the objective lens. (The lens appears to glow when the beam is on.) Also, when power is first applied to the optical circuits, the objective lens moves up and down two or three times to focus the beam on the CD, as described in Sec. 8.7.2. So, if you see the objective lens move when power is first applied, it is reasonable to assume that the laser is on and producing enough power to operate the optics.

Of course, this brings up some obvious problems. First, on most players, if you open the CD compartment and gain access to see the lens, you must override at least one interlock. Next, many players have some provision for shutting down the player optics if there is no CD in place (Sec. 8.7.2), so you must override this feature.

Most important, never, never look directly into the objective lens with power applied, and keep your eye at least 12 in from the lens. The purpose of the lens is to focus the beam sharply onto the CD. The lens can do the same job for your eye.

The service literature for early model CD players sometimes recommends monitoring the laser with a light meter. However, it is more practical (and much easier) to adjust the laser diode output until you get an EFM signal of correct amplitude. This not only checks the laser but also checks the photodiodes and IC amplifiers following the photodiodes.

Figure 8.11*a* is the diagram for testing and adjusting the laser diode using the EFM signal. Before you make the adjustment, set R_{629} to minimum and then increase the setting as required.

Note that chuck switch S_3 must be in the closed (tray in) position before power is applied to Q_{601} and the laser. You must override S_3 manually during adjustment. If S_3 is in the tray-open position, the laser has no power and IC_{301} receives a 5-V signal to shut the system down.

1. Connect the scope as shown in Fig. 8.11*a*. With this connection you are monitoring the EFM signal (after the photodetector output is preamplified). As discussed, the EFM signal (at this test point) is also applied to the tracking, focus, and pickup-motor servos, as well as to the signal processing circuits.
2. Load a CD in the player and select play. The EFM signal should appear on the scope and produce a waveform similar to that of Fig. 8.11*a*.
3. Adjust R_{629} until the EFM signal level is 0.7 V (or as specified in the service literature, typically 0.5 to 0.9 V).

Be aware that laser diodes can be damaged by current surges (as can any semiconductor). Typically, the lasers used in CD players have drive-current limits in the 40- to 70-mA range, possibly 100 mA. Generally, 150 mA is sufficient to damage (if not destroy) any CD laser diode.

Current limitations can present a problem since laser diodes may require more drive current to produce the required light as the diode ages. Some service literature spells out "safe" limits of laser drive current. The simplest way to check laser drive current is to measure the voltage across a resistor in series with the

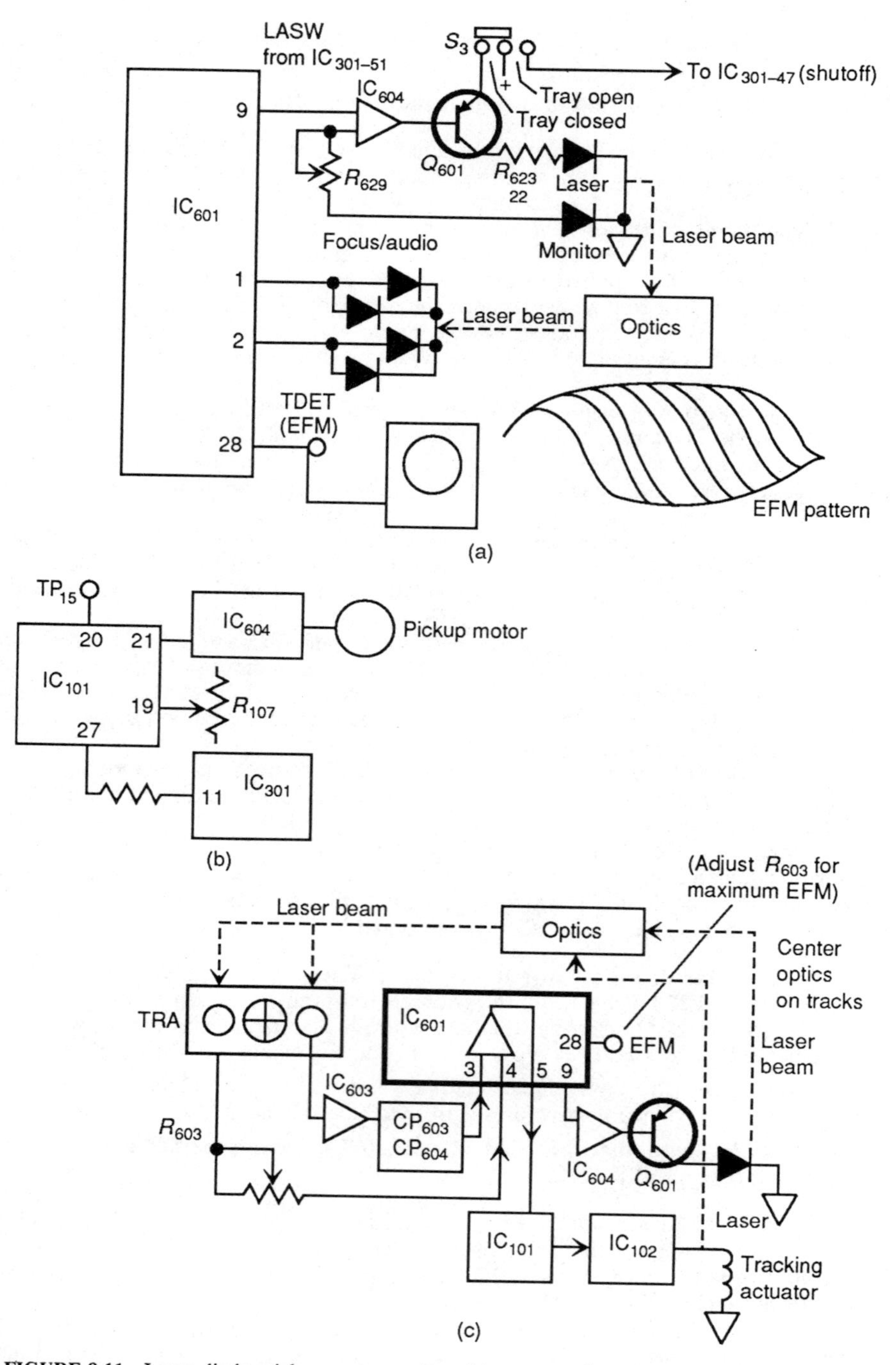

FIGURE 8.11 Laser diode, pickup-motor, and tracking-servo adjustments.

diode, such as R_{623} in Fig. 8.11*a*, and then calculate the drive current. For example, if the recommended laser diode current is 40 to 70 mA and the series resistance is 22 Ω, the voltage should be between 0.88 and 1.54 V. You can make this check before adjustment of the diode, and you should make the check after adjustment (to make sure that you have not exceeded the safe drive limits).

8.10.2 Pickup-Motor Adjustment

This adjustment is not available on all CD players. When available, the adjustment sets the point where the pickup accesses the beginning of the CD (the directory). If the adjustment is not correct, the program information may not be read properly.

Note that the adjustment controls the pickup-motor servo and is not to be confused with the inner-limit switch (S_1, Fig. 8.4). However, the two adjustments are interrelated. For example, if you set the switch so that the pickup motor cannot reach the inner limit, the servo cannot be adjusted to access the full CD directory.

1. Monitor the voltage at test point TP_{15} as shown in Fig. 8.11*b*. With this connection, you are monitoring the motor gain output from servo IC_{101}.
2. Load a CD in the player and select play mode.
3. While the CD is playing, connect pin 11 of IC_{301} to ground. This simulates a low TSW signal to IC_{301} and prevents pin 11 from going high (which would cause IC_{301} to shut the system down).
4. Set the player to stop. After about 10 s, measure the dc level at TP_{15}, and adjust R_{107} so that the reading is 0 V ± 50 mV. Adjust R_{107} in small increments and wait for the voltage level to stabilize before continuing to adjust. (Make sure to remove the ground from pin 11 of IC_{301} when adjustment is complete.)

8.10.3 Tracking Adjustment

Figure 8.11*c* is the adjustment diagram. With this setup, you are monitoring the EFM signal and adjusting the optical pickup (through the servo and tracking actuator) so that the laser beam is properly centered on the tracks (as indicated when the EFM is maximum). Note that R_{603} sets the offset of the two tracking photodiodes but not the four remaining focus and audio photodiodes.

1. Load a CD in the player and select play mode. The EFM should produce a waveform on the scope (similar to that shown in Fig. 8.11*a*).
2. Adjust R_{603} until the EFM is maximum (optical pickup centered on the tracks).

On some players, the display may become erratic and the audio may mute after this adjustment is made. If the display is erratic, set the player to stop and then go back to play. This should eliminate the erratic display.

8.10.4 Focus Adjustment

Figure 8.12*a* is the adjustment diagram. With this setup, you are again monitoring the EFM, but you are now adjusting the optical pickup (through the servo and focus actuator) so that the laser beam is properly focused on the tracks (as indicated by maximum EFM). Note that R_{116} sets the offset of the four focus and audio photodiodes but not the two remaining tracking photodiodes.

1. Load a CD in the player and select play mode. The EFM should produce a waveform on the scope (similar to that shown in Fig. 8.11*a*).
2. Adjust R_{116} until the EFM is maximum (optical pickup focused on the CD tracks).

Again, if the display becomes erratic after this adjustment, stop and restart the player.

8.10.5 Turntable Motor Adjustment

Figure 8.12*b* is the adjustment diagram. With this setup, you are monitoring the drive signals to both coils A and B of the turntable motor (from motor drive IC_{201}).

1. Load a CD in the player and select play mode.
2. Adjust R_{201} so that the output levels at DMCA and DMCB are equal. Usually, DMCA and DMCB are about 2 V p-p.

8.10.6 Dropout Sample/Hold Adjustment

Figure 8.12*c* is the adjustment diagram. This adjustment, not available on all players, is not to be confused with the sample/hold (S/H) audio circuits (Sec. 8.6). The S/H circuits shown in Fig. 8.12*c* are located in pickup servo IC_{101}, and they control the TER signals (Sec. 8.8).

With the setup in Fig. 8.12*c*, play a CD with a simulated defect, and adjust TER signals to produce the best response (minimum audio dropout). The effect is simulated by placing a black (nonreflective) tape on the mirror side of the CD. Then monitor the EFM and adjust for minimum dropout (ideally there should be no dropout).

You can make this adjustment by ear. The simulated defect produces a chattering or ticking in the audio. Adjust for minimum noise. The scope is generally more accurate (or you can monitor both ways). Do not turn up the volume with a simulated defect; the noise is unbearable.

1. Load a CD in the player and select play mode.
2. Adjust R_{103} for minimum audio dropout on the EFM display or for minimum chattering in the audio or both.

Note that with such a defect, a portion of the EFM display is cut out (typically a notch or wedge, starting from the top, as shown in Fig. 8.12*c*). However, you should be able to eliminate all (or most) of the audio dropout (as indicated by a

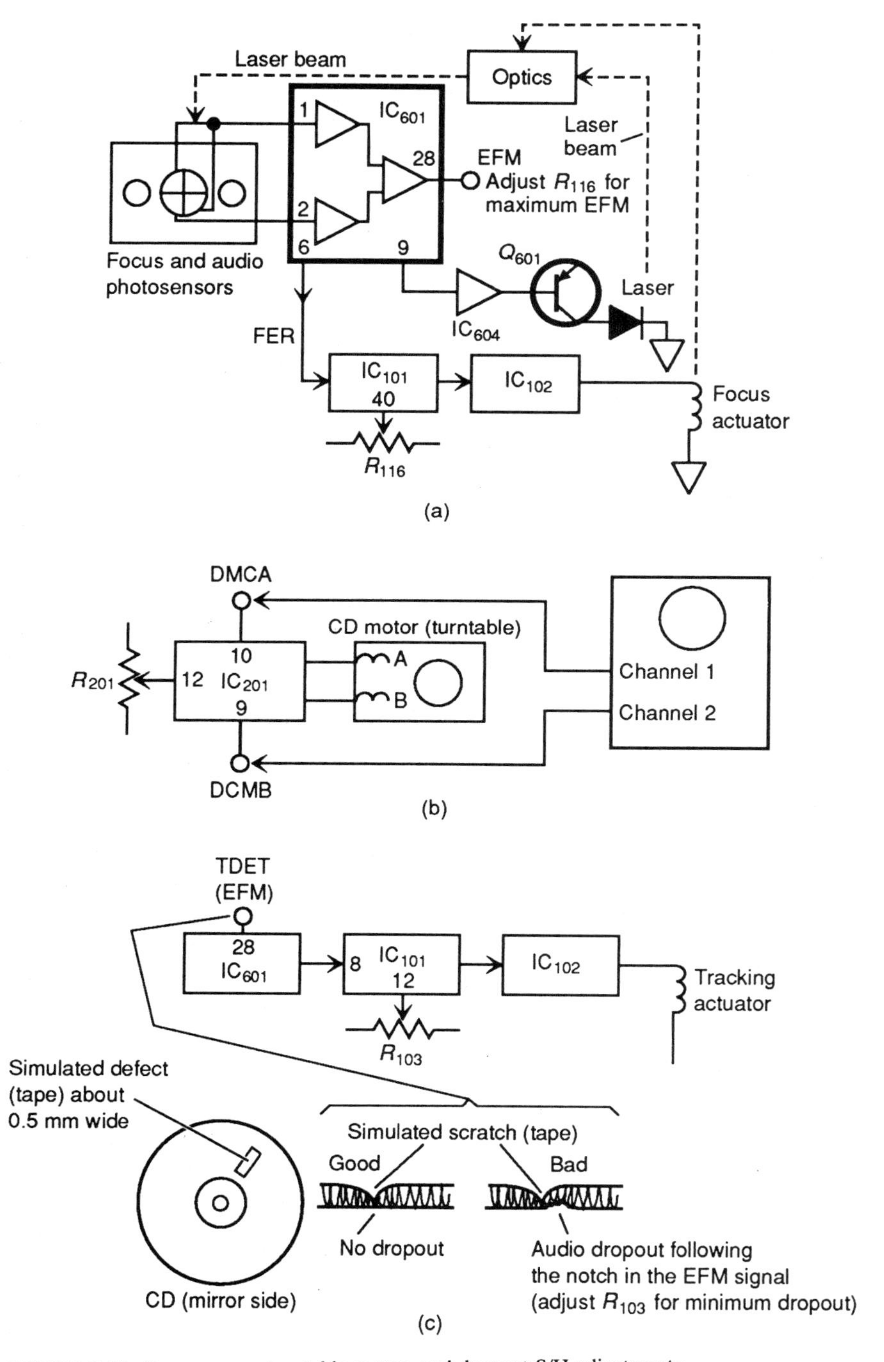

FIGURE 8.12 Focus-servo, turntable motor, and dropout S/H adjustments.

cutout at the bottom of the EFM display). If you get considerable dropout at all R_{103} settings, IC_{101} may be defective.

8.11 PRELIMINARY TROUBLESHOOTING

Here are the checks to be made before going into the CD player circuits:

If the CD player operates manually but not by remote, check the remote-unit batteries and the control cable. Then try resetting the power circuits by pressing the front-panel power button on and off.

If the left or right speaker is dead when the CD player is used in a system, try playing another component connected to the audio system line (AM/FM tuner or cassette deck). If operation is normal with the other component, suspect the CD player. Also check for a loose cable between the CD player and other components. Then temporarily reverse the left and right speaker leads. If the same speaker remains dead, the speaker is at fault.

If the left or right speaker is dead when the CD player is used in a nonsystem configuration, temporarily reverse the left and right cable connectors at the amplifier input from the CD player. If the same speaker remains dead, the amplifier (or speaker) is probably at fault. If the other speaker goes dead, suspect the CD player or the player output cable.

If there is no sound from either speaker, check for the following. Both speaker selector switches may be turned off. The wrong speaker terminals may have been selected (on amplifiers with two sets of speakers). The speaker cables may be disconnected. The CD player cables may be disconnected or improperly connected. In a nonsystem configuration, the output level may be too low (control too far counterclockwise), or the amplifier output selector may be set for the wrong source. Try playing the amplifier with a different input.

If the sound is distorted, the output level may be too high, or the CD player output may be connected to the phono input of the amplifier.

If there is hum or noise (only when the CD player is used), check for the following. The shield of the audio cable from the CD player may be broken, or the connector may not be firmly seated in the jacks. The CD player may be too close to the amplifier. The magnetic fields produced by the amplifier may induce hum into the player circuits (not likely, but possible).

If the player does not start, check the following. Make certain that there is a CD in the tray, that the CD is properly loaded (not upside down; the label should be up), and that the CD is firmly seated on the supports. Also check to see if the CD is very dirty, scratched, or warped. It is also possible that moisture has condensed on the CD or the objective lens.

If the sound cuts or repeats at some point, there may be a very dirty spot on the CD. Clean the CD with a soft cloth and mild detergent. (Do not use any commercial cleaners unless they are specifically recommended for CDs.) It is also possible that there is a scratch on the CD. Try skipping the point where the sound cuts or repeats.

CHAPTER 9
GRAPHIC EQUALIZER AND LP TURNTABLE AUDIO

This chapter is devoted to graphic equalizers and LP turntables. Although the LP is being replaced by the CD in rapid order, there are still many LP turntables in use. The following sections describe a cross section of graphic equalizer and LP turntable circuits.

9.1 GRAPHIC EQUALIZERS

Figure 9.1 shows the operating controls for a typical graphic equalizer and preamp (the Alpine 3311). Note that both equalizer and preamp functions are usually combined. Figure 9.2 shows the block diagram. Before we get into the circuits, let us discuss the purposes and functions of equalizers.

9.1.1 Equalizer Functions

Equalizers (EQs) are used to "custom tailor" a stereo system to the taste of the listener and to compensate for acoustic problems in the audio environment (listening room or car). A problem common to both room and car stereo systems is poor frequency response because of less-than-optimum speaker placement. In the case of a car, speaker placement is usually dictated by the car manufacturer. In room installations, there may be only certain places where the speakers can be placed (as dictated by a stubborn customer).

An EQ takes the concept of bass and treble adjustment (Sec. 5.7) an important step further. With EQ, the audio spectrum is divided into *bands or ranges of adjustment*. There are two basic types of EQs: parametric and graphic.

Parametric EQs offer the ultimate in system equalization and provide the installer with a tool to fine tune the system. The specific bands of adjustment are determined by selecting a *center frequency for each band*. Then the installer decides on the width of the adjustment band by assigning a Q, or slope, to the band. The slope can be gradual (producing many adjacent frequencies) or narrow (for a more focused adjustment), as shown in Fig. 9.3*a*.

Once center frequency and slope are selected, the installer adjusts the system by boosting (with amplifier gain) or cutting (with attenuation) the signal in a par-

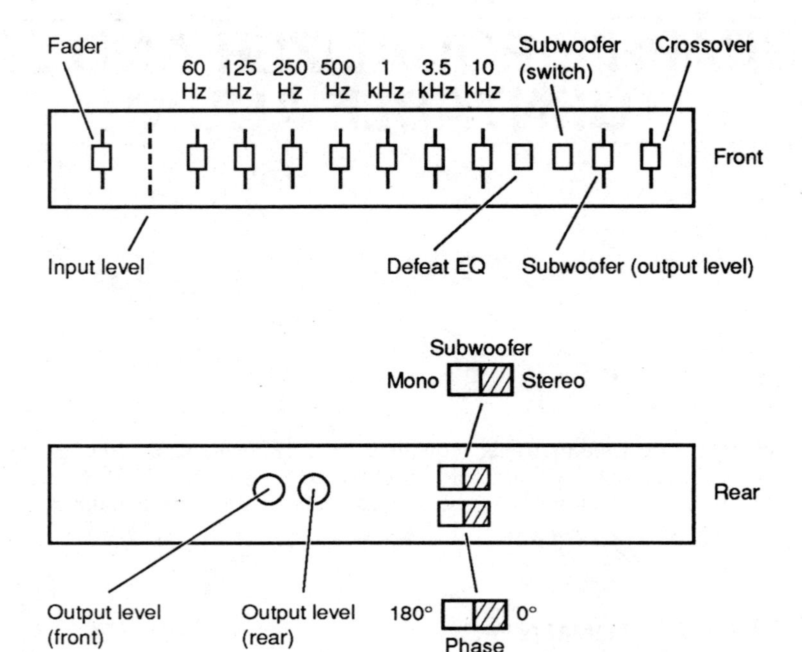

FIGURE 9.1 Operating controls and indicators for a typical graphic equalizer.

ticular section of audio bandwidth. A parametric EQ allows the installer to improve system frequency response by filling the "holes" or smoothing out the "peaks" that produce poor audio reproduction, as shown in Fig. 9.3*b*.

Proper installation of a parametric EQ usually requires a real time analyzer (RTA), at least for the initial set up. Then the installer can "tweak" the EQ controls to the customer's satisfaction (ideally). The RTA allows the installer (and customer) to see the effects of adjustment on the system frequency response. Usually, a parametric EQ is not located where the customer can make constant adjustments (fortunately) and is designed to be adjusted on a one-time basis (often only at initial installation).

Graphic EQs (unlike parametric) are usually located for easy access to the customer (typically on the stereo rack front panel or on the car's dashboard), and the center frequencies are predetermined. The center frequency and slope are assigned to provide adjustment of the *most common problem areas* of the audio bandwidth. The listeners are able to cut or boost the levels of the assigned band to compensate for any problems in the listening area and to customize the frequency response to their own tastes.

The more elaborate (and expensive) graphic EQs have more bands (usually 11 bands at most). Our graphic EQ has seven bands and includes a defeat switch which cuts the EQ functions out. This permits the audio produced by an equalized system to be compared with an unequalized or "flat" system. Some graphic EQs have power-level indicators (a series of differently colored LEDs) that provide visual confirmation of the output level.

Our graphic EQ also includes a *subwoofer output*. This output is produced by an active dividing network that provides a low-pass output with three choices for

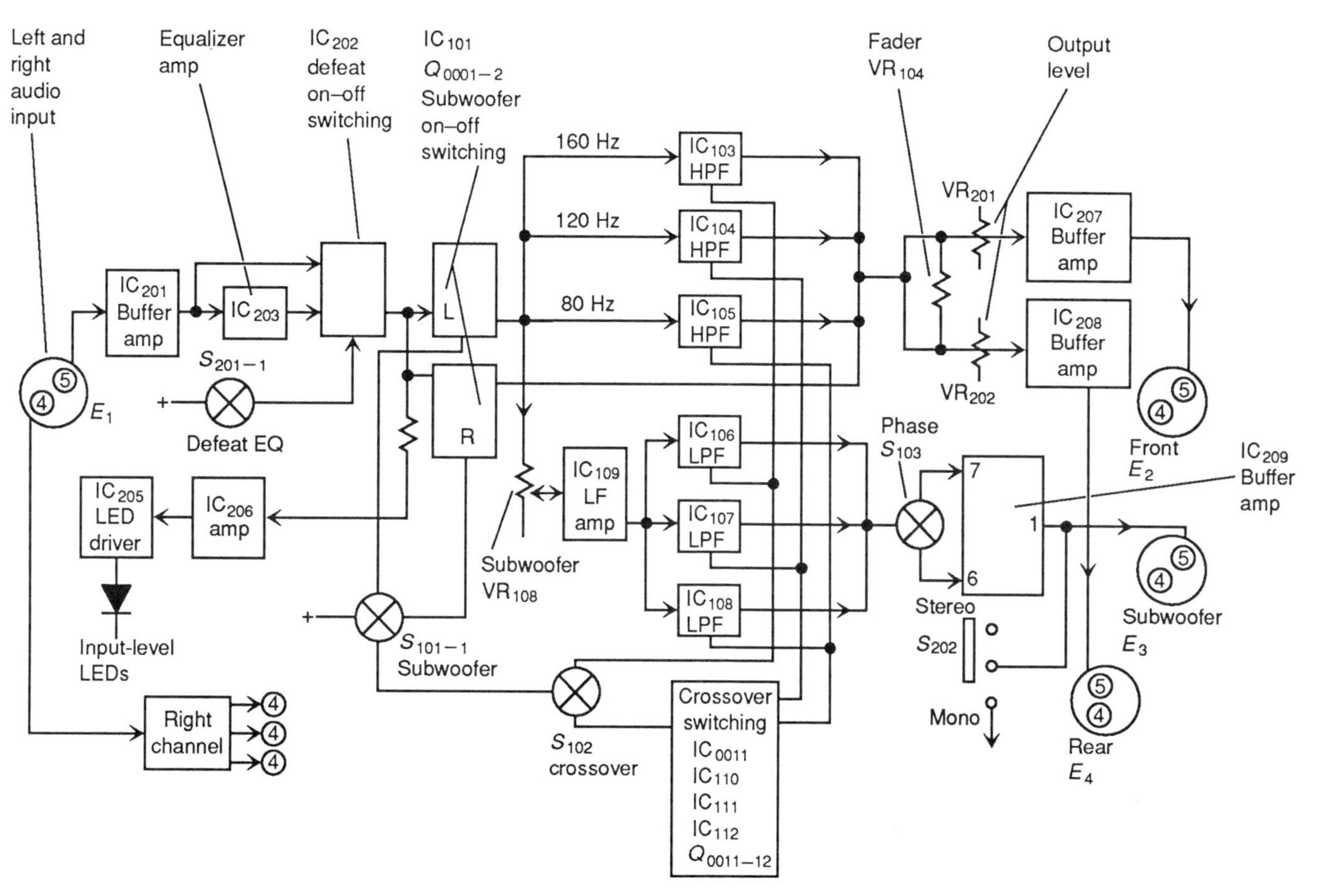

FIGURE 9.2 Graphic equalizer block diagram.

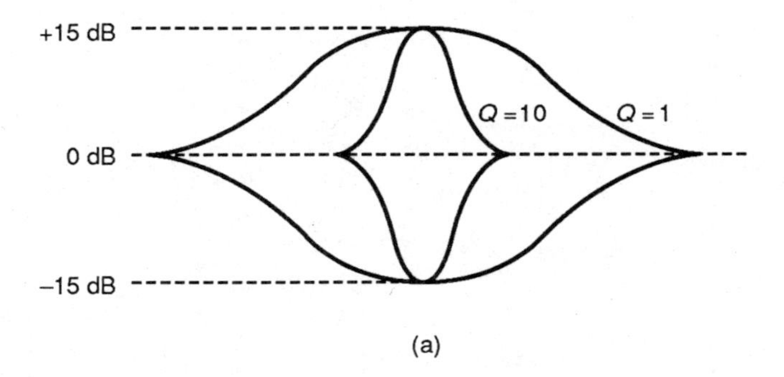

(a)

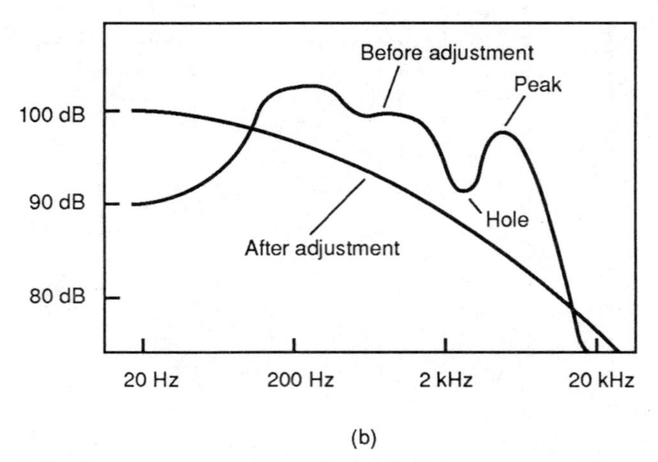

(b)

FIGURE 9.3 Graphic equalizer band-slope and sound-curve adjustment characteristics.

the crossover point: 80, 120, or 160 Hz. When the subwoofer output is on, the front and rear outputs become high pass above the selected crossover frequency. There is a *subwoofer level control* to adjust the amount of subbass in the system. Also included are a *stereo/mono switch* (to choose a stereo subwoofer system or a system with a summed mono output) and a *subwoofer phase switch* to correct phase problems that may occur in bass response.

9.1.2 Equalizer Features

Our graphic EQ is designed for car installation and has seven slide controls for individual control of seven frequency bands. LED indicators on all slide controls illuminate and display control positions. The crossover selector allows selection of the three frequency-crossover points (80, 120, and 160 Hz). Five LED input-level indicators reflect signal input levels (which are adjustable by the volume control of the tuner or tape deck).

9.1.3 Basic EQ Operating Procedures

The following is a summary of the basic operating procedures for the graphic EQ shown in Fig. 9.1. Most EQs have similar procedures.

The graphic EQ is turned on and off automatically when the tuner or tape deck is switched on or off. When the EQ is on and the defeat EQ switch is in the off (signals through the EQ) position, the defeat EQ switch illuminates in green. The LED indicators on all EQ controls are on.

With the graphic EQ on, adjust the volume control on the tuner or tape deck, the fader control, and the seven equalizer controls to suit listening preference. Figure 9.3*b* shows a typical audio sound curve with a hole and a peak, both before and after adjustment with a graphic EQ.

When the defeat EQ switch is in the on (EQ defeated) position, defeat EQ illuminates in amber and the LED indicators on all EQ controls are off. Under these conditions, adjust the volume *and tone* controls of the tuner or tape deck and the fader control of the EQ to suit listening preference.

Regardless of the defeat EQ switch position, the five input-level indicators light up in response to the input signal levels (even though the input level is controlled by the tuner or tape-deck volume control).

When the five input-level indicators stay on, the input signal is at the maximum that the EQ will accept. Generally, you get the best sound when the input is lower than maximum (all amplifiers and networks not overdriven).

9.1.4 Subwoofer Operation

With the graphic EQ turned on and the subwoofer switch set to off, the subwoofer switch illuminates in green. Under these conditions, there is no subwoofer output, and a full range of signals is passed to front and rear speakers.

With the subwoofer switch set to on, the subwoofer switch illuminates in amber, and the LED on the subwoofer level control turns on. High-pass signals go to the front and rear, and low-pass signals go to the subwoofer output.

With the subwoofer function on, select 80-, 120-, or 160-Hz crossover frequency with the crossover switch. Then adjust the output levels (front and rear) and select subwoofer mono or stereo, as well as phase (180° or 0°), using the following guidelines.

Phase. The acoustical characteristics of an automobile can vary from car to car. With certain installations, the characteristics may cancel out some frequencies. In such cases, inverting the subwoofer phase may restore proper response. Likewise, the type of speaker can throw the sound out of phase, making otherwise good sound appear distorted. Without going into any elaborate measurements, simply listen to the sound while moving the phase switch between 0° and 180°. Leave the phase switch in the position where there is *more bass* from the subwoofer speaker.

Subwoofer. If there is one subwoofer in the system, set the (rear) subwoofer switch to mono. If there are two or more subwoofers, use either the mono or stereo positions. You *gain sound separation* in the stereo position and *gain volume* (instead of separation) in the mono position.

Output Level. The output level for both the front and rear are factory-set for maximum, and no further adjustment is usually required. However, if there are differences in output levels between amplifiers used with the EQ or if there is a difference in speaker efficiency, the fader control may not be able to make up for the unbalanced level between front and rear speakers. In such a case, adjust the output level by setting the fader to the center position, and determine which output level (front or rear) is greater. Then use the output-level controls to balance the sound volume between front and rear speakers. (Since both controls are probably at maximum, always reduce the volume that is greater.)

9.1.5 Defeat EQ Circuits

During the defeat on mode, the left- and right-channel audio signals appearing on the input buffer amplifier IC_{201} are amplified and coupled to switch IC_{202}. Since the defeat EQ switch $S_{201\text{-}1}$ is set to on, the signals are switched directly to the output of IC_{202}. The equalized signals from IC_{203} are also available but are not switched to the output of IC_{202} because the corresponding (equalized) contacts of IC_{202} are open.

The switches in IC_{202} are controlled by S_{201}, IC_{0001}, Q_{0001}, and Q_{0002}. When $S_{201\text{-}1}$ is set to on, B+ is applied to Q_{0002}, turning Q_{0002} on. This applies a negative to IC_{0001}. The negative voltage is inverted and amplified twice before application to IC_{202}. At the same time, a positive (from Q_{0001} and IC_{0001}) is applied to the opposite input on IC_{202}. With this combination of voltages, IC_{202} switches the unequalized input signals from IC_{201} through to IC_{101}. A portion of the L and R input signals at IC_{101} are applied to the input-level LEDs through IC_{206} and IC_{205}.

During the defeat off mode, the equalized audio signals from IC_{203} are applied to IC_{202} and then to IC_{205} and IC_{101}. The unequalized signals from IC_{201} are also available but are not switched to the output of IC_{202} because the corresponding (unequalized) contacts of IC_{202} are open.

9.1.6 Subwoofer, Filter, and Crossover Circuits

The equalized or unequalized signals appear at the input of IC_{101}. The output from IC_{101} is determined by the combined action of subwoofer switch $S_{101\text{-}1}$, the portion of IC_{0001} that controls subwoofer switching, and Q_{0001}.

With $S_{101\text{-}1}$ set to on, B+ is removed from the base of Q_{0001}, shutting Q_{0001} off. The collector of Q_{0001} then drives an input of IC_{0001} positive. This closes IC_{101} switch sections, permitting signals to pass.

The left-channel signal is coupled to IC_{103}, IC_{104}, IC_{105}, and IC_{109}, while the right-channel signal is coupled to IC_{106}, IC_{107}, IC_{108}, and IC_{109}. The signal outputs of the high-pass and low-pass circuits are selected by a function of the crossover circuits. For example, if crossover switch S_{102} is set to 160 Hz, only IC_{103} and IC_{106} pass the IC_{101} signals. (Crossover frequencies are determined by *RC* networks connected at the filter circuits.)

9.1.7 Output Circuits

After the signals pass through IC_{110} through IC_{112}, the signals are applied to the appropriate buffer amplifiers for distribution to the output connectors. The L and

R channels are coupled through fader controls VR_{104} and VR_{109} (right channel), output-level controls VR_{201} and VR_{202}, and buffer amplifiers IC_{207} and IC_{208}.

The subwoofer channel is coupled through phase switch S_{103} and buffer amplifier IC_{209}. If only one amplifier and speaker is used for subwoofer operation, subwoofer switch S_{202} is set to mono. This connects both outputs from IC_{209} to a common line, thus providing a single subwoofer signal. When more than one subwoofer amplifier and speaker are used, subwoofer switch S_{202} is set to stereo. This connects each output from IC_{209} to a separate line (at pins 4 and 5 of the E_3 subwoofer output connector).

9.2 LP TURNTABLES

Figure 9.4 shows the operating controls and connections for a typical turntable. This particular turntable is a linear-tracking device. For those not familiar with turntables, the major difference between linear-tracking and conventional turntables is in the operation of the tonearm. On a linear-tracking turntable, the tonearm is driven by a motor, separate from the platter-drive motor.

9.2.1 Shipping Devices

Many turntables have clamps or screws that hold the platter in place during shipment. Make sure that such devices are removed or loosened before you apply power to the turntable. Most conventional turntables have a clamp or screw that

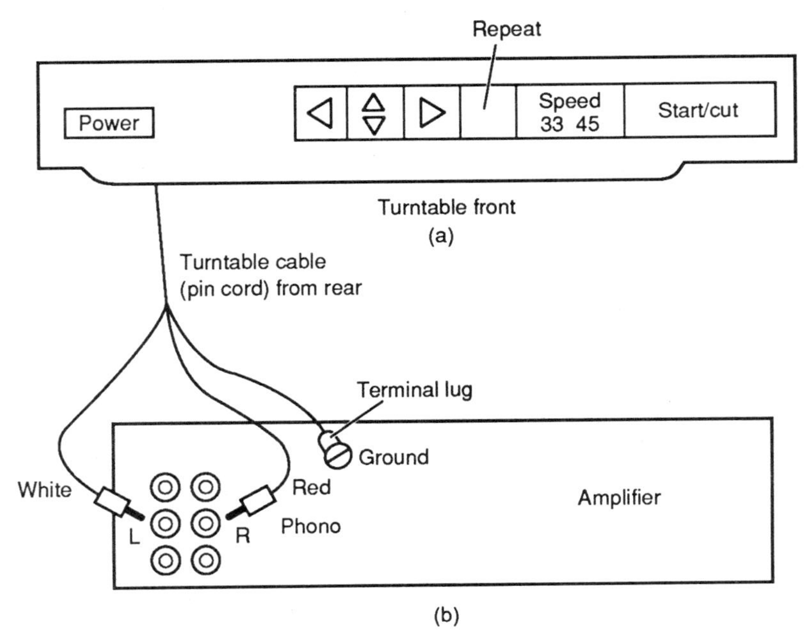

FIGURE 9.4 Operating controls, indicators, and connections for a typical LP turntable.

holds the tonearm in an at-rest position when not in use. Generally, linear-tracking turntables do not have such an arrangement since the tonearm is moved along a travel guide by means of a cord and pulley. In most cases, a linear-tracking tonearm cannot move unless the pulley is driven by the tonearm motor. However, look for any tonearm restraints (which must be removed before use and should be replaced before shipment or when being carried back by the customer).

9.2.2 Controls and Indicators

Figure 9.4*a* shows the operating controls and indicators for our turntable. Compare the following with the controls and indicators of the turntable you are servicing:

Press power to turn the turntable on or off.

Press tonearm in (*left arrow*) to move the tonearm toward the center of the record.

Press tonearm up/down (*up/down arrows*) to move the tonearm up or down.

Press tonearm out (*right arrow*) to move the tonearm away from the center of the record.

Press repeat to repeat the entire record.

Press speed to set the turntable speed at either 33⅓ or 45 rpm.

Press start/cut to either start or stop playing a record.

9.2.3 Connections

Figure 9.4*b* shows typical connections for a turntable. External connections between the turntable and amplifier are made from the back of the turntable. In most cases, a pin cord is supplied with the turntable (or amplifier) for connection between the L and R stereo outputs and the corresponding inputs on the amplifier. Although the connections are very simple, certain precautions must be observed (for all turntables).

Look for any color coding on the pin cord. Typically, red is used for the right channel and white is used for the left channel (but do not count on it). Also, look for any *ground terminals or leads that are part of the pin cord.* Often, the pin cord has a third lead (with a terminal lug) to connect the turntable chassis with the amplifier chassis.

Do not connect the turntable to the CD, aux, or tape inputs of an amplifier. Instead, always use the phono input (or whatever the turntable input is called). Generally, no damage is done if you connect the turntable to the wrong input. However, the turntable output is very low in relation to other audio components (about 5 mV for a turntable, 500 mV for a tape deck, and 2 V for a CD player), so you must have some preamplification for a turntable output signal.

9.2.4 Keyboard and Audio-Mute Circuits

Figure 9.5 shows the keyboard and audio-mute circuits. The front-panel control buttons operate momentary-contact push switches S_1 through S_7.

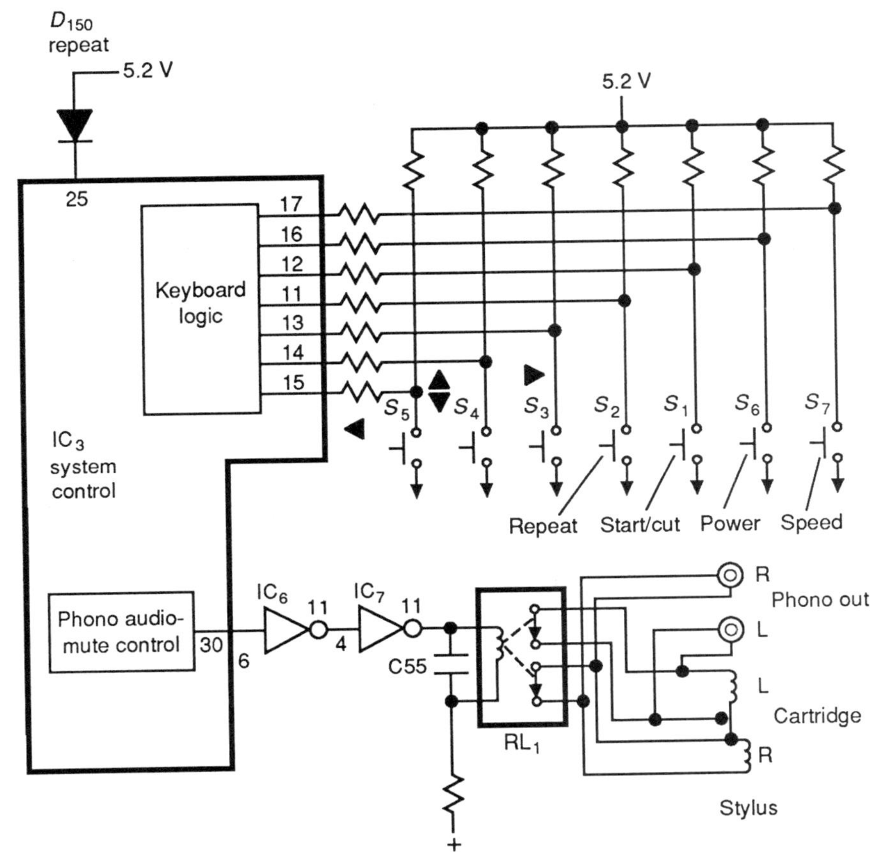

FIGURE 9.5 LP turntable keyboard and audio-mute circuits.

Note that IC_3 accepts commands from the front-panel switches in the form of 0 V, or ground, when the corresponding switch button is pressed and sends high or low command signals to the appropriate circuit.

For example, when repeat switch S_2 is pressed, pin 11 of IC_3 is grounded and IC_3 produces all commands necessary to repeat play of the record (up to 16 times on our turntable). When this occurs, pin 25 of IC_3 goes low, and repeat indicator D_{150} (an LED) turns on.

All of these functions are programmed into IC_3 and cannot be changed, nor can you repair any circuit within IC_3. You must replace IC_3 if *any one* function is absent or abnormal (as is the case with all microprocessors). However, from a practical standpoint, you must understand and check *all results* of a particular command applied to IC_3 before you pull IC_3 from the board (a difficult and tedious job, at best).

For example, if S_2 is operated, pin 25 of IC_3 should change state, and the repeat LED D_{150} should turn on. Simultaneously, the turntable should repeat play. So when any command is applied to IC_3, check that all commands from IC_3 are made properly and that only those commands are made.

The audio-mute circuits shown in Fig. 9.5 eliminate noise caused by the stylus making contact with the record whenever the tonearm is raised or lowered by operation of the cueing solenoid (Sec. 9.2.8), in response to commands from IC_3.

When the tonearm is not down, pin 30 of IC_3 is high (about 5 V). The high is inverted to a low by IC_6 and back to a high by IC_7. This deenergizes RL_1 and shorts the windings of the pickup cartridge as well as the phono output connections. The audio output of the turntable appears as a short to the amplifier input, and the audio is muted.

When the stylus is lowered onto the record by IC_3 and the cueing solenoid, pin 30 of IC_3 is made low. The low is inverted twice by IC_6 and IC_7 and applied to RL_1. This energizes RL_1, removing the short from the cartridge and phono output. Whatever audio is present on the cartridge is applied to the amplifier.

Troubleshooting. If the turntable turns on and the tonearm moves *but there is no audio* (with audio from other components), the *cartridge* is the most likely suspect. You can try replacing the cartridge as the first step, but there are some other simple checks to make if a cartridge is not readily available.

Start by measuring both the left and right phono outputs with a known-good record playing. Typically, the output is in the range of 3 to 5 mV for a dual-magnet cartridge. The actual output should be substantially the same for both channels. If not, this usually indicates a failure of one cartridge coil (unless the record happens to be recorded in that way).

Next, check the audio-mute circuits. RL_1 must be energized for audio to pass from the cartridge to the phono output connectors. RL_1 is energized when pin 30 of IC_3 is made low during normal play. This low is inverted twice by IC_6 and IC_7 and applied to the open end of RL_1.

If pin 30 is not low, suspect IC_3. If pin 30 is low, check for a low from IC_7. If the open end of RL_1 is low but RL_1 is not energized, suspect RL_1 or the associated power circuit.

We do not go into cartridge replacement here (consult the service literature). However, because cartridge and stylus replacement is so simple (usually), many technicians prefer to replace the cartridge first for a "no audio output" symptom, after the muting circuits are checked.

If the repeat function cannot be selected, first check that pin 11 of IC_3 goes low when repeat switch S_2 is pressed. If it does not, suspect S_2 and the associated wiring.

Next check that pin 25 of IC_3 goes low and that repeat LED D_{150} turns on when S_2 is pressed. If pin 25 does not go low, suspect IC_3. If pin 25 goes low but D_{150} does not turn on, suspect D_{150}. Make certain that D_{150} is receiving power (about 5.2 V).

Remember that IC_3 should issue the same commands during repeat as during play (tonearm up/down, platter on, etc.). So do not pull IC_3 to correct a "no repeat" symptom until you are sure that normal play is available.

9.2.5 Automatic Size and Speed Control

Figure 9.6 shows the automatic size and speed control circuits. These circuits detect record size and select the appropriate playing speed (33⅓ or 45 rpm). The circuits are optical (using LEDs and phototransistors) and are in operation only when standard-size records are played.

When the turntable dustcover is open, interlock S_8 is closed, grounding pin 21

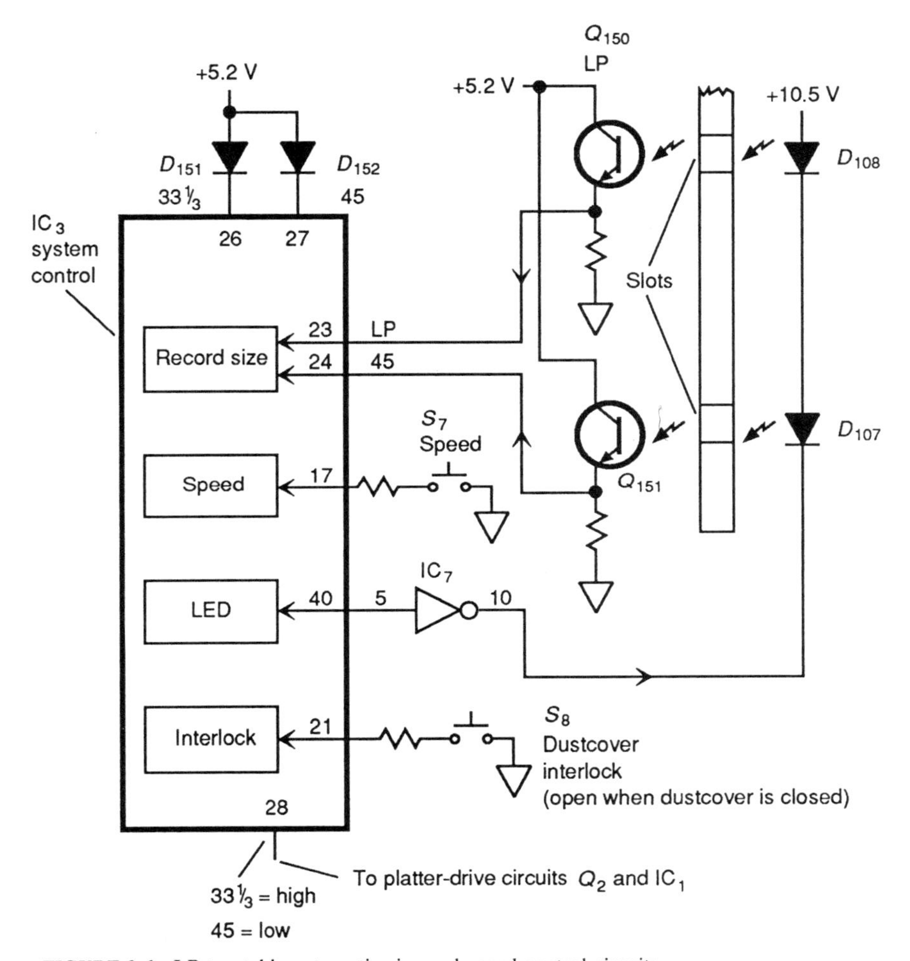

FIGURE 9.6 LP turntable automatic size and speed control circuits.

of IC_3. This places the turntable in the stop mode. The platter stops spinning, and the tonearm moves to the rest position (outside the record circumference).

When the dustcover is closed, S_8 opens, removing the ground from $IC_{3\text{-}21}$ and causing pin 40 of IC_3 to go high. The high is inverted to a low by IC_7 and applied to LEDs D_{107} and D_{108}, causing D_{107} and D_{108} to turn on.

The turntable platter contains slots at about 3 and 5 in from the center of the platter. D_{107} and D_{108} are positioned over the 3- and 5-in slots, respectively. LP-detect Q_{150} is positioned beneath the platter in line with D_{108}; 45-detect Q_{151} is positioned beneath D_{107}.

With no record (or a transparent record) on the platter, both Q_{150} and Q_{151} conduct, and a high is applied to both pins 23 and 24 of IC_3. This is detected by IC_3 as a no-record-present condition. IC_3 turns off power to the platter drive and tonearm circuits. This automatically protects the stylus from possible damage caused by lowering the tonearm to the platter when there is no record present. (With nonstandard or transparent records, the turntable must be operated manually.)

When a standard 7-in 45-rpm record is on the platter, the light from D_{107} to Q_{151} is blocked, and a low is applied to pin 24 of IC_3 (but pin 23 remains high). IC_3 senses this as a 45-rpm record condition and applies appropriate commands to the platter and tonearm circuits.

When a standard 12-in LP record is on the platter, the light from D_{107} to Q_{151} and D_{108} to Q_{150} is blocked. Both pins 23 and 24 of IC_3 go low. IC_3 senses that as a 33⅓-rpm LP record condition and applies appropriate commands to the platter and tonearm circuits.

When nonstandard (not 7- or 12-in, transparent, etc.) records are on the platter, the playing speed must be selected with front-panel speed switch S_7. When S_7 is pressed, pin 17 of IC_3 is grounded, and IC_3 produces all commands necessary to change speeds. When this occurs, pins 26 and 27 change states, as to the 33⅓- and 45-rpm speed-indicator LEDs D_{151} and D_{152}.

Troubleshooting. If speed can be selected manually *but there is no automatic speed selection*, the fault is probably in the record size-detection circuits, Q_{150} and Q_{151} and/or D_{107} and D_{108}, although the problem can be in IC_3.

First, with no record installed, check that both D_{107} and D_{108} are turned on by a high at pin 40 of IC_3. This high is inverted to a low by IC_7. Then check that both pins 23 and 24 are high. If either pin 23 or 24 is low, check the corresponding size-detection circuit (Q_{150} and D_{108} for pin 23, Q_{151} and D_{107} for pin 24).

Next, check that both the 33⅓- and 45-rpm speed indicator LEDs (D_{151} and D_{152}) are on (with both pins 23 and 24 of IC_3 high). If they are not, check for a low at pins 26 and 27. If pins 26 and 27 are low but the corresponding LED is not on, suspect the LED. If pins 26 and 27 are not low, with pins 23 and 24 high, suspect IC_3.

Next, check that pin 28 is high when 33⅓ rpm is selected (pin 23 high, pin 24 low) and goes low when 45 rpm is selected (pin 24 high, pin 23 low). If it does not, suspect IC_3.

If pin 28 of IC_3 goes to the correct state when pins 23 and 24 are high or low but the correct speed is not selected, suspect Q_2 or IC_1. When 33⅓ rpm is selected, pin 28 of IC_3 goes high. This turns Q_2 on and makes pin 7 of IC_1 low.

If speed cannot be selected manually, check to see if speed switch S_7 controls the status at pin 28 of IC_3. First, check that pin 17 of IC_3 goes low when S_7 is pressed and returns to high when S_7 is released. If it does not, suspect S_7. Pin 28 of IC_3 should change to the opposite state each time S_7 is pressed. If it does not, suspect IC_3.

After making any change in the record size and detect circuit, or after replacement of IC_3, always check for correct platter speed and phase adjustment as described in the service literature.

9.2.6 Platter Motor-Control Circuits

Figure 9.7 shows the platter motor-control circuits. These circuits are essentially a PLL, and they use Hall-effect elements (as is typical for most present-day turntables).

The platter motor is controlled by a single PLL IC_1, which receives turntable

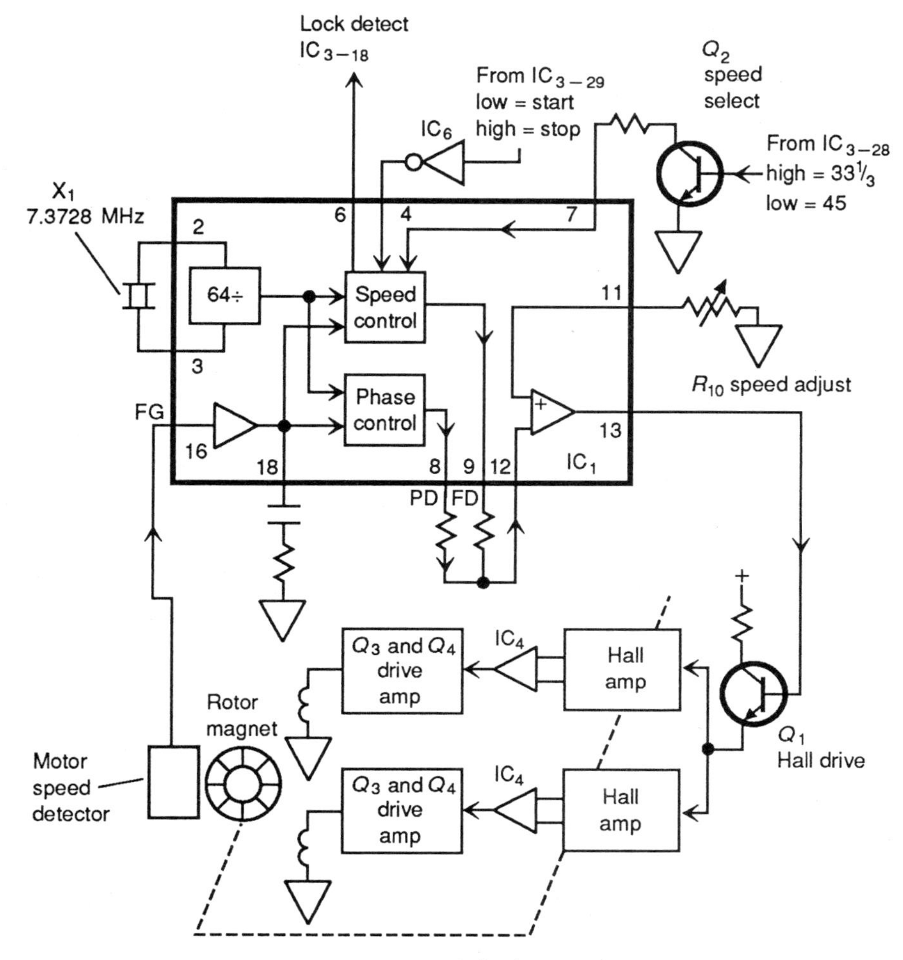

FIGURE 9.7 LP turntable platter motor-control circuits.

start-stop and speed-select information from IC_3. IC_1 also receives platter-speed information from the Hall-effect elements.

Note that IC_1 is controlled by 7.3728-MHz clock crystal X_1. As a troubleshooting hint, if all functions of IC_1 appear to be absent (turntable does not turn, speed cannot be controlled), check for 7.3728-MHz signals at pins 2 and 3 of IC_1. If the signals are absent or abnormal, it is possible that X_1 or the associated crystal circuits are defective.

The signals generated by X_1 are divided by a circuit within IC_1 and serve as a reference for both speed-control and phase-control portions of the PLL. During normal operation, the motor speed detector (an integral part of the platter motor) generates frequency-generator (FG) signals. These FG signals are applied to IC_1 at pin 16 and inform both the speed-control and phase-control circuits within IC_1 of the actual motor speed.

The speed-control circuits of IC_1 compare the reference signals (from X_1) and the FG signals to produce a frequency-difference (FD) signal at pin 9 of IC_1. The phase-control circuits of IC_1 compare the reference signals and the FG signals to produce a phase-difference (PD) signal at pin 8 of IC_1.

The FD and PD signals are added together (in a network outside IC_1) and returned to an op-amp within IC_1 (through pin 12) at the inverting input. The noninverting input is connected to R_{10}. The output of the op-amp is applied to Q_1 through pin 13 of IC_1. If the motor speeds up, the drive signal to Q_1 is altered to slow the motor down, and vice versa. The motor can be locked to the desired speed by adjustment of R_{10}.

Transistor Q_1 drives two switching-type Hall amplifiers. Note that the Hall amplifiers are part of the platter motor (as are the motor speed-detection elements and the rotor magnet) and are not serviceable. Each Hall amplifier supplies inputs to differential amplifiers within IC_4 which, in turn, drive corresponding amplifiers Q_3 and Q_4 and Q_5 and Q_6 (Sec. 9.2.7).

When the turntable is in the stop mode, pressing either the start/cut or tonearm-in buttons causes IC_3 to produce a low, or start, signal at pin 29 of IC_3. This low is inverted to a high by IC_6 and applied through pin 4 of IC_1 to the speed-control circuits within IC_1. The speed-control circuits recognize this as a start command and cause pins 8 and 9 of IC_1 to go low for an initial start period. During this period, pin 13 of IC_1 goes high, turning Q_1 and the Hall amplifiers on to start the platter motor. Once the platter motor approaches the desired operating speed, the platter motor drive circuits take over to control motor speed, as described in Sec. 9.2.7.

Troubleshooting. If the *platter motor does not rotate*, apply power, close the dustcover, press start/cut, and check for a high at pin 4 of IC_1. If it is absent, check for a low at pin 29 of IC_3. If it is absent, suspect IC_3 or start/cut switch S_1 (Fig. 9.5).

Try pressing the tonearm-in switch S_5. If this starts the turntable but pressing S_1 does not, IC_3 is most likely at fault (although it could be the S_1 circuit).

If the platter does not rotate with either S_1 or S_5 pressed, check that pin 21 of IC_3 is low (Fig. 9.6). If it is not and the dustcover is definitely closed, suspect S_8.

If pin 4 of IC_3 is high and X_1 is operating normally, check that pins 8 and 9 of IC_1 are near 0 V initially and rise to about 2.4 V when S_1 or S_5 are actuated. If not, suspect IC_1.

Check that pin 13 of IC_1 is about 6 V. If it is not, suspect the *RC* network at pins 8, 9, and 12 of IC_1 or possibly IC_1 itself.

Check that the base of Q_1 is about 3.5 V. If it is not, suspect Q_1 or the associated circuits.

Finally, check for motor-drive signals from Q_1 through the platter circuits as described in Sec. 9.2.7.

If the platter rotates at a very high or very low speed, no matter what speed is selected, disconnect the power cord, short the base of Q_1 to ground, and then reconnect the power cord. Turn the platter by hand and check for an ac output (FG signals) at pin 18 of IC_1.

If there is no output at pin 18 of IC_1 with the platter being rotated, suspect the speed detector in the platter motor or the summing amplifier in IC_1. You may be able to pin down which of the two components is at fault if you monitor the signal at pin 16 of IC_1.

If there is an ac signal from the speed detector to pin 16 of IC_1 but not at pin 18 of IC_1, it is reasonable to assume that IC_1 is at fault. However, the output from the speed detector is very low (even when the motor is being driven normally) and can be difficult to monitor.

If you get an output at pin 18 of IC_1, check that the output at pins 8 and 9 of IC_1 start at 0 V when the platter is rotated slowly and then increase to about 2.4 V when the platter is rotated quickly (by hand). If not, suspect IC_1.

If you get correct outputs at pins 8 and 9 of IC_1, check that the output at pin 13 of IC_1 is about 6 V when the platter is rotated slowly and then drops to about 0 V when the platter is rotated quickly. If not, suspect IC_1 (or possibly the *RC* networks at pins 8, 9, and 12 of IC_1).

9.2.7 Platter Motor-Drive Circuits

Figure 9.8 shows the platter motor-drive circuits. As discussed in Sec. 9.2.6, Q_1 receives input drive from pin 13 of PLL IC_1. In turn Q_1 supplies drive bias to both Hall amplifiers within the platter motor.

The Hall amplifiers are magnetic-sensitive devices coupled to the platter-motor rotor (an 8-pole magnet). Each Hall amplifier has two outputs of opposite polarity. As the magnetic fields of the rotor pass over the Hall amplifiers, the

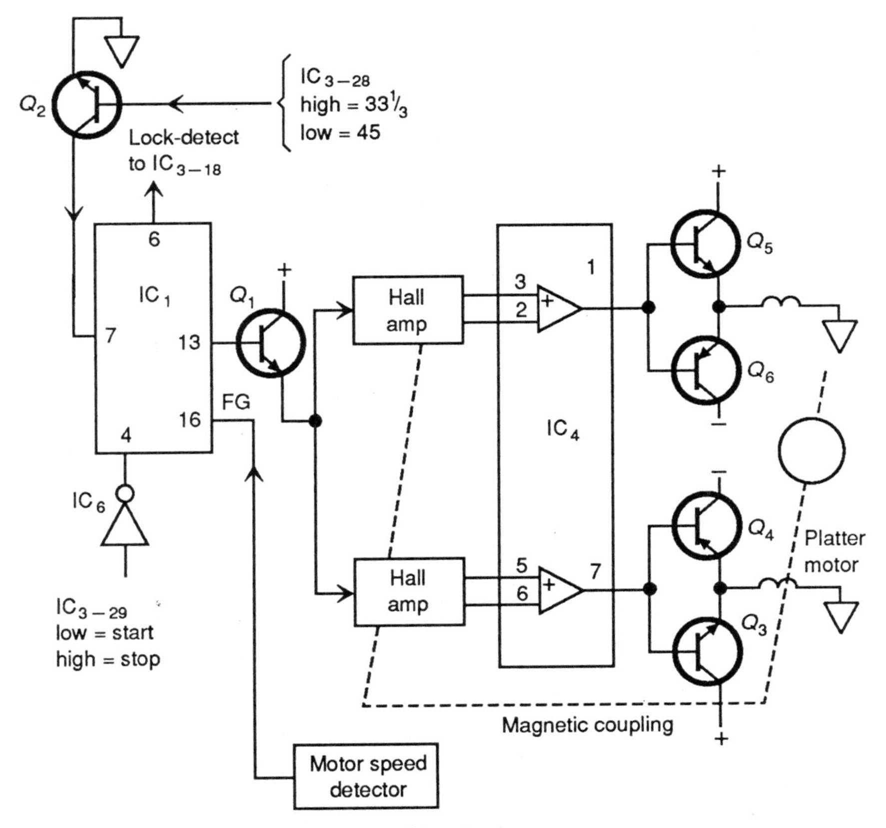

FIGURE 9.8 LP turntable platter motor-drive circuits.

outputs change polarity. The outputs of the Hall amplifiers are applied to the inputs of two differential amplifiers within IC_4.

A positive-going signal is applied at pin 3 of IC_4. At the same time, a negative-going signal is applied to pin 2 of IC_4. The resulting output at pin 1 of IC_4 is positive-going and is applied to the bases of Q_5 and Q_6. A positive signal on the base of Q_6 (PNP) turns Q_6 off. A positive signal on the base of Q_5 (NPN) turns Q_5 on to conduct through the motor winding.

The current through Q_5 and the motor winding causes the motor to turn. As the motor rotates, the output polarities of the Hall amplifiers reverse and are applied to pins 2 and 3 of IC_4. The result is a negative output signal at pin 1 of IC_4. This negative signal turns Q_5 off and Q_6 on, reversing the current flow through the motor winding. In effect, this action produces an ac motor drive signal. The ac frequency is 2.2 Hz at 33⅓ and 3 Hz at 45 rpm.

Note that the two windings of the platter motor are positioned such that the respective drive signals are about 90° out of phase. This sustains continuous rotation of the platter motor (in one direction) in the conventional manner.

Troubleshooting. If the *platter does not rotate* but there is a drive signal at the base of Q_1, trace the drive signals from Q_1 through to the platter motor. Keep the following points in mind.

The outputs of the two switching amplifiers in the platter motor should have substantially the same output. If they do not, suspect the platter motor. Note that on most turntables, the entire platter motor assembly must be replaced as a package.

The outputs of IC_4 should also be the same. If they are not, suspect IC_4. Similarly, the outputs from Q_3 and Q_4 and Q_5 and Q_6 should be the same in amplitude (although shifted in phase by 90°). So if there are outputs to both motor windings and the outputs are substantially equal in amplitude but the motor does not operate, suspect the platter motor.

Note that if the platter is locked mechanically, there are no FG pulses to pin 16 of IC_1. This causes pin 6 of IC_1 to go low, applying a lock-detect low to pin 18 of IC_3. The drive at pin 13 of IC_1 is removed and the platter motor is turned off. So always look for anything that might jam the platter or prevent the motor from turning.

Check the service literature for any lubrication instructions. In some cases, the platter motor requires lubrication (such as a drop or two of silicone oil on the drive shaft). *Do not lubricate* unless the literature so recommends.

Note that the status at pin 7 of IC_1 determines the speed of the platter motor. When pin 28 of IC_3 goes low, Q_2 is cut off, causing pin 7 of IC_1 to go high. This selects the 45-rpm speed. However, the platter motor should run no matter what the status at pin 7 of IC_1.

9.2.8 Tonearm Control

Figure 9.9 shows the control circuits for the tonearm of a linear-tracking turntable. In a nonlinear turntable, the tonearm motor, tracking circuit, and position-detect circuits (shown in blocks) are not used.

In a linear-tracking turntable, IC_3 receives tonearm-position information from the position-detect circuits and the tonearm-rest switch S_9. IC_3 evaluates this information to keep track of exactly where the tonearm is in relation to the start and end of a record. IC_3 controls movement of the tonearm with the tracking circuits that drive the tonearm motor.

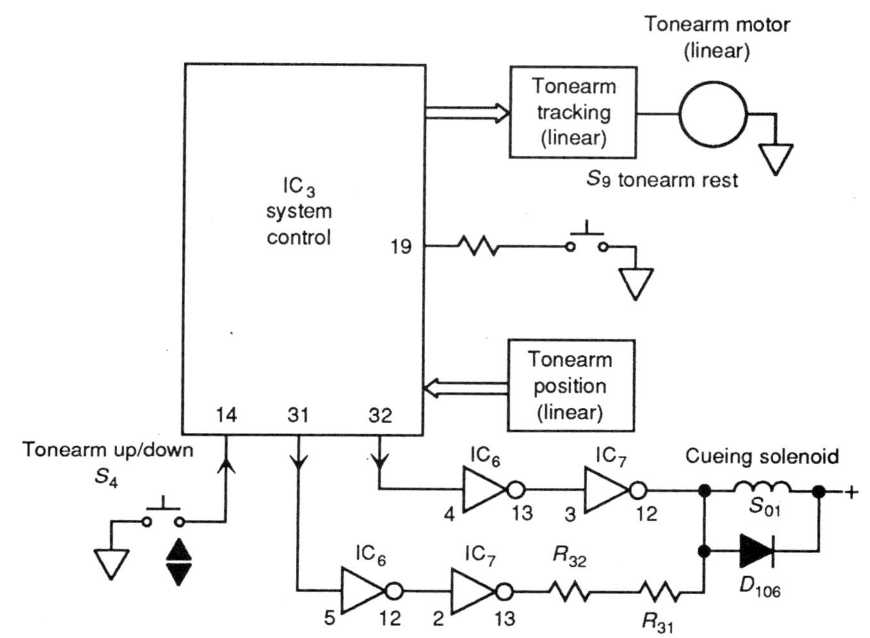

FIGURE 9.9 LP turntable tonearm-control circuits.

In our turntable, the tonearm is raised and lowered by cueing solenoid S_{01}. When the tonearm up/down button is pressed, S_4 is closed and pin 14 of IC_3 is grounded. This causes pins 31 and 32 of IC_3 to change state. In turn, this causes cueing solenoid S_{01} to be energized (or deenergized) to move the tonearm up or down. Note that IC_6 and IC_7 invert the outputs from pins 31 and 32 twice, thus restoring the original state.

When S_4 is first pressed to lower the tonearm, pins 31 and 32 both go low, causing the junction of R_{31} and S_{01} to be grounded. This lowers the tonearm.

When S_4 is released, pin 32 returns to high, but pin 31 remains low, grounding the junction of R_{32} and pin 13 of IC_7. This allows enough current through S_{01} to hold the solenoid and tonearm down.

When S_4 is pressed again to raise the tonearm, pin 31 returns to high, removing current through S_{01} and allowing the tonearm to be raised (by the tonearm mechanism).

Note that the tonearm is also moved up and down automatically by IC_3 during normal play, repeat play, and so on. This is done by changing the states at pins 31 and 32 of IC_3.

Troubleshooting. If the *tonearm does not move up or down*, check that pin 14 of IC_3 goes low when tonearm up/down switch S_4 is pressed. If it does not, suspect S_4.

Next check that pins 31 and 32 of IC_3 are about 3.5 V (initially, with the tonearm up). If they are not, suspect IC_3.

Check that pins 12 and 13 of IC_6 are inverted from pins 4 and 5. If they are not, suspect IC_6. Also check that pins 12 and 13 of IC_7 are inverted from pins 2 and 3. If they are not, suspect IC_7.

Check that pins 12 and 13 of IC_7 both go low when S_4 is pressed (pin 14 of IC_3 low) and that pin 12 of IC_7 goes high (but pin 13 of IC_7 remains low) when S_4 is released. Then check that both pins 12 and 13 of IC_7 return to high when S_4 is pressed again. If it does not, suspect IC_3 or IC_6 and IC_7.

If pin 12 of IC_7 is low but the tonearm does not lower onto the record, suspect S_{01} or D_{106}. If pin 13 of IC_7 is low but the tonearm does not stay lowered, suspect R_{31} and R_{32} or possibly S_{01} and/or D_{106}.

Generally, solenoid S_{01} cannot be replaced as an individual component. The entire tonearm assembly must be replaced as a package. However, always check the parts list in the service literature for the turntable you are servicing.

9.2.9 Preliminary Troubleshooting

Here are the checks to be made before going into the turntable circuits:

If the turntable operates manually but not by remote, check the remote-unit batteries and the control cable. Then try resetting the power circuits by pressing the front-panel power button on and off.

If the left or right speaker is dead when the turntable is used in a system, try playing another component (tuner, cassette deck, CD player). If operation is normal with the other component, suspect the turntable. Also check for a loose cable between the turntable and amplifier. Temporarily reverse the left and right cable connectors at the amplifier input from the turntable. If the same speaker remains dead, the amplifier (or speaker) is probably at fault. If the other speaker goes dead, suspect the turntable (probably the cartridge) or the turntable output cable. Remember that the turntable is generally played through a separate *phono preamp* in the amplifier.

If there is hum or noise (only when the turntable is used), check for the following. The shield of the audio cable from the turntable may be broken, or the connector may not be firmly seated in the jacks. The *turntable ground wire* (Fig. 9.4*b*) may be disconnected from the amplifier.

CHAPTER 10
STEREO TV AND SURROUND-SOUND AUDIO

This chapter is devoted to stereo TV and surround-sound audio circuits. Although there are many so-called stereo VCRs and TV sets, not all can decode and play stereo TV broadcasts with *multichannel television sound* (MTS or MCS, whichever term you prefer). Likewise, there are many "surround sound" or "sound enhancers" but not all are capable of decoding and reproducing video tape (or TV broadcasts) recorded in true *Dolby Surround*. We discuss both systems here.

10.1 STEREO TV BASICS

Before we get into the circuits, let us review the basic stereo TV system.

10.1.1 Basic Signal Flow in Stereo TV

Figure 10.1 shows the TV transmitter and receiver for the MTS/MCS system which delivers audio for a stereophonic TV program and separate audio program (SAP). The MTS/MCS stereo TV broadcasts are made in accordance with the *dbx Noise Reduction (NR) System* and the *Zenith Transmission System*. Figure 10.2 shows the multichannel signal spectrum used in the MTS/MCS system.

The main-channel signal is composed of the sum (L + R) signal of L (left) and R (right) signals and is the same as in the conventional TV sound specification. As a result, the same TV sound as in conventional broadcasting can be received by an ordinary TV set (or VCR) when stereo TV broadcasting is in effect.

When the difference between the L and R signals (L − R) is transmitted, the L and R signals are reproduced from the main-channel signal (L + R) and the stereo signal (L − R) by the TV set. The same signals can be recorded and played back by a VCR equipped for stereo TV.

The L − R signal is produced by amplitude modulating a subcarrier with a frequency double the horizontal scanning frequency (fH) using the double-sideband suppressed carrier (DSB-SC) system. As a result, the frequency division of the main-carrier signal is double (50 kHz) that of the main-channel signal (L + R).

The carrier of the L − R signal is suppressed, so it is necessary to transmit a

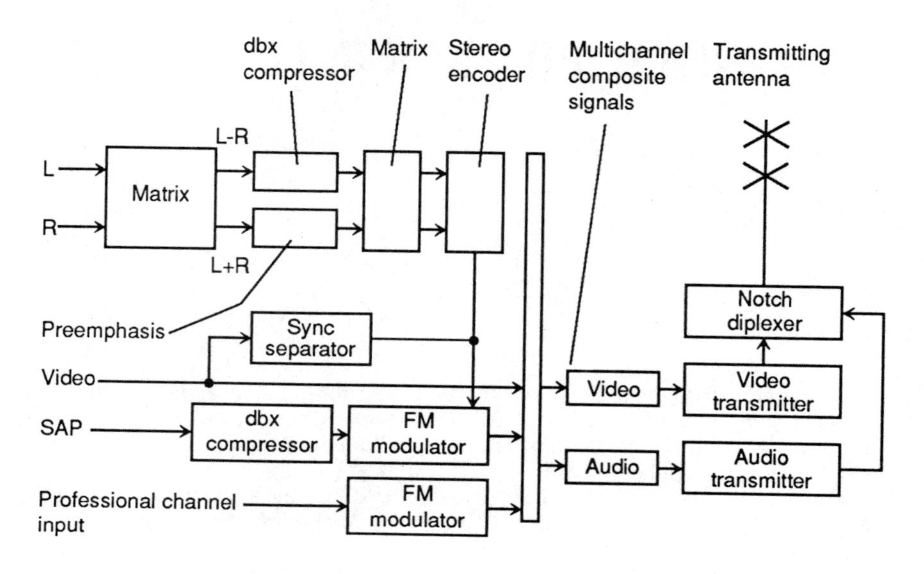

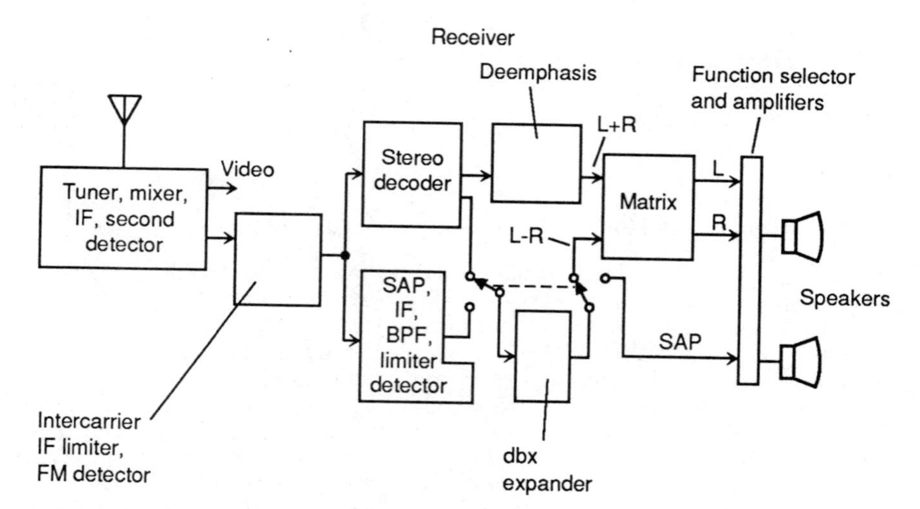

FIGURE 10.1 TV transmitter and receiver for the MTS/MCS system.

reference signal to demodulate the L − R signal correctly in the TV set or VCR. To do this, a signal with a frequency equivalent to the horizontal scanning frequency (fH), called the *pilot signal*, is inserted between the main-channel signal (L + R) and the stereo signal (L − R). The pilot signal is also used to indicate the presence or absence of the stereo signal (no pilot, no stereo).

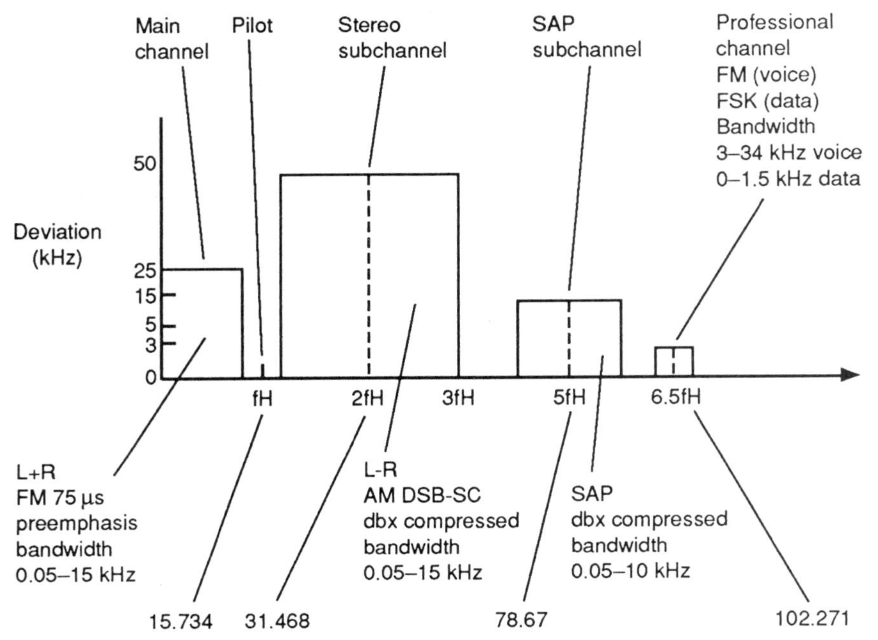

FIGURE 10.2 MTS/MCS multichannel signal spectrum.

The SAP-channel signal frequency modulates a subcarrier with a frequency of 5 fH and is locked to 5 fH during nonmodulation.

The *nonpublic channel* (also called the *professional channel*) signal frequency modulates the subcarrier with a frequency of 6.5 fH. This channel is scheduled to be used for business, not for TV broadcasting.

The sum (L + R), difference (L − R), pilot, SAP-channel, and the nonpublic-channel signals are added to produce the *multichannel composite signal* which, in turn, frequency modulates the transmitted TV broadcast signal. A *noise-reduction system* (NR system), which specifically encodes and transmits the L − R and SAP signals, is used to reduce noise in the TV set and VCR.

10.1.2 dbx NR System

The dbx noise reduction is an essential part of the stereo TV system. In the simplest of terms, portions of the transmitted signal are compressed (or encoded) at the TV transmitting station. These same signals are expanded (or decoded) at the TV set or VCR. The process is known as *companding* (or compandoring in some literature).

The following is a brief discussion of the companding process, covering both the why and how. Figure 10.3 shows simplified block diagrams of both compressor and expander circuits.

Stereo TV Transmission System. As discussed, the stereo TV system transmits the sum of the left and right stereo audio signals (L + R) in the spectrum space

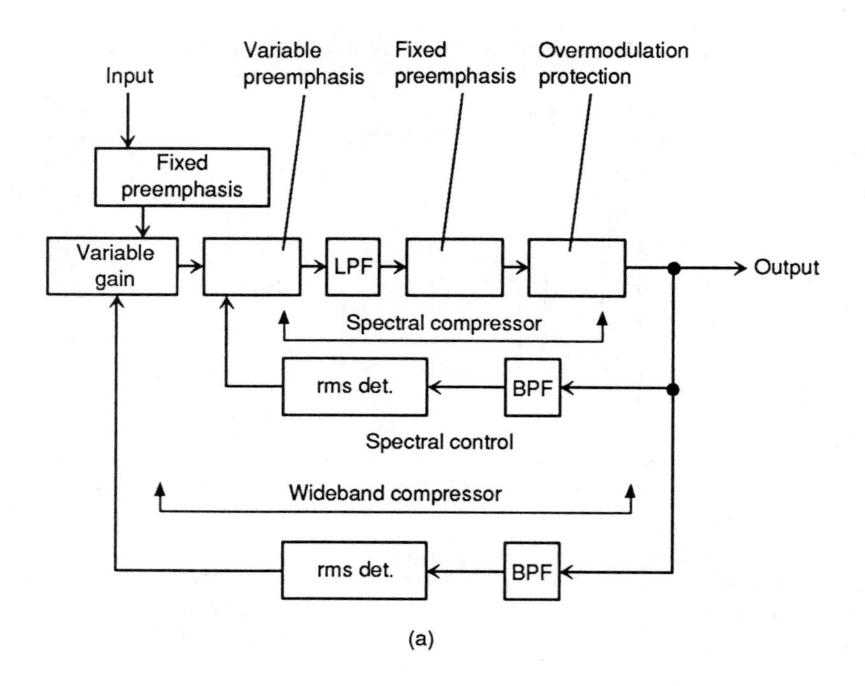

(a)

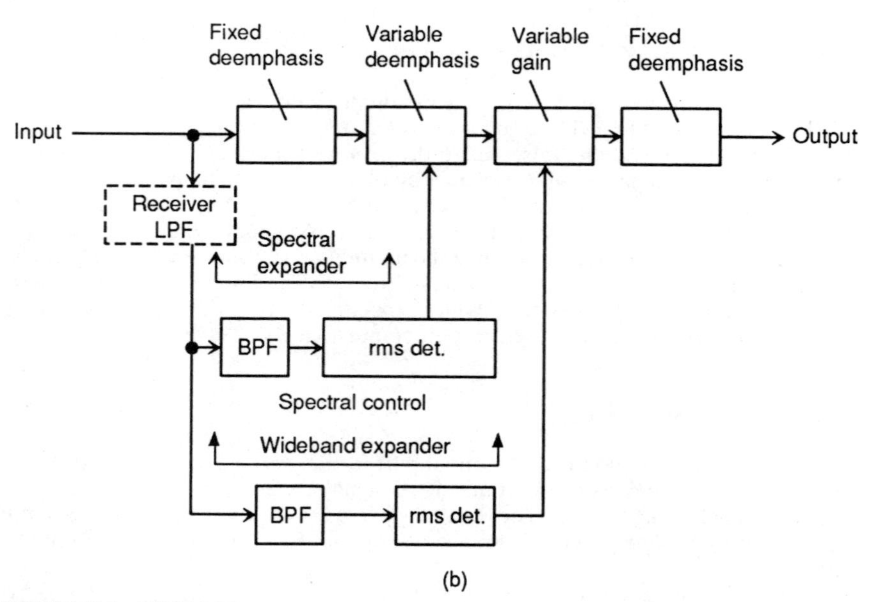

(b)

FIGURE 10.3 MTS/MCS compressor and expander circuit.

usually occupied by the conventional or mono TV audio signal (Fig. 10.2). The stereo information is encoded by subtracting the right audio signal from the left (L − R) and transmitting the difference via an AM subcarrier located at 31.468 kHz (twice the video horizontal-scanning frequency of 15.734 kHz), superimposed on the conventional FM audio carrier.

Noise Reduction. Because the L − R or stereo subcarrier is at a higher frequency than the L + R signal, stereo reception is about 15-dB noisier than mono reception, even under ideal conditions. Typically, the stereo S/N ratio is about 50 dB. The SAP subcarrier (at an even higher frequency of 78.67 kHz) has a typical S/N ratio of 33 dB. The practical effect of this noise is to reduce the coverage area for both stereo and SAP as compared to the mono coverage area (under identical conditions).

The dbx noise-reduction system is designed to improve the S/N ratios. (Theoretically, the goal is to eliminate any noise increase when going from mono to stereo and to make SAP listenable.)

Note that noise reduction is added only to the L − R and SAP channels, leaving the mono L + R signal unchanged. This ensures compatibility with conventional TV broadcasts. (If you have a conventional TV set or VCR, you will not know that stereo TV is being broadcast.)

Identical companding is used in both the L − R and SAP channels. This allows a single noise-reduction circuit to be switched between the L − R and SAP channels in the TV set or VCR. Operation of the noise-reduction system involves preemphasis, deemphasis, spectral companding, and wideband-amplitude companding.

Preemphasis and Deemphasis. The dbx noise-reduction system uses preemphasis (at the TV broadcast transmitter) and deemphasis (at the TV set or VCR). The preemphasis circuits alter the frequency-response curve of the transmitted audio to overcome noise. The deemphasis circuits restore tonal balance to the program material and reduce hiss picked up in transmission.

Spectral Companding. The dbx noise-reduction system includes a stage where preemphasis is varied to suit the signal. The function is called *spectral companding*.

When very little high-frequency information is present in the audio, the spectral compressor (Fig. 10.3*a*) provides large high-frequency preemphasis. When strong high frequencies are present, the spectral compressor provides deemphasis, thus reducing the potential for high-frequency overload. The transmitted signal is dynamically adjusted to have high-frequency content to provide good masking.

In the TV set or VCR (Fig. 10.3*b*), the *spectral compander* restores high-frequency signals to the proper amplitude. The expander also attenuates high-frequency noise when little high-frequency is present. When strong high-frequency signals are present, the signal itself masks the noise.

Wideband-Amplitude Companding. The third stage or function of the dbx noise-reduction system is the wideband-amplitude companding, which keeps the signal level in the transmission channel high at all times.

The compressor (Fig. 10.3*a*) reduces the dynamic range of input signals by a factor of 2:1 (in dB). The transmitted signal level tends toward about 14 percent modulation, which allows transient peaks to overshoot without causing overmodulation.

During reception (Fig. 10.3*b*), the wideband-amplitude expander restores signal levels to the proper amplitude. When the signal is low in amplitude, the decoder attenuates channel noise toward the point of inaudibility. During high-amplitude passages, the signal masks the noise.

10.1.3 Stereo TV Generators

A special-purpose stereo TV generator is required to test and adjust the stereo circuits of a TV set or VCR. There are inexpensive stereo TV generators (such as the B&K Model 2009) that provide all of the signals necessary for service and troubleshooting. Such generators use *spot modulation* to simulate dbx encoding *at specific frequencies* (rather than providing encoding across the entire TV frequency spectrum, as do expensive broadcast studio instruments).

The 2009 generates an FM-modulated RF carrier on TV channel 3 or 4 and at the standard TV IF carrier, as well as on a standard 4.5-MHz audio carrier. A mode selector permits four combinations of internal modulation: left-channel only (L), right-channel only (R), baseband (L + R), and subband (L − R). A 15.734-kHz pilot signal is generated and combined with the composite audio. The pilot may be switched on and off when desired, making it possible to test the pilot detector circuits in the decoder. A SAP mode is also selectable, providing a subband signal centered at 78.67 kHz, to test SAP operation.

Stereo TV Generator Operating Notes. The following notes supplement the specific test and adjustment procedures described in Sec. 10.3.

The L, R, L + R, and L − R modes should produce *equal audio outputs* in the TV or VCR. That is, the audio levels in the left and right channels should be equal for all modulation modes. The L + R or L − R mode should produce equal audio level in each channel. Similarly, the L or R mode should produce the same level in the active channel as in the L + R or L − R modes. This can be used in troubleshooting.

For example, under a given set of conditions (same measurement points, same audio frequencies) the L and R channels should produce the same audio output voltage. If they do not, an unbalance condition is indicated. (This unbalance condition is not necessarily limited to the stereo-decoder circuits but could as likely be in the audio amplifier circuits.)

The SAP function should produce equal SAP audio-output levels at all frequencies. The SAP function is monaural in our generator but has dbx encoding. This is not necessarily the case with a TV station signal, where both stereo and SAP may be broadcast simultaneously (even though you can listen to only one at a time).

10.2 STEREO TV CIRCUITS

Figure 10.4 shows the basic audio circuits for a VCR with stereo TV (and hifi, Chap. 11) capability.

There are two basic audio paths. One supplies signals to the conventional sta-

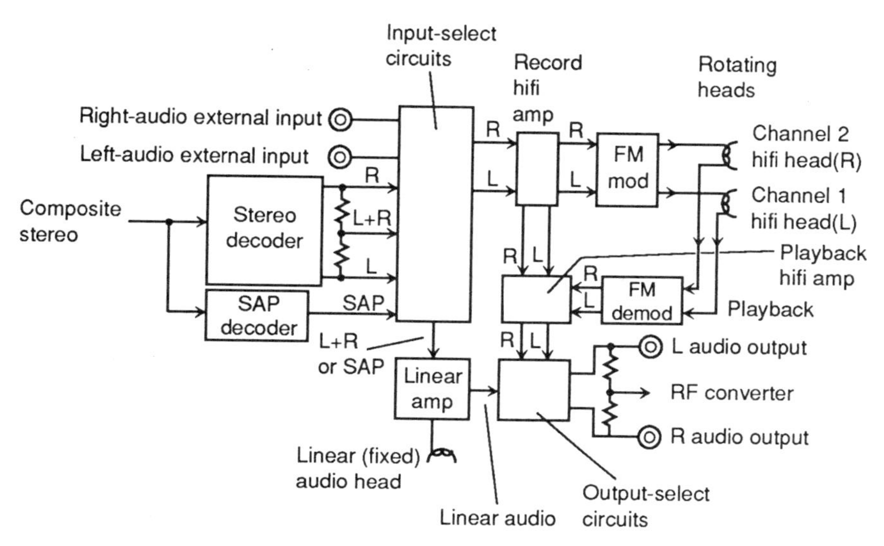

FIGURE 10.4 Audio circuits for hifi VCR with stereo TV capability.

tionary audio head (the linear head); the second supplies signals to the circuits used by the two rotating hifi heads. Signals to be recorded on the rotating hifi heads may be taken from two sources: the external audio inputs or the TV broadcast signal.

The RF and IF circuits between the VCR tuner and the stereo and SAP decoders are essentially the same as for a conventional VCR. The output of the RF and IF circuits is the usual TV signal (composite audio and video), with the audio taken from a 4.5-MHz detector.

The TV signal from the 4.5-MHz detector (shown as composite stereo in Fig. 10.4) may assume one of two forms: normal TV monaural sound or composite stereo sound. The TV signal may or may not contain a SAP signal. The signal from the 4.5-MHz detector is directed to the stereo-decoder circuits. If stereo is being broadcast, the signal is decoded, and the individual right and left audio signals are applied to the input-select circuit. If the TV sound transmission is mono, the stereo decoder produces identical mono signals at both the right- and left-channel inputs to the input-select circuit.

The input-select circuit selects either TV sound or external audio from the external inputs and directs the selected sound to the record hifi amplifier. The sound signal is subjected to both fixed and dynamic preemphasis (to improve the S/N ratio), as well as amplification, and is applied to the FM modulator. A sample of the signal from the record hifi amplifier is also applied to the playback hifi amplifier. The sample signal is then applied to the audio output jacks and the RF converter through the output-select circuit.

The right and left audio signals modulate two individual FM carriers in the FM modulator. The resulting FM signals are combined and applied to the hifi heads (rotating heads) for recording on tape. Since both audio signals used in hifi VCR recording are FM, there is no need for a bias oscillator.

During playback, the hifi heads pick up the previously recorded audio. The two audio signals are combined, amplified, frequency separated, and directed to the FM demodulator circuits. After demodulation, the right and left audio signals are applied to the playback hifi amplifier.

The playback amplifier amplifies both audio signals, as well as performing both fixed and dynamic deemphasis on them. This returns the audio to the original form, with a hifi-quality S/N ratio. The restored audio is directed to the audio output jacks and to the RF converter through the output-select circuit.

There are also two possible signal sources for the linear audio head (stationary or fixed head). Portions of the left and right outputs of the stereo decoder are added together to form a mono L + R signal for the linear head. If SAP is broadcast, the output of the SAP decoder may also be directed to the linear head.

The composite stereo signal applied to the stereo decoder is also applied to the SAP decoder. If a SAP signal is present, the SAP signal is decoded, and SAP audio is applied to the input-select circuit.

Either mono L + R audio or SAP audio, as selected by the input select circuit, is applied to the linear head through the linear amplifier. Note that the linear amplifier is essentially the same as that used in conventional nonstereo TV and VCRs.

During playback, the audio signal is picked up by the linear audio head (either L + R or SAP), processed by the linear amplifier, and directed to the output-select circuit. If desired, the output-select circuit directs audio from the linear head to the audio output jacks and the RF converter.

10.2.1 Overall Stereo TV Decoder

Figure 10.5 shows the combined L + R and L − R signal paths, while Fig. 10.6 shows the SAP signal paths. Before we get into the circuit details, let us consider the overall stereo and SAP functions shown in Fig. 10.7.

Individual components of the composite stereo signal may take any of three different paths. Which path is determined by low-pass and band-pass filters (LPFs and BPFs) in the input circuit.

The L + R audio is selected by an LPF designed to pass only those L + R signals below 15 kHz. The L + R (mono) signal is amplified and applied to the audio matrix.

The second LPF passes all frequencies below 46 kHz, which includes the L + R, L − R sidebands, and the pilot signal. These signals are all applied to the L − R decoder.

The L + R signal (because of the low frequency) is rejected by the L − R decoder. (Only the L − R sidebands and pilot signal are used in the L − R decoder.) The pilot signal is used to synchronize an oscillator within the L − R decoder. The oscillator produces a suppressed 31.468-kHz L − R subcarrier, thus permitting subsequent decoding of the L − R sidebands.

The output of the L − R decoder is applied to the audio matrix through a dbx noise-reduction circuit. The L − R and L + R signals are combined in the matrix to form individual left and right audio channels. The left- and right-channel audio signals are applied to the audio input-select circuits and are then selected for recording on tape by the rotating hifi audio heads.

Samples of each audio channel are added to form the L + R mono audio signal. This signal is recorded by the rotating hifi heads if no stereo is broadcast. The L + R signal is also applied to the stationary or linear head when SAP is not broadcast (or selected).

If SAP is present, the SAP audio is passed through the BPF. This BPF passes

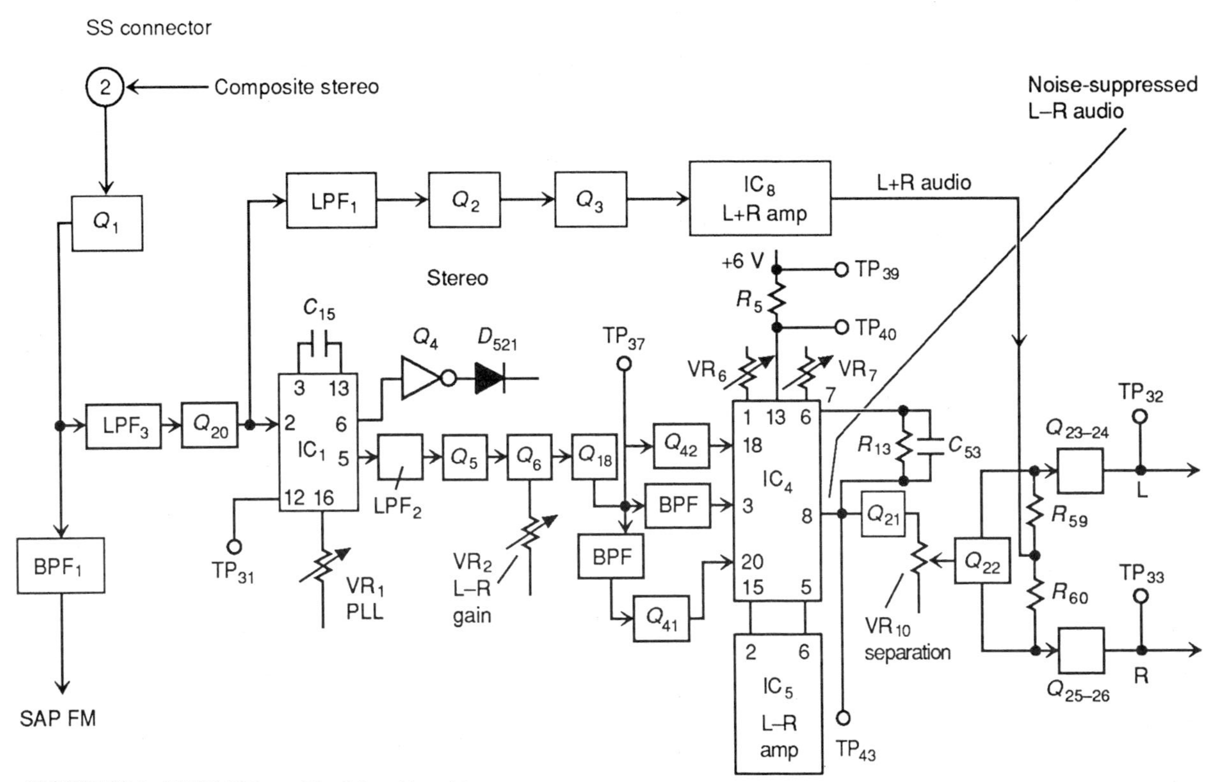

FIGURE 10.5 MTS/MCS combined L + R and L − R signal paths.

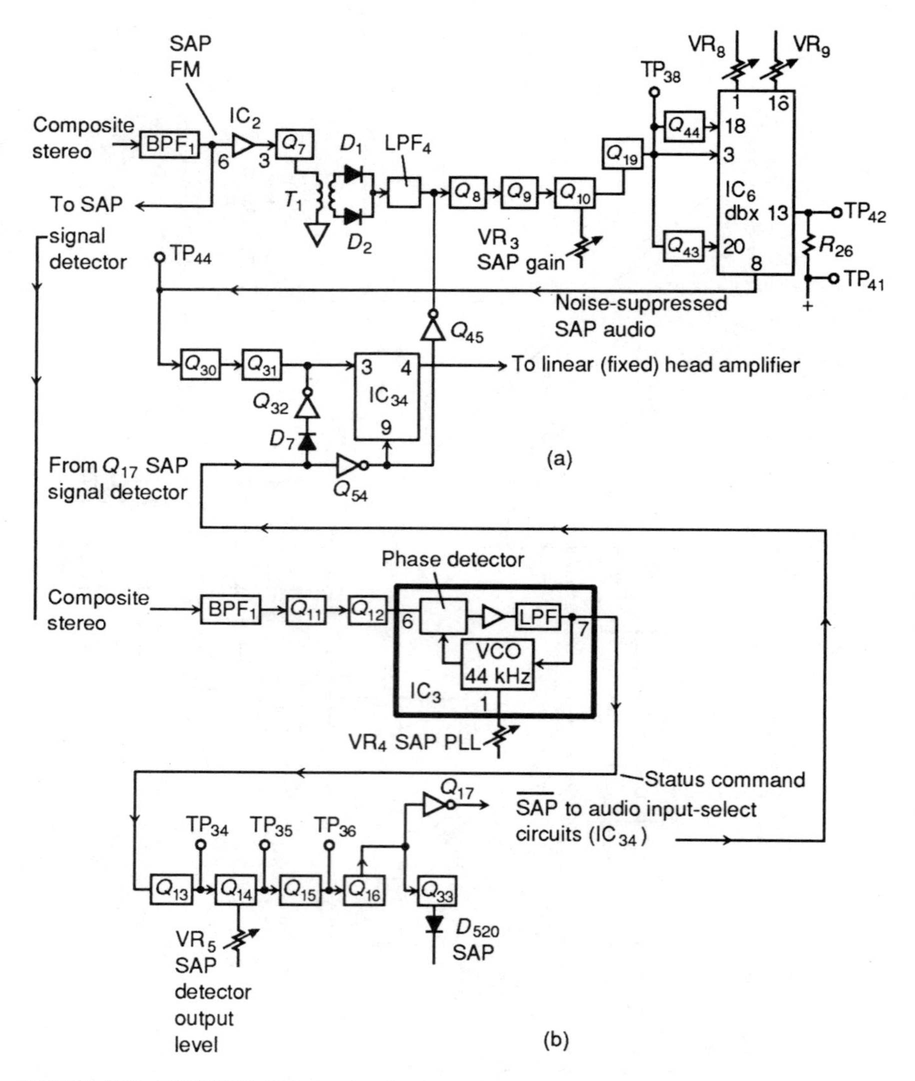

FIGURE 10.6 MTS/MCS SAP signal paths.

signals from 65 to 95 kHz and thus rejects the L + R and L − R signals. The SAP signal is decoded by the SAP decoder and applied to the audio input-select circuit through a separate dbx noise-reduction circuit. The SAP signal can be recorded on the stationary linear head, in place of the L + R, when so selected (on many VCRs).

10.2.2 L + R Circuits

As shown in Fig. 10.5, the composite stereo signal is amplified by Q_1 and applied through filters to the stereo (L − R) and SAP signal paths. The L + R and L − R

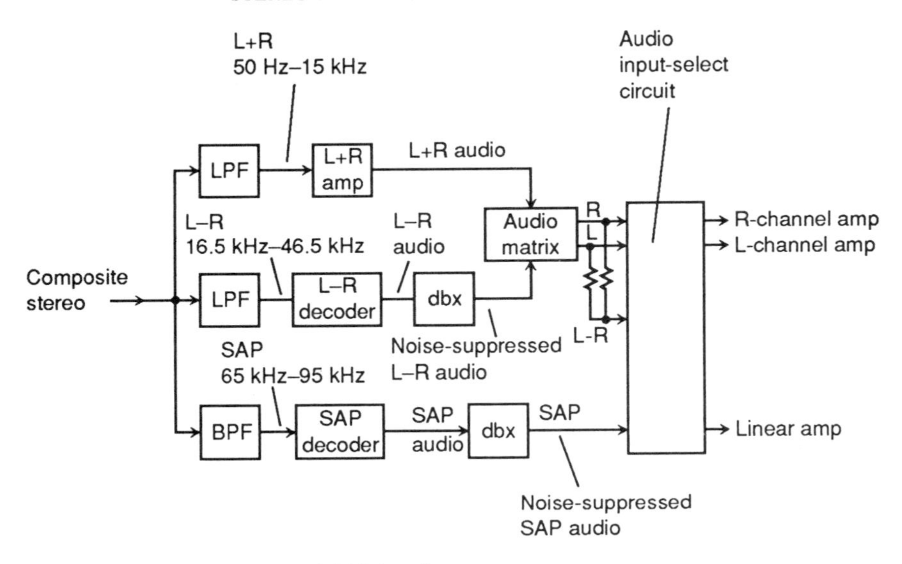

FIGURE 10.7 Overall stereo and SAP functions.

signals are rejected by BPF_1 but are passed to Q_{20}, where both signals are amplified. The L + R signal is applied to amplifier IC_8 through LPF_1, Q_2, and Q_3.

The L + R audio from IC_8 is applied to the junction of resistors R_{59} and R_{60}. Both resistors are part of a matrix circuit used to combine the L + R and L − R signals.

10.2.3 L − R Decoder

Most of the L − R functions are performed within IC_1. These functions include detecting the presence of a stereo signal (a signal with pilot) and decoding the L − R sidebands to produce L − R audio. If the broadcast is in stereo (pilot present), IC_1 produces a command at pin 6 to turn on the stereo indicator D_{521}.

The L − R sidebands and the pilot are applied to pin 2 of IC_1. The L −R sidebands pass to an L − R decoder within in IC_1, while the pilot is amplified by a preamp. The amplified pilot signal is applied to two comparators through C_{15}.

A VCO in IC_1 operates at 4 times the pilot frequency, or 62.936 kHz. The VCO output is divided twice by 2, resulting in a 31.468-kHz output (applied to the L − R decoder as a substitute carrier) and a 15.734-kHz output (applied to both comparators as a second input). When a stereo signal is transmitted, the pilot and the divided-comparison signals from the VCO are compared in both comparators.

The PCL comparator (phase comparator lamp) generates a signal that is filtered by an LPF and amplified by a lamp driver to produce a low at pin 6 of IC_1. If there is no pilot, pin 6 of IC_1 remains high.

A low at pin 6 of IC_1 turns on Q_4. In turn, Q_4 applies 12 V to stereo indicator D_{521}. As a result, D_{521} turns on when a pilot signal is present (indicating a stereo broadcast to the listener or viewer). If pin 6 is high (no pilot, no stereo), Q_4 and D_{521} remain off.

The PCV comparator (phase comparator VCO) develops a correction voltage if a frequency or phase error exists between the pilot signal and the VCO comparison signal. The correction voltage is filtered, amplified, and applied to the VCO to correct any frequency or phase error. VR_1, at pin 16 of IC_1, is used to set the free-running frequency of the VCO to 4 times that of the pilot frequency.

The output of the VCO is applied to a divide-by-2 circuit, reproducing a phase- and frequency-corrected duplicate of the original L − R subcarrier. The resulting 31.468-kHz signal is applied to the L − R decoder, along with the L − R sidebands from pin 2 of IC_1.

The L − R sidebands are decoded, and the resulting L − R audio is available at pin 5 of IC_1. The L − R audio is passed through LPF_2, where all unwanted signals above 15 kHz are removed. The L − R audio is then amplified by Q_5, Q_6, and Q_{18} and applied to the dbx circuit.

10.2.4 dbx Circuits

Although the dbx noise-reduction format used in stereo TV is the most complex part of the system, virtually all of the dbx circuits are contained within IC_4. Figure 10.8 shows these circuits in block form.

A discrete-component BPF passes only those L − R signals within the spectral band (4 to 15 kHz) and applies signals to the spectral rms detector (which detects the signal amplitude). Simultaneously, a second discrete-component BPF passes those L − R signals within the wideband (100 Hz to 4 kHz) and applies the signals within this band to the wideband rms detector (which also detects the signal amplitude).

The complete L − R audio signal is deemphasized by a fixed 73-μs deemphasis network. The deemphasized L − R signal is then applied to the spectral voltage-controlled amplifier (VCA). Note that in some dbx ICs the VCAs are called *current*-controlled amplifiers (CCAs) just to confuse you.

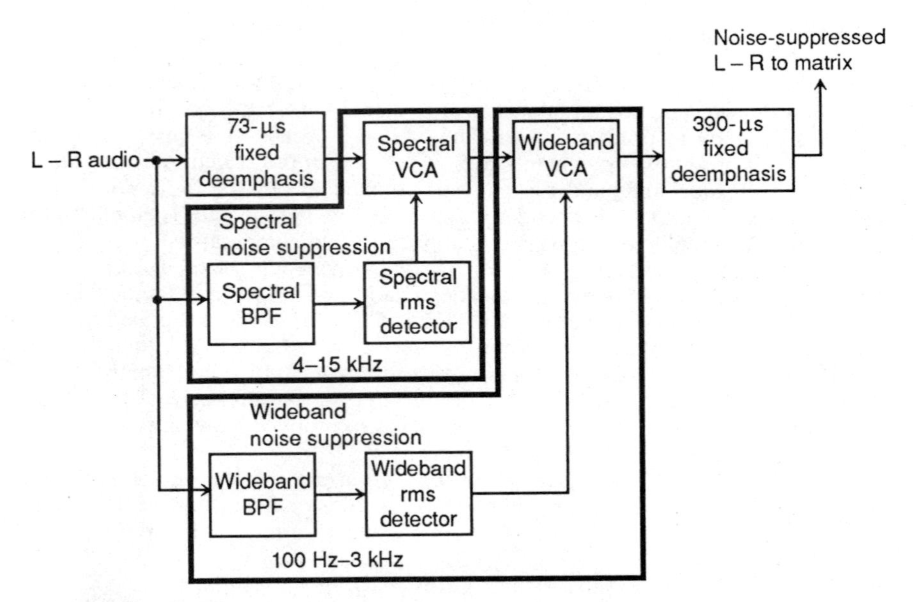

FIGURE 10.8 The dbx noise-reduction circuits.

The gain of the spectral VCA is controlled by the detected level of the spectral rms detector. The signals from 4 to 15 kHz are either compressed or expanded (depending on the respective amplitude) by varying the gain of the spectral VCA. The L − R signal with spectral compression is then directed to the wideband VCA.

The gain of the wideband VCA is controlled by the wideband rms detector. The signals from 100 Hz to 3 kHz are compressed by varying the gain to the wideband VCA. The L − R signal from the wideband VCA is deemphasized by a fixed 390-μs deemphasis network and is applied to the matrix circuit for mixing with the L + R audio signal.

dbx Deemphasis and BPF. As shown in Fig. 10.5, the L − R audio from Q_{18} is applied to the spectral VCA at pin 18 of IC_4. Q_{42} and the associated parts form the fixed 73-μs deemphasis network.

Q_{41} and the associated parts form the spectral BPF. L − R signals in the range from 4 to 15 kHz are applied to the spectral rms detector at pin 20 of IC_4.

A discrete-component BPF forms the wideband BPF. L − R signals in the range from 100 Hz to 3 kHz are applied to the wideband rms detector at pin 3 of IC_4.

dbx Processing. Two rms detectors within IC_4 receive power from a constant-current generator (also in IC_4). The output of the generator is adjustable by VR_6 (the so-called L − R timing control), connected at pin 1 of IC_4. The setting of VR_6 determines the amount of control output from the rms detectors for a given signal amplitude.

Spectral processing is performed by the spectral VCA at pin 18 of IC_4, controlled by the output of the spectral rms detector. The L − R signal is then applied to an L − R amplifier IC_5 and an op-amp within IC_4. VR_7 sets the gain of the op-amp and is sometimes called the variable deemphasis (VD) control.

The amplified L − R signal is then applied to another op-amp within IC_4. This op-amp output at pin 8 of IC_4 is applied to Q_{21} in the matrix circuit. C_{53} and R_{13}, between pins 7 and 8 of IC_4, produce the required 390-μs fixed deemphasis.

10.2.5 L + R and L − R Troubleshooting

Here are some thoughts on troubleshooting the L + R and L − R circuits shown in Fig. 10.5.

The L − R gain is set by adjustment of VR_2 in the emitter of Q_6, ahead of the dbx circuits. This is not to be confused with the separation adjustment VR_{10} after the dbx circuits, although both controls set the L − R signal level.

The dbx functions are set by adjustment of VR_6 and VR_7. L − R timing control VR_6 sets the amount of control output from the rms detectors within IC_4 for a given signal amplitude. Variable deemphasis control VR_7 sets the gain of the spectral op-amp in IC_4.

If there is no stereo TV but mono operation is good, the problem is in the L − R path, with IC_1 and IC_4 being the likely suspects. The problem could also be improper adjustment of VR_1, VR_2, VR_6, VR_7, or VR_{10}. Signal tracing (or signal injection) through the entire L − R path should lead to the problem.

If there is excessive background noise, suspect the dbx noise-reduction circuits. IC_4 is the most likely suspect, although the BPFs at the input to IC_4 are also suspect since the BPFs separate the frequencies to be processed by IC_4.

If there is no sound during a mono broadcast and distorted sound during a stereo broadcast, suspect the L + R audio path. During a stereo broadcast, the L − R signal produces audio. However, if the left- and right-channel audio paths are identical, the L − R signal falls to zero at that instant. As a result, the reproduced sound is erratic and/or distorted since there is no L + R signal at the matrix.

10.2.6 SAP Circuits

Figure 10.6*a* shows the SAP demodulation signal path that decodes the SAP audio. This audio is applied to the audio input-select circuit through the SAP dbx processing. Figure 10.6*b* shows the SAP signal-detect path that turns on the front-panel SAP indicator and produces a SAP-present signal (also applied to the audio input-select circuit).

SAP Demodulation Path. The composite stereo signal is applied to BPF_1, which passes only those signals with frequencies between 65 and 95 kHz. The output of BPF_1 is applied to the SAP signal-detect path and to pin 6 of IC_2, which acts as a limiter. The SAP FM from pin 3 of IC_2 is applied to the SAP demodulator (consisting of T_1, D_1, and D_2) through Q_7.

SAP audio from the SAP demodulator is applied to the SAP dbx processing circuits (at Q_{44}) through LPF_4, Q_8, Q_9, Q_{10}, and Q_{19}. Note that VR_3 in the emitter of Q_{10} sets the level of the SAP audio signal.

SAP Demodulation Troubleshooting. What may appear to be a loss of SAP audio and a defect in the SAP demodulation path can be caused by a problem in the SAP signal-detect path (discussed next). For example, if the SAP signal-detect circuit does not apply the SAP-present signal to the audio input-select circuit, the SAP signal cannot be directed to the linear audio head.

SAP Signal-Detect Path. As shown in Fig. 10.6*b*, the heart of the SAP signal-detect circuit is PLL IC_3. The SAP FM signal is applied at pin 6 of IC_3 through BPF_1, Q_{11}, and Q_{12}.

The VCO in IC_3 does not oscillate at the SAP carrier frequency of 78.67 kHz. Instead, the VCO oscillates at 44 kHz. If the input to IC_3 is 44 kHz, the output of the phase detector is zero. However, since the SAP signal extends from 46 to 95 kHz, the SAP signal is always higher than the VCO frequency.

The phase detector produces a negative voltage when the incoming signal is higher than the VCO. This negative voltage, generated only when SAP is present, is amplified and filtered. The negative or low output from IC_3 at pin 7 is applied through Q_{13}, Q_{14}, Q_{15}, and Q_{16} and appears as a high at the collector of Q_{16}. The high drives Q_{33} into conduction, turning the SAP indicator D_{520} on and informing the user that a SAP signal is being broadcast.

The high at Q_{16} is also applied to Q_{17} and drives the output of Q_{17} low. This output is a status command (called $\overline{SAP}$) applied to the audio input-select circuit (Sec. 10.2.7). If the $\overline{SAP}$ line does not go low, the output of the SAP demodulator is not applied to the linear audio head, making it appear as though no SAP is present. This point should be remembered when troubleshooting for a loss of SAP.

The high at Q_{17} is inverted to a low by Q_{32} and applied to pin 3 of IC_{34}, muting the SAP audio. The Q_{17} high is also inverted by Q_{45} to a low at the base of Q_8, also muting SAP audio.

10.2.7 Audio Input-Select Circuits

Figure 10.9 shows typical audio input-select circuits in simplified form. The circuit is responsible for selecting the audio signals to be directed to the rotating hifi heads and to the stationary linear head. The user controls involved are the input select switch S_1 and the normal audio switch S_3.

The bulk of the audio input-select circuits are within IC_{34}. Left-channel audio is output from pin 14 of IC_{34}, while right-channel audio is taken from pin 15. The linear-head amplifier is supplied from pin 4 of IC_{34} and may be either L + R (mono) or SAP audio.

The right and left outputs of the circuit are controlled by the logic at pins 10 and 11 of IC_{34}, taken from S_1. Both the ext. audio and external positions of S_1 select the external inputs of the VCR. In either position, S_1 opens the path to the 9-V supply, and the logic at pins 10 and 11 goes low. This causes the switches within IC_{34} to connect pins 2 and 15, as well as pins 12 and 14, and passes the external audio to the hifi amplifier inputs.

In the TV position of S_1, the logic at pins 10 and 11 of IC_{34} goes high (+9 V), opening the analog switches to the external audio inputs and closing the switches to the left and right outputs of the stereo decoder.

Switch S_3 determines the audio signal to be applied to the linear audio head. If SAP is not present, L + R is applied automatically. (SAP is applied at pin 3 of IC_{34}, and L + R is applied at pin 5 from Q_{52} and Q_{53}.) The switches within IC_{34} (for the linear audio head) are controlled by the logic at pin 9.

A high at pin 9 of IC_{34} connects the SAP input to the linear-head output at pin 4, while a low at pin 9 connects the L + R input to pin 4. When S_3 is in the L + R position, pin 9 is grounded through S_3. As a result, only L + R can be applied to the linear head. When S_3 is in SAP, the logic at pin 9 of IC_{34} is determined by the

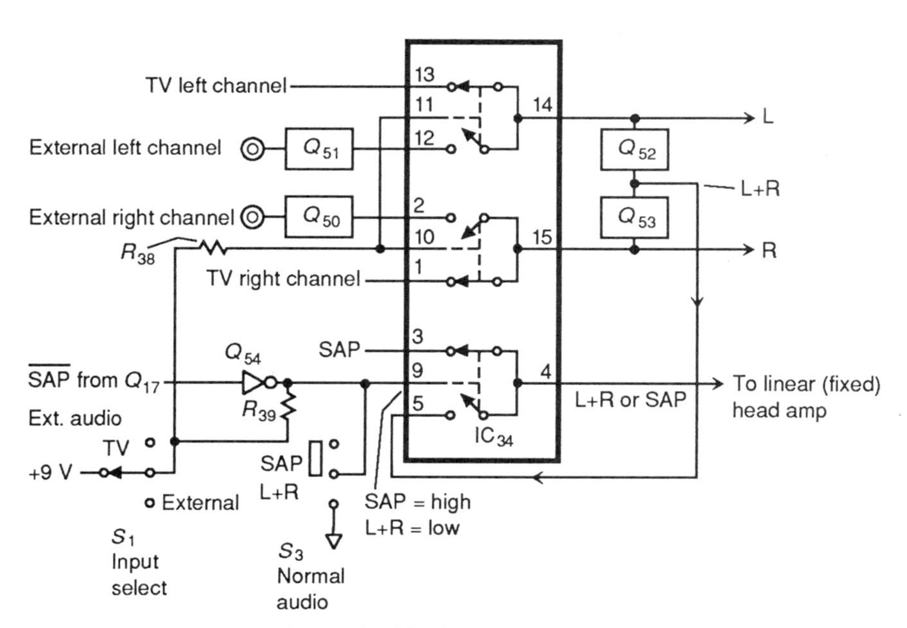

FIGURE 10.9 Typical audio input-select circuits.

$\overline{\text{SAP}}$ status command from the SAP signal-detect circuits (Sec. 10.2.6). The $\overline{\text{SAP}}$ status line is low when SAP is present and high when SAP is not being transmitted.

When SAP is present, the collector of Q_{17} is low. This low is inverted to a high by Q_{54} and applied to pin 9 of IC_{34}. This connects pins 3 and 4 of IC_{34} and applies SAP to the linear-head amplifier.

When no SAP is being transmitted, the collector of Q_{17} is high, and the high is inverted by Q_{54}. With pin 9 of IC_{34} low, pins 4 and 5 are connected, and L + R is applied to the linear-head amplifier.

10.3 STEREO TV TESTS AND ADJUSTMENTS

Compare the following adjustment procedures with those shown in the service literature for the stereo TV or VCR being serviced. The test and adjustment points are shown in Figs. 10.5 and 10.6.

10.3.1 L − R Adjustment

The purpose of the L − R IC_1 is to reinsert a carrier into the AM stereo L − R sidebands at pin 2 and produce corresponding audio output at pin 5. The missing L − R carrier is at 31.468 kHz and is produced by a VCO within the L − R decoder. Since there is no L − R carrier at the decoder output, the VCO is usually locked to the 15.734-kHz pilot (or a multiple).

In the circuit of Fig. 10.5, the VCO can be set by adjustment of PLL adjust VR_1 at pin 16 of IC_1 and can be monitored at TP_{31}. The VCO signal at TP_{31} is 15.734 kHz, even though the VCO operates at a different frequency (typically higher).

1. Ground the input to the stereo decoder circuits at pin 2 of the SS connector. This removes all signals to the L − R decoder input, including the 15.734-kHz pilot. If the pilot is present, it is possible that the VCO will lock onto the incoming signal even though the VCO is not exactly on frequency when free-running. Check that stereo indicator D_{521} is off.
2. Connect a frequency counter to TP_{31}, and adjust VR_1 for a reading of 15.734 kHz. Remove the ground from pin 2 of the SS connector. Check that the TP_{31} reading remains at 15.743 kHz.

10.3.2 L − R Gain Adjustment

The purpose of this adjustment is to set the level of the stereo L − R signal before dbx noise-reduction processing. This is not to be confused with the L − R separation adjustment that sets the L − R level after dbx processing (Sec. 10.3.10).

1. Apply a modulated L − R signal to pin 2 of the SS connector (using a stereo generator such as described in Sec. 10.1.3). The stereo indicator D_{521} should turn on. Use a modulation frequency of 300 Hz (unless otherwise specified in the service literature).
2. Monitor TP_{37} for the 300-Hz signal, using an audio voltmeter or scope. Adjust

VR_2 for the correct voltage level at TP_{37}. Always use the values specified in service literature (for both input amplitude at pin 2 of the SS connector and output at the test point).

10.3.3 SAP Level Adjustment

The purpose of this adjustment is to set the level of the SAP signal (before dbx processing) in relation to the stereo L − R signal. If the SAP and L − R signals are not approximately equal at this stage in the audio path, the user audio output or volume control must be reset each time when switching between stereo and SAP.

1. Apply a modulated SAP signal to pin 2 of the SS connector. The SAP indicator D_{520} should turn on. Use a modulation frequency of 300 Hz unless otherwise specified in the service literature.
2. Monitor TP_{38} for the 300-Hz signal. Adjust VR_3 for the correct voltage level at TP_{38}. Always use the values specified in the service literature (for both the input amplitude at pin 2 of the SS connector and output at the test point). Note that the SAP level adjustment is usually performed after adjustment of the SAP detector (but check the service literature for the preferred sequence).

10.3.4 SAP Detector Adjustment

The purpose of the SAP signal detector IC_3 is to produce a low at output pin 7 when the 78.67-kHz SAP carrier is present at input pin 6. (Output pin 7 remains high when the SAP carrier is not present.) This is done by comparing the incoming SAP carrier with the IC_3 decoder VCO. When both signals are present and locked in frequency and phase, pin 7 goes low. The VCO is set by adjustment of SAP PLL adjust VR_4 at pin 1 of IC_3.

1. Apply a 78.67-kHz signal to pin 2 of the SS connector. Monitor the voltage across TP_{35} and TP_{36}. Set VR_4 to full counterclockwise. The SAP indicator D_{520} should be off. The voltmeter should read about −1.0 V.
2. Adjust VR_4 clockwise until the voltage at TP_{35} and TP_{36} changes to about +1.0 V. This indicates that pin 7 of IC_3 is switched to low. You could measure at pin 7 of IC_3 or at TP_{34}, but the change in signal level is much more difficult to detect. Also, by checking at TP_{35} and TP_{36}, you also confirm operation of Q_{13}, Q_{14}, and Q_{15} simultaneously.
3. With the SAP carrier still applied at the circuit input (pin 2 of the SS connector), check that SAP indicator D_{520} has turned on when the TP_{35} and TP_{36} voltage indication changes from −1.0 to +1.0 V. This confirms operation of both D_{520} and Q_{33}.

10.3.5 SAP Detector Signal Output-Level Adjustment

The purpose of VR_5 is to set the level of the SAP detect signal. (This signal appears when the SAP carrier is present, and pin 7 of IC_3 goes low.)

1. Make certain that no SAP carrier signal is present. (On some circuits, it may be necessary to ground the SAP circuit input at pin 2 of the SS connector.) Monitor the voltage across TP_{35} and TP_{36}. Adjust VR_5 until the voltage at TP_{35} and TP_{36} is −1.0 V.

10.3.6 Stereo Noise-Reduction Time-Constant Adjustment

The purpose of VR_6 is to set the amount of control output from dbx noise-reduction IC_4 for a given L − R signal amplitude. Connect a digital voltmeter to TP_{39} and TP_{40}. Adjust VR_6 for the correct voltage across R_5. Use the service-literature values.

Note that although you are measuring voltage across R_5, VR_6 is set for a given current through circuits within IC_4 (the more current, the more control). Typically, R_6 is set for a current of 15 mA at pin 13 of IC_4. Since R_5 is 1000 Ω, the voltage reading should be 15 mV.

10.3.7 Stereo-Reduction VD Adjustment

The purpose of VR_7 is to set the L − R gain, after spectral processing by IC_4 but before wideband processing, to produce the desired variable deemphasis (VD). (This is sometimes called the *wideband* or *high-band* VD adjustment.)

1. Make certain that no L − R signal is present. On some circuits, this can be done by grounding the input at pin 2 of the SS connector. In other circuits, it is necessary to disable the L − R audio path (such as connecting the emitter of Q_{18} to +9 or +12 V).
2. Apply a 300-Hz signal to TP_{37}, using the correct level specified in the service literature. A typical input level at TP_{37} is −24 dB.
3. Monitor the 300-Hz signal (after noise reduction) at TP_{43}. Make certain that the level at TP_{43} is within limits specified by the service literature. The typical 300-Hz output level at TP_{43} should be between −23 and −35 dB. Note the actual level at TP_{43}.
4. Change the frequency of the audio signal applied at TP_{37} from 300 Hz to 8 kHz. Set the amplitude of the 8-kHz signal as specified in service literature. A typical input level at TP_{37} (with the 8-kHz audio) is −17 dB.
5. Adjust VR_7 so that the 8-kHz signal (after noise reduction) at TP_{43} is as specified in the service literature. A typical 8-kHz output level at TP_{43} is the actual 300-Hz level (as measured in step 3) less −11 dB. Remember that these noise-reduction adjustments are *critical* to proper operation of the stereo TV circuits. Also, the service literature generally recommends that the VD adjustment described in Sec. 10.3.6 be performed before the separation adjustment described in Sec. 10.3.10.

10.3.8 SAP Noise-Reduction Time-Constant Adjustment

The purpose of VR_8 is to set the amount of control output from dbx noise-reduction IC_6 for a given SAP signal amplitude. Connect a digital multimeter to TP_{41} and TP_{42}. Adjust VR_8 for the correct voltage level across R_{26} (at TP_{41} and TP_{42}). Use the service-literature values.

Note that although you are measuring across R_{26}, VR_8 is set for a given current through the circuits within IC_6 (the more current, the more control). Typically, VR_8 is set for a current of 15 mA at pin 13 of IC_6. Since R_{26} is 1000 Ω, the voltage reading should be 15 mV.

10.3.9 SAP Noise-Reduction VD Adjustment

The purpose of VR_9 is to set the SAP gain, after spectral processing by IC_6 but before wideband processing, to produce the desired variable deemphasis. The procedure is the same as for adjustment of VR_7 (Sec. 10.3.7), except that TP_{38} and TP_{44} are used instead of TP_{37} and TP_{43}.

10.3.10 L − R Separation Adjustment

The purpose of this adjustment is to set the level of the stereo L − R signal in relation to the mono L + R signal. Both signals are combined in the Q_{22} matrix (R_{59} and R_{60}). If the L − R signal is low in relation to L + R, you will hear only mono. If the L − R signal is high in relation to L + R, you will hear both signals, but there will be poor separation between the left and right audio (the audio sounds like mono even though stereo is present).

1. Apply a modulated L − R signal to pin 2 of the SS connector. The stereo indicator D_{521} should turn on. Use a modulation frequency of 300 Hz unless otherwise specified in the service literature.
2. Monitor TP_{32} and TP_{33} for the 300-Hz signal. Adjust VR_{10} for the correct voltage level at TP_{32} and TP_{33}. Always use the values specified in the service literature (for both input amplitude and output at the test points). Remember that this adjustment can be *critical* in producing good stereo sound.
3. If you cannot find a separation adjustment procedure in the service literature, use the following as an *emergency procedure only*.
4. Set VR_{10} so that L − R is zero. (Generally, this means setting VR_{10} full counterclockwise.) Apply an L + R signal with 300-Hz modulation at pin 2 of the SS connector. Note the voltmeter reading at TP_{32} and TP_{33}. This is the mono L + R signal level.
5. Remove the L + R signal and apply L − R with 300-Hz modulation at pin 2 of the SS connector.
6. Adjust VR_{10} until the readings at TP_{32} and TP_{33} are the same as with the L + R (or just below L + R) in step 4.

For a final test, measure stereo separation as described in Sec. 10.3.11. (As a practical matter, some technicians recommend adjusting VR_{10} for a given amount of separation or for maximum separation at the speakers, rather than for a given reading in the decoder circuits.) A stereo separation of 60 dB or better is possible on some circuits.

10.3.11 Stereo Separation Tests

Virtually all stereo TV decoders require a stereo separation test and a stereo indicator (pilot) test. Compare the following procedures with those shown in (or omitted from) the service literature.

1. A pilot signal must be present for all stereo tests. Operate the generator controls as necessary for a pilot signal, and check that the stereo indicator turns on.
2. Before making any stereo tests, make certain that the stereo function is selected on the TV set or VCR.
3. Select the desired audio modulating frequency of 300 Hz, 1 kHz, or 8 kHz, using the appropriate generator controls.
4. Select the desired form of modulation (L, R, L − R, or L + R), using the corresponding generator controls. Note that many stereo tests are conducted using L (left), followed by a repeat of the test with R (right) modulation. Often, this is followed with a test using L − R modulation.

On some generators, the controls are interlocked, so you can select only one of the four modulation signals at a time. No matter what controls are involved, remember that if you select L + R, there is no stereo operation at the decoder, even though the pilot is present and the stereo indicator is on.

To measure stereo separation, select L (with the pilot on). The stereo indicator should turn on and sound should be heard from the left speaker. Monitor the left-channel speaker (or left-channel audio output, whichever is convenient). Measure the left-channel output voltage. Select R (with a pilot). The stereo indicator should remain on, and sound should be heard from the right speaker.

Measure the right-channel output voltage, and compare the right-channel voltage to the left-channel voltage. The *ratio* of the left-channel output to that of the right-channel is the *stereo separation* and is usually expressed in dB.

Generally, the actual voltages are not critical; it is the *voltage ratio* that counts. Of course, both voltages (L and R) must be measured at the same point in the audio path (typically at the L and R speakers or outputs), and both must be measured using the same input voltage level and frequency.

Note that when the pilot is present (stereo on), the channel outputs may (or may not) be equal. However, with the pilot off (stereo off), the channel outputs should be substantially the same, no matter what modulating signal is used. Compare the measured stereo separation to that given in the TV or VCR specifications.

10.4 STEREO TV TROUBLESHOOTING APPROACH

In each of the following trouble symptoms, it is assumed that an active TV channel is tuned in, that both stereo and SAP are (supposedly) being broadcast (in addition to a mono broadcast), and that video can be recorded and played back (in the case of a VCR). If you do not have a good picture, the problem is probably in the sections ahead of the stereo TV circuits (RF, IF, etc.). It is also possible that the problem is in the TV set used to monitor the VCR output (in the case of a VCR).

In the case of a VCR being serviced, a good preliminary troubleshooting step is to play back a known-good tape using a known-good TV set. (As an alternative to a TV set, you can monitor the audio output of a VCR at the headphone jack and/or audio-output jacks.)

If a VCR is capable of hifi operation (Chap. 11), the test tape should have both hifi audio (recorded on the rotating-head tracks) and normal audio (recorded on the linear-head track).

If the audio playback is good on all tracks, it is reasonable to assume that the heads and playback amplifiers are good (as are any audio amplifiers that follow the playback amps).

If playback is good on all but one track, you have isolated the problem to either the hifi heads and amplifiers or to the normal-audio head and amplifiers.

If there is no playback from any of the tracks or if playback is abnormal for all tracks, the problem is probably in the playback or audio amps following the stereo TV circuits.

Once you have established that audio playback is good, the next step is to record and play back broadcast audio (in the case of a VCR). Check to see if any audio is available at the audio output jacks (or headphone jack), indicating that a mono broadcast is available. Then check to see if the stereo and SAP indicators are turned on, indicating stereo and SAP broadcast signals.

10.4.1 No Audio Available at the Audio-Output Jacks

If no audio is available at the audio-output jacks but the picture is good and either the stereo or SAP indicator is on, the problem is probably in the mono L + R audio path or in the audio amplifiers that follow the stereo TV circuits. The fact that either indicator is on makes it likely (but not absolutely certain) that the sections ahead of the stereo TV circuits are good.

Before you plunge into any circuits, make certain that all switches are in the correct positions. For example, even if the SAP indicator is on, do not expect to record SAP unless the normal audio switch S_3 (Fig. 10.9) is set to SAP.

Start by applying a composite signal (with L + R) at the input of the stereo TV circuits (Fig. 10.5). Listen for an audio signal at the speakers (or monitor at output jacks, whichever is convenient).

If the modulated signal is heard with L + R applied, it is fair to assume that the audio path through the stereo TV circuits and audio amplifiers is good. The problem is probably ahead of the stereo TV circuits, but it could be in the audio input-select circuits (Fig. 10.9).

If the modulated signal is not heard with L + R applied, trace through the audio path using signal injection or signal tracing, whichever is most convenient. The L + R path (up to the audio input-select circuits) is shown in Fig. 10.5.

10.4.2 No Stereo Operation; Mono Operation Is Good

With these symptoms and a good picture, check that the stereo indicator is on (indicating that stereo is being broadcast). If stereo is on, the problem is probably in the L − R audio path shown in Fig. 10.5. If stereo is off, the problem may be that no stereo is being broadcast. Try injecting an L − R signal into the stereo TV circuit input.

If stereo turns on and you get stereo at the output, with L − R injected, the problem is likely to be ahead of the stereo TV section. In that case, proceed as described in Sec. 10.4.1.

If stereo does not turn on and there is no stereo at the output, with L − R injected, the problem is in the L − R path (Fig. 10.5). The same is true if stereo is on, but there is no stereo at the output. There are several points to consider when tracing through the L − R path. First, try correcting the problem by adjustment of the L − R path (VR_1, VR_2, VR_6, VR_7, and VR_{10}) as described in Sec. 10.3. These adjustments could cure the problem. If they don't, simply making the adjustments might lead to the defect.

For example, if there is no 15.734-kHz at TP_{31}, no matter how VR_1 is adjusted, IC_1 is suspect. Similarly, if pin 6 of IC_1 does not go low when a pilot is injected at the stereo TV input, IC_1 is again suspect.

10.4.3 No SAP Operation; Mono and Stereo Operation Are Good

With these symptoms and a good picture, check to see that normal audio switch S_3 (Fig. 10.9) is in SAP and that the SAP indicator is on (indicating that SAP is being broadcast).

If SAP is off, the problem may be that no SAP is being broadcast. Similarly, on a VCR, if normal audio S_3 is set to L + R, SAP will not be recorded.

If SAP is on and S_3 is set to SAP, the problem is likely to be in the SAP audio path (Fig. 10.6*a*), the SAP signal-detect path (Fig. 10.6*b*), or possibly the audio input-select path (Fig. 10.9). Try injecting a SAP signal at the input of the stereo TV circuits.

If SAP turns on and you get SAP at pins 3 and 4 of IC_{34}, the problem is probably ahead of the stereo TV circuits. Proceed as described in Sec. 10.4.1.

If SAP does not turn on and there is no SAP at IC_{34}, the problem is in the SAP paths (Fig. 10.6). The same is true if SAP is on, but there is no SAP at IC_{34}.

There are several points to consider when tracing through the SAP paths. First, try correcting problems in the SAP paths by the adjustments described in Sec. 10.3 (VR_3, VR_4, VR_5, VR_8, and VR_9). These adjustments could cure the problem. If they do not, simply making the adjustments might lead to the defect. For example, if pin 7 of IC_3 does not go low when there is a SAP signal applied at the input (Fig. 10.6*b*), no matter how VR_4 is adjusted, IC_3 is suspect. (IC_3 is also suspect if pin 7 remains low when the SAP input is removed.)

If pin 7 of IC_3 goes low when SAP is applied, check that the high appears on the base of Q_{33}. If it does so and SAP indicator D_{520} does not turn on, D_{520} and Q_{33} are suspect. If the base of Q_{33} does not go high where there is a SAP input (and pin 7 of IC_3 is low), suspect Q_{13} through Q_{16}. It is also possible that VR_5 is not properly adjusted (Sec. 10.3.4).

If the SAP indicator D_{520} is turned on, check for a high at pin 9 of IC_{34} (Fig. 10.9). If pin 9 is high, check that the audio signals at pins 3 and 4 of IC_{34} are substantially the same. If pin 9 is not high, suspect Q_{17} and Q_{54}.

If there is no audio at pin 3 of IC_{34}, trace through the SAP audio path (Fig. 10.6*a*). Remember that pin 3 of IC_{34} should be high when SAP is present. When SAP is removed, the output of Q_{17} goes high. This high is inverted by Q_{32} to a low at pin 3 of IC_{34} (to mute the SAP audio). The high is also inverted by Q_{54} to a low at pin 9 of IC_{34} (to pass L + R).

10.4.5 Audio Input-Select Problems

Problems in the audio input-select circuits (Fig. 10.9) can produce trouble symptoms that appear to be in other circuits. For example, if there is no mono or stereo, it is possible that IC_{34} has not received a command that selects pins 1 and 13. This could be the fault of S_1 or the associated circuit. Similarly, IC_{34} may receive the correct command but not respond properly. This is the fault of IC_{34}.

Before you condemn the stereo TV circuit ahead of the audio input network, check all signals to and from IC_{34}. As an example, set S_1 to TV, and check that pins 9, 10, and 11 are high. If they are not, suspect S_1, R_{38}, and R_{39}.

If pins 10 and 11 are high, check that the signal at pin 14 of IC_{34} is substantially the same as at pin 13, while the signal at pin 15 is the same as at pin 1. If they are not, suspect IC_{34}. If pin 9 is high, check that the signal at pin 4 is substantially the same as at pin 3. If it is not, suspect IC_{34}.

Now set S_3 to L + R and check that pin 9 of IC_{34} is low (no matter what the position of S_1). If it is not, suspect S_3. If pin 9 is low, check that the signal at pin 4 is substantially the same as at pin 5. If it is not, suspect IC_{34}.

Note that the L + R signal is applied to IC_{34} through Q_{52} and Q_{53}. So if it is possible to record and play back SAP on the linear head but not L + R (even though pin 9 of IC_{34} is low and pin 4 is connected to pin 5), suspect Q_{52} and Q_{53}.

10.5 SURROUND SOUND

Figure 10.10 shows some typical surround-sound configurations available to present-day stereo, TV, and VCR listener and/or viewers. With surround sound (which is not directly related to stereo TV), it is possible to simulate the sound of motion-picture theaters in the home.

The following paragraphs describe three of the most popular surround-sound configurations: *matrix surround, Hall matrix*, and *Dolby Surround*. These configurations may be called by other names (music surround, mono enhance, simulated stereo, time-delay surround, etc.) and can be provided by *built-in circuits* (in TV or stereo receivers) or by external *surround processors*. Although all of the configurations produce a surround effect, only the Dolby systems (Dolby Surround or Dolby Prologic) are capable of decoding Dolby surround sound recorded on tape and film (and occasionally broadcast by TV stations).

10.5.1 Matrix Surround

Figure 10.11*a* shows a basic matrix-surround circuit. Such a circuit extracts "ambience" information from a pair of stereo signals by finding the difference between the signals. When the speaker A switch is on, normal stereo sound is heard from both speakers connected to the A terminals. When speaker B switch is on, the sound is also available from the speakers connected to the B terminals. When both switches are on, the A speakers reproduce the normal stereo signal, but the B speakers produce the *difference* between the left- and right-chanel signals (essentially the same as L − R found in stereo TV).

If the B speakers are placed behind the listener (Fig. 10.10) *ambience is synthesized*. Many matrix systems also include amplification. In some matrix-surround configurations (with amplification), 2L + R is produced from the front left speaker, 2R + L from the front right speaker, 2L − R from the rear left speaker, and 2R − L from the rear right speaker.

10.5.2 Hall Surround or Hall Matrix

Hall surround or Hall matrix (Fig. 10.11*b*) is similar to the basic matrix, except that a *time delay* is added to the rear (surround) speakers. This gives the feeling

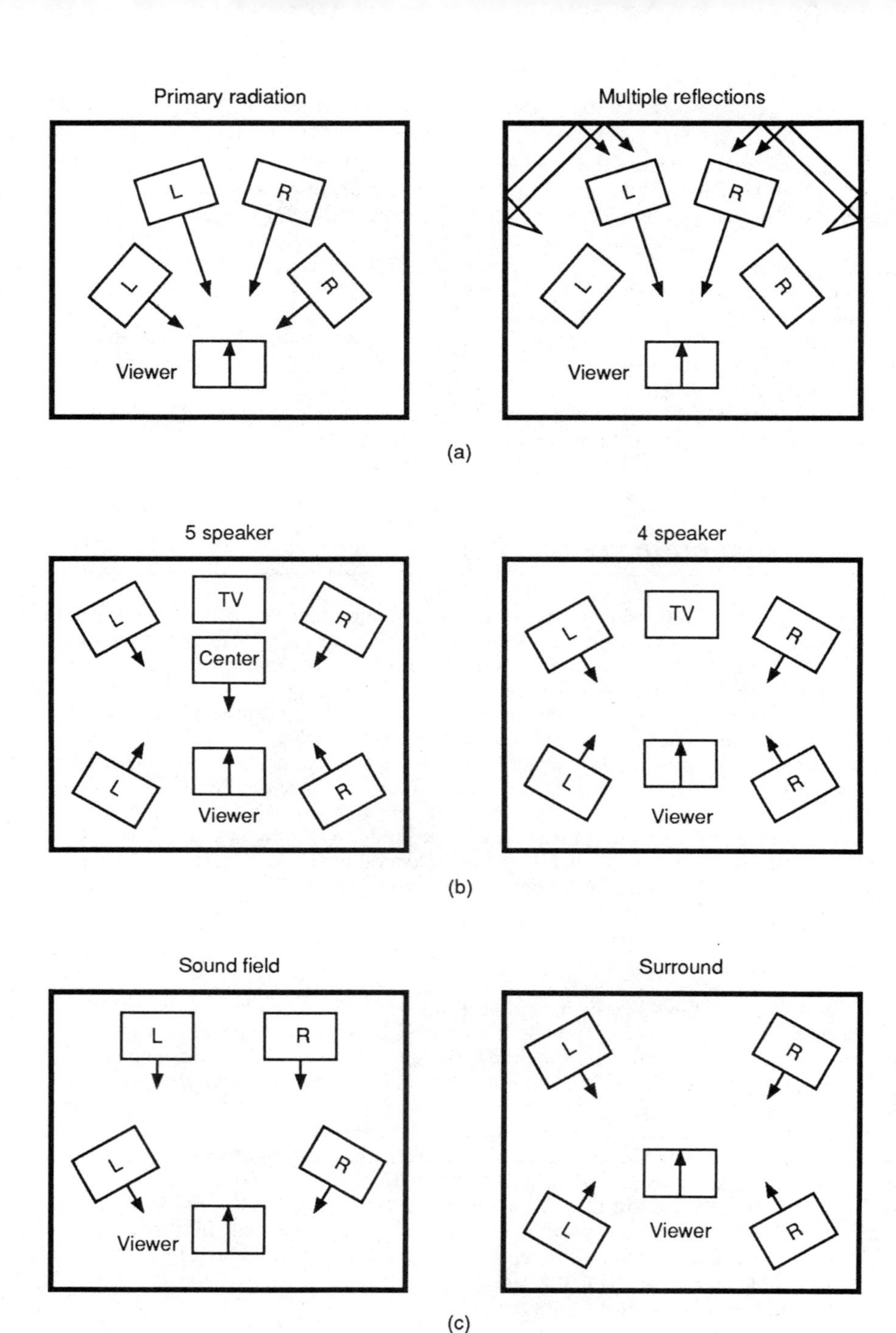

FIGURE 10.10 Typical surround-sound configurations. (*a*) Time delay on Hall; (*b*) Dolby; (*c*) matrix or simulated.

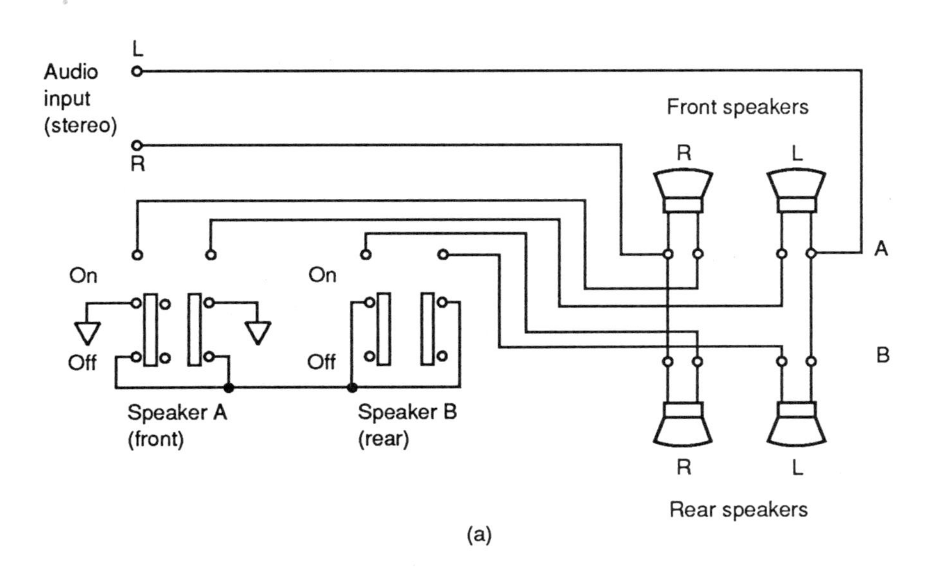

(a)

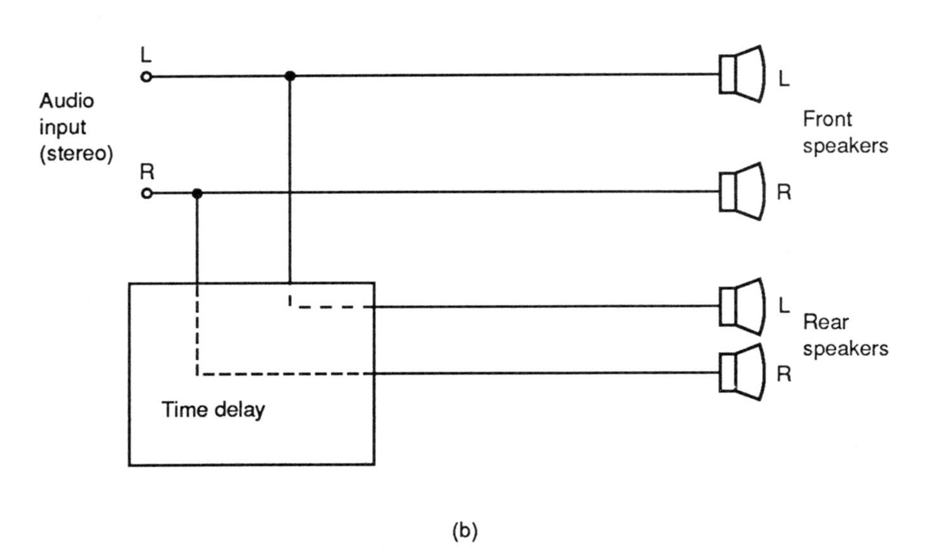

(b)

FIGURE 10.11 Matrix and Hall-surround circuits.

(or illusion) of spaciousness (such as being in a concert hall, thus the name *Hall surround*). The time delay may also produce sound reverberations (in some listening areas).

10.5.3 Dolby Surround Processing

To understand either form of Dolby Surround (usually found in surround processors but not built into TV), it is necessary to understand the basics of the theatrical Dolby stereo encoding and decoding system. The basic Dolby Surround decoding circuit is shown in Fig. 10.12. Note that four channels of sound are encoded on only two sound tracks.

The basic Dolby decoder derives surround information by subtracting the right channel from the left (L − R). Unlike either Hall surround or matrix surround, Dolby also produces a *center channel* by summing the left and right channels (L + R). The mono center channel prevents a center "hole" when the L and R speakers are widely separated.

Like the Hall matrix, the basic Dolby decoder adds time delay to the surround output. The delay (of 15 and 30 ms) is added to take advantage of the Haas effect. (The Haas effect causes the mind to identify the sound source as that from which the sound is first heard and to ignore the same sound arriving later at the ear.) This "first arrival" effect ensures that front-channel sounds are clearly identified as originating from the front, rather than from the rear.

Unlike Hall matrix, the Dolby decoder cuts off the surround sound at 7 kHz and adds a modified form of Dolby B (5 dB of Dolby processing, instead of the usual 10 dB found in normal Dolby B, Sec. 7.5). In addition to providing noise

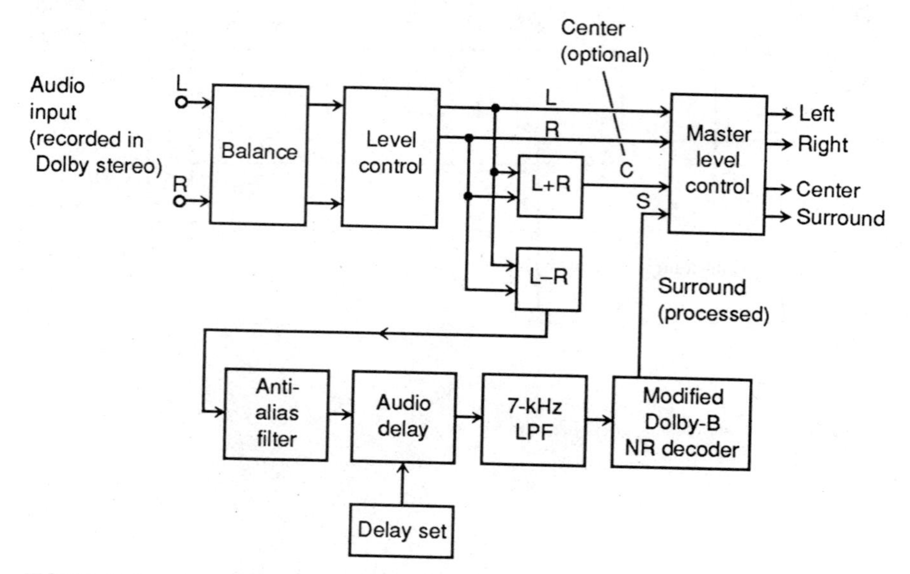

FIGURE 10.12 Basic Dolby Surround decoding circuit.

reduction, the modified Dolby prevents the (rear) surround signal from altering the (front) L and R audio.

10.5.4 Dolby Prologic

The ultimate Dolby surround system is the Dolby Prologic, which contains active circuits that provide steering logic for the recorded audio. As shown in Fig. 10.13, this steering logic is part of an adaptive matrix (also called a directional-emphasis circuit and found in IC form). The steering logic senses the direction of *soundtrack dominance* (from which direction the loudest sound on the track seems to originate) and generates control signals that increase the gain in the appropriate (left, right, center, surround) combinations of channels.

By comparing the left, right, center, and surround signal pairs and taking the logarithms of the values, *a pair of bipolar control signals* is generated. (Logs are used, in part, because human hearing works in a logarithmic manner.) The bipolar control signals adjust gain of eight voltage controlled amplifiers (VCAs), four VCAs for each channel. The output of the VCAs, together with the L and R signals, produces a total of 10 control signals.

When the control signals are applied to the four output channels, a total of 40 summed directional-audio components are available. Separation between any pair of channels, adjacent or opposite, is 30 dB. This compares to Dolby Surround's 3 dB of adjacent separation and 40 dB of opposite direction.

Prologic decoders are two-speed devices. When only one sound source dominates, the circuits operate in a slow mode. When there are two distinct sound sources, the circuits go into a fast time-division multiplexing mode, where the circuits control one source and then the other. The decoder switches back and forth between the two sources so quickly that the effect is unnoticed by the listener.

10.5.5 Surround-Sound Troubleshooting

Troubleshooting for the surround-sound circuits is essentially a matter of signal tracing, even for the more complex Dolby Prologic. A possible exception is where the circuits must be adjusted. Of course, the service-literature procedures must be used for adjustment (and there are no general or typical procedures for surround-sound adjustment). Note that some circuits (particularly those in surround processors with Dolby Prologic) have built-in signal-generator circuits to produce simulated Dolby recordings during test.

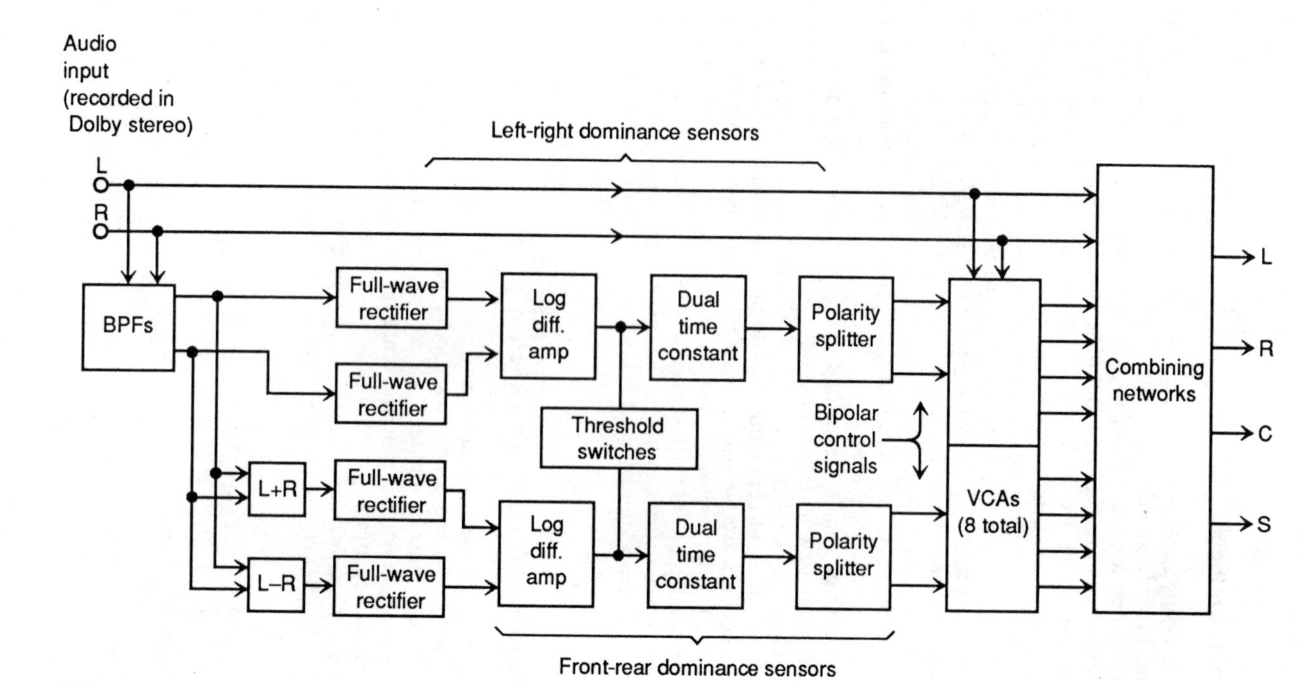

FIGURE 10.13 Dolby Prologic circuits.

CHAPTER 11
HIFI AUDIO-TAPE (VHS, BETA, 8 MM, DAT) EQUIPMENT

This chapter is devoted to the formats and circuits found in hifi audio-tape equipment. The first three formats (VHS, Beta, and 8 mm) are used in both VCRs and camcorders. The DAT format is used in digital audio-tape players and recorders. All four formats are related in that *both audio and video* are recorded on rotating heads (to get the bandwidth necessary for hifi reproduction). However, the circuits and format details are different for the four formats (and, of course, DAT does not include video). We start with the most popular format (VHS hifi), which forms a basis for understanding the remaining formats.

11.1 VHS HIFI

With a hifi VCR, the audio is recorded and played back through stereo heads located on the cylinder, as shown in Fig. 11.1. This is the same cylinder (also called the drum or scanner) used to rotate the video heads. As in the case of video, the hifi audio is recorded and played back in FM. This is different from conventional VCRs, where the audio is recorded and played back in AM. The combination of high-speed recording and playback and FM produces the high-fidelity audio with a frequency response and quality not possible in conventional VCRs.

In Beta hifi and Super Beta, the same heads are used for both audio and video. With either Beta or VHS, sound is also recorded as an AM (mono or stereo) signal with the conventional fixed audio head. This assures complete audio compatibility with nonhifi VCRs. Likewise, the normal audio-erase and full-erase heads operate in the normal manner.

In VHS hifi, the audio signal is used to frequency-modulate left- and right-channel carrier frequencies; 1.3 MHz is used for the left channel, and 1.7 MHz is used for the right channel, as shown in Fig. 11.2. These carriers are mixed and output is to audio heads mounted on the cylinder. During playback, the FM signals are picked up by the rotating audio heads and routed to the hifi playback circuits, where the left- and right-channel signals are separated and demodulated.

Normal audio is picked up by the normal audio head and amplified by the normal audio circuits (as in any VCR). Either the hifi or normal audio may be selected by the front-panel audio controls and applied to the audio output jacks, headphones, and RF modulator. The audio-erase head is used to erase previously

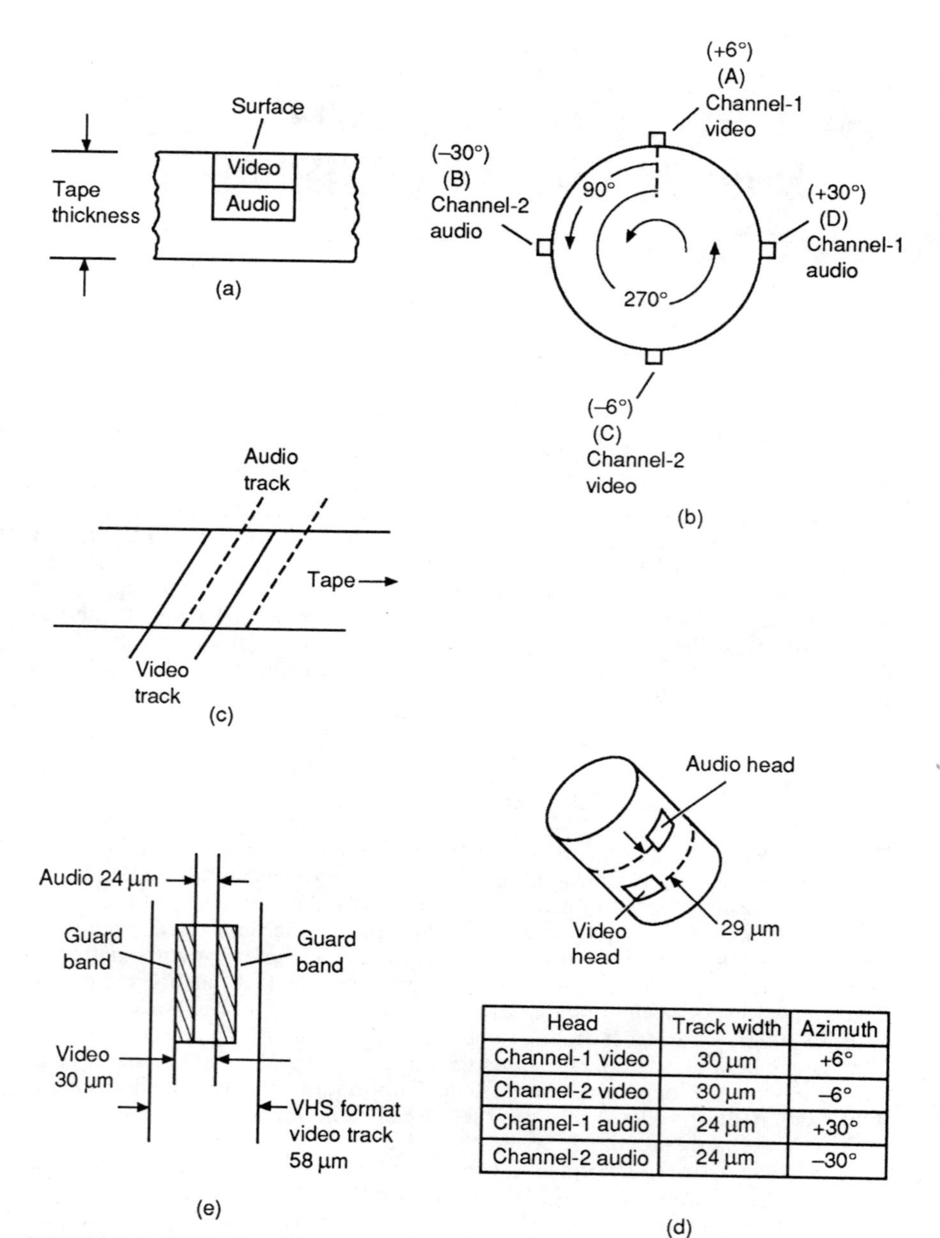

Head	Track width	Azimuth
Channel-1 video	30 μm	+6°
Channel-2 video	30 μm	–6°
Channel-1 audio	24 μm	+30°
Channel-2 audio	24 μm	–30°

FIGURE 11.1 Hifi audio-tape recording characteristics.

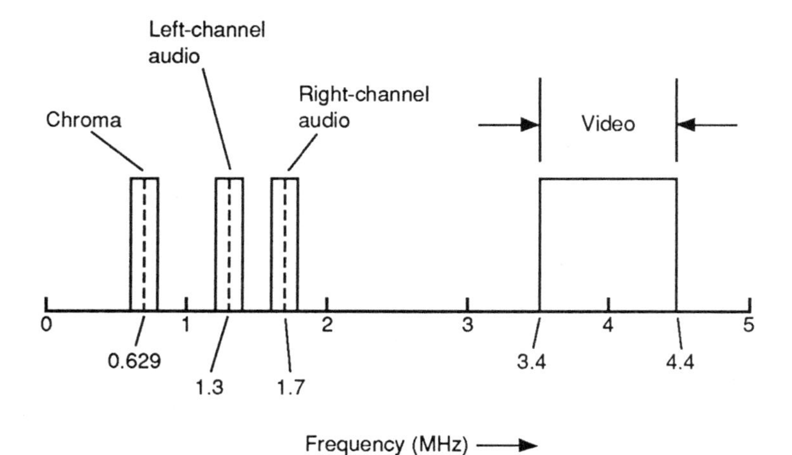

FIGURE 11.2 VHS hifi audio signal-conversion spectrum.

recorded normal audio when doing an audio dub. The full-erase head is used to erase both video and audio in the normal manner.

11.1.1 Hifi Recording Techniques

In addition to FM and rotating heads, a special recording process (described as depth-layer or multiple-layer recording) is also required for hifi. Figure 11.1 shows the basic principles involved.

The techniques shown in Fig. 11.1*a* permit both video and audio information to be deposited on a single recorded track. Initially, the audio is recorded deep into the tape's magnetic medium. The video signal is then recorded on the surface of the tape. In effect, the video is superimposed above the audio signal.

As is the case with VCR video, an azimuth recording technique is used for VHS hifi audio; both the audio and video signals on a given track are recorded at opposing azimuths. With azimuth recording, any interaction between the two signals during playback is at a minimum.

Figure 11.1*b* shows the relative positions of the video and audio heads mounted on the rotating drum of a VHS VCR. As dictated by the VHS format, the two video heads are spaced 180° apart, and each audio head is positioned 90° from the corresponding video head.

As shown in Fig. 11.1*c*, the video and audio tracks could not be superimposed without some form of compensation, because of the movement of the tape and the 90° differential between the video and audio heads. If this condition were allowed to exist, the succeeding audio track would erase most of the previously recorded video track.

To prevent this condition, the video and audio heads are displaced, as shown in Fig. 11.1*d*. Each audio head (which records first) is mounted on the cylinder 29 μm higher than the corresponding video head. Such displacement assures that the audio and video tracks align properly, one on top of the other.

The table in Fig. 11.1*d* specifies the azimuth and recording track width for each of the individual recording heads. As shown, the video-head azimuths remain identical to those of conventional VHS VCRs and camcorders (+6° and −6°). The azimuths of the audio heads are expanded considerably (+30° and −30°).

Figure 11.1*e* shows the hifi recording technique used in a 2H speed mode. The channel 1 video track is recorded on top of the channel 2 audio track. Since the VHS format allows a 58-μm video track width in the 2H mode and the widest track produced by a given head is equal to only 30 μm, guard bands are produced.

Note that, in the 2H mode, corresponding video and audio signals are recorded at opposing azimuths (+6° for the video track and −30° for the corresponding audio track, and vice versa). This produces a total azimuth difference of 36°. As a result, there is little or no interaction between the two signals.

During the 2H hifi recording mode, the video signal recorded on top of a given audio track is supplied by the adjacent head. For example, referring to Fig. 11.1*b*, when audio head B records channel 2 audio, video head A records channel 1 video, superimposing the video track onto the audio track. Then audio head D records channel 1 audio, with video head C recording channel 2 video.

Because of tape movement and the 90° differential between head-mounting positions, the video track to be recorded appears to fall on top of one-half the breadth of the audio track. This is prevented by raising the head 29 μm, as discussed. As a result, video 1 track is properly superimposed above the channel 2 track, and video 2 track is superimposed directly above the channel 1 audio track.

11.1.2 Hifi Recording with FM Conversion

Figure 11.2 shows the audio signal-conversion spectrum for VHS hifi. As in the case of the video signal, the audio signal to be recorded is converted to an FM signal at a relatively high frequency. The separate audio signals are developed, one for the right channel and one for the left channel. The two signals are then combined and directed to the audio heads.

The right-channel FM center frequency is 1.7 MHz, and the left-channel frequency is at 1.3 MHz. Nominal deviation for both channels is + 50 kHz, with a maximum allowable deviation of +150 kHz. The selection of these audio frequencies places the audio FM signal between the down-converted VHS chroma signal (629 kHz) and the VHS video signal (3.4 to 4.4 MHz), as shown in Fig. 11.2.

In addition to normal FM recording techniques, the signal is processed with *dbx noise reduction*, similarly to that for stereo TV circuits (described in Chap. 10). The noise-reduction feature improves the signal-to-noise ratio of the reproduced audio.

11.1.3 Hifi and Nonhifi Compatibility

To assure compatibility with existing VHS VCRs and existing tapes recorded for use with such VCRs, a hifi VCR also has the conventional *fixed audio head*. Such heads (also called the *normal heads or linear heads*) record a separate audio track along the upper edge of the video tape (as does any conventional VHS VCR). If you are not familiar with basic VCR formats and circuits, you can refer to *Lenk's Video Handbook*.

11.2 BASIC HIFI VCR AUDIO CIRCUITS

Figure 11.3 shows the basic audio circuits for a hifi VHS VCR. The audio input from either the tuner or line is applied to the fixed linear head and to the hifi heads through processing circuits. The linear processing circuits are the same as for any VHS VCR. Operation of the hifi processing circuits is as follows.

11.2.1 Record Noise Reduction and Preemphasis

The left and right input signals are first applied to a noise-reduction (NR) system. Noise reduction is done by compressing the dynamic range of the audio signal during the record process and then expanding the dynamic range during playback. As a result of the compression during record, the input signal is compressed to about one-half the dynamic range. Large transistions are decreased and small transitions are increased, resulting in a lower overall dynamic range.

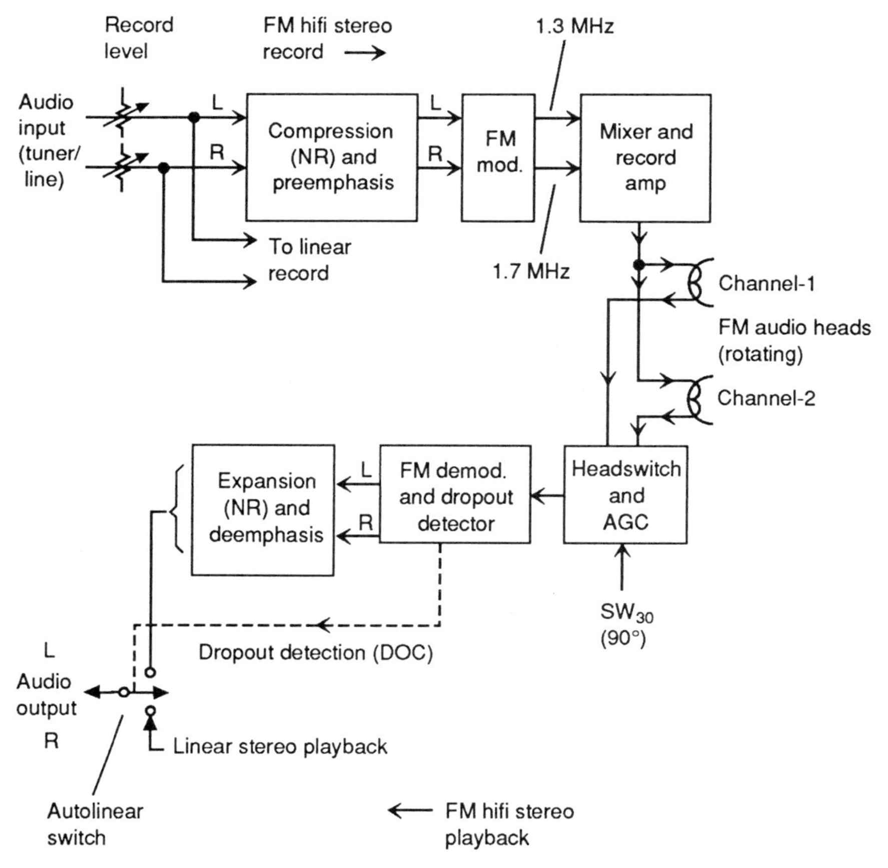

FIGURE 11.3 Basic audio circuits for a hifi VHS VCR.

During playback, the opposite must be performed on the audio signal to prevent distortion (similarly to dbx, discussed in Chap. 10).

The compressed right and left audio signals are also applied to a preemphasis network, which accentuates the high-frequency signals. After compression and preemphasis, the left and right signals are then applied to the audio FM modulator circuits.

11.2.2 Record FM Modulation and Mixer and Amplifier

Within the audio FM modulator, the left-channel audio signal modulates a 1.3-MHz RF carrier, while the right-channel audio signal modulates a 1.7-MHz audio carrier. The resultant two FM carrier signals are then applied to a mixer and record amplifier where the signals are mixed and applied to the rotating FM audio heads.

Note that each audio head records both the 1.3- and 1.7-MHz signal. As is the case in the video record and playback system, the audio signals are applied to both heads at all times during record.

11.2.3 Playback FM Head-Switching and Demodulation and Dropout Compensation

Playback operation of the FM audio signal is very similar to that of the FM video playback signal. The output signal of the two audio heads is applied to a head-switching circuit much like a video head-switching system. The head-switching circuit switches the input to the AGC circuit from one head to the other at a 30-Hz rate (the familiar SW_{30} signal). However, because of the 90° phase difference between the audio and video heads, it is necessary to delay the SW_{30} signal by 90°. The audio FM carrier output from the head-switching and AGC circuits (both 1.3 and 1.7 MHz) is then applied to the FM demodulator and dropout compensation circuits.

11.2.4 Playback Noise Reduction and Deemphasis

The left- and right-channel audio signals, after being demodulated, are applied to the expander and deemphasis circuits. The expander circuits expand the audio signal to the same dynamic range that was present before compression. As a result, after expansion and deemphasis, the audio signal (applied to the audio output through the autolinear switch) has the same characteristics that the audio provides to the input during record.

11.2.5 Autolinear Switching

The autolinear switch selects between the hifi stereo signal and the linear stereo signal from the fixed linear head. Note that even though the VCR records the linear track on a fixed head, the linear record/playback can be in stereo (on most hifi VCRs) but not in hifi.

The autolinear switch is also activated by an output from the dropout compensation (DOC) circuits associated with the FM demodulator (Sec. 11.2.3). If a dropout of longer than 2 s occurs, output from the dropout detector automatically switches the audio source from the FM hifi system to the linear playback system. This is done to prevent noise from occurring in the event that audio FM carriers are lost.

11.3 BETA HIFI AND SUPER BETA AUDIO

Figure 11.4 shows the frequency spectrum of both Super Beta and Beta hifi (which are essentially the Beta versions of S-VHS and VHS hifi). The following paragraphs summarize operation of Super Beta and Beta hifi audio circuits.

11.3.1 Beta Hifi

Beta hifi permits the VCR to record left- and right-channel audio (stereo), using the video heads, in addition to the linear head (also called the audio/control, or A/C, head). As in the case of VHS hifi, this provides much higher fidelity audio.

Carrier	Frequency MHz	Video track	Audio channel
F_1	1.380682 (87.75 fH)	A	L
F_2	1.530157 (97.25 fH)	B	L
F_3	1.679633 (106.75 fH)	A	R
F_4	1.829108 (116.25 fH)	B	R

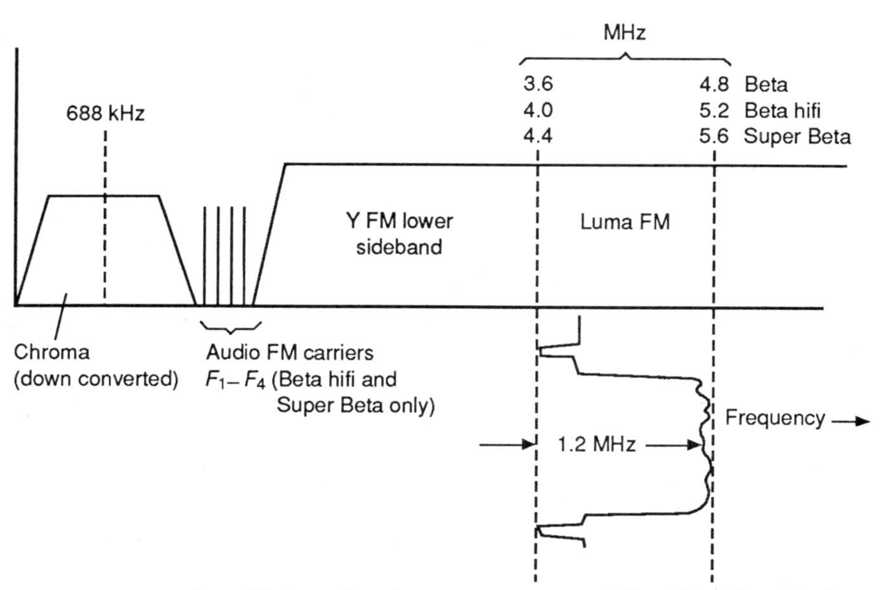

FIGURE 11.4 Beta, Beta hifi, Super Beta frequency spectrums and Beta hifi audio-carrier frequencies.

In Beta hifi, the left- and right-channel audio signals are FM modulated and divided into four pilot audio carriers. Four carriers (two for each channel, separated by 150 kHz) are required to maintain separation between the left- and right-channel audio, as well as to reduce the crosstalk between adjacent tracks of the video information.

Unlike VHS, which uses separate audio heads for video and audio, Beta uses the same heads for video and audio. The four audio carriers are centered about 1.5 MHz and are mixed with the video information to be recorded by the video heads on tape. (Figure 11.4 shows the frequencies of the four audio carriers.) During playback, the four carriers are reconstructed into left- and right-channel audio by alternately switching the A and B tracks in sync with the video heads.

The addition of the four audio FM carriers requires that the FM luma signal be shifted upward by 0.4 MHz (from Beta) to make room between the chroma and luma information. The high-frequency limitations of the video heads result in a loss of some FM sidebands (because of the shift in luma frequency). This causes a slight reduction of resolution in the picture produced by some Beta hifi VCRs.

11.3.2 Super Beta

Super Beta overcomes this reduction of resolution by narrowing the video-head gaps and shifting the FM luma carrier up by 0.8 MHz (from Beta), resulting in a larger luma FM sideband (shown as the Y FM lower sideband in Fig. 11.4). The increased bandwidth of the total luma signal results in resolution greater than achieved by both Beta hifi and conventional Beta VCRs.

11.3.3 Beta Audio Record

Figure 11.5 shows the audio record sequence in a typical Beta hifi VCR. The Super Beta audio-record sequence is similar.

The audio to be recorded is first filtered by a 20-kHz LPF to remove any stereo TV pilot signals (Chap. 10). If a stereo TV transmission is recorded, the front-panel MPX switch can be engaged to lower the LPF frequency (from 20 to 15 kHz), removing the 15.734-kHz pilot.

With the stereo TV pilot (if any) removed, the audio is preemphasized by dynamic-emphasis circuits in IC_1. In effect, the high-frequency components of the audio are boosted. The emphasized signal is then amplified by a record amp in IC_1 and passed to a limiter stage in the FM-modulator circuit of IC_3.

A record-compensate circuit (Q_{401} through Q_{404}) adds 30-Hz pulses (RF-switching pulses) to the recorded audio signal. This shifts the modulation frequency slightly at the beginning of each tape pass (or swipe). The pulses are added both during record and playback. A fixed dc value is added during pause operation of the VCR. (Note that if this circuit fails, *there is a 30-Hz buzz in both audio channels*.)

A limiter stage within IC_3 limits the maximum signal amplitude to prevent unwanted signals from the FM modulator. The modulated FM carrier is filtered by

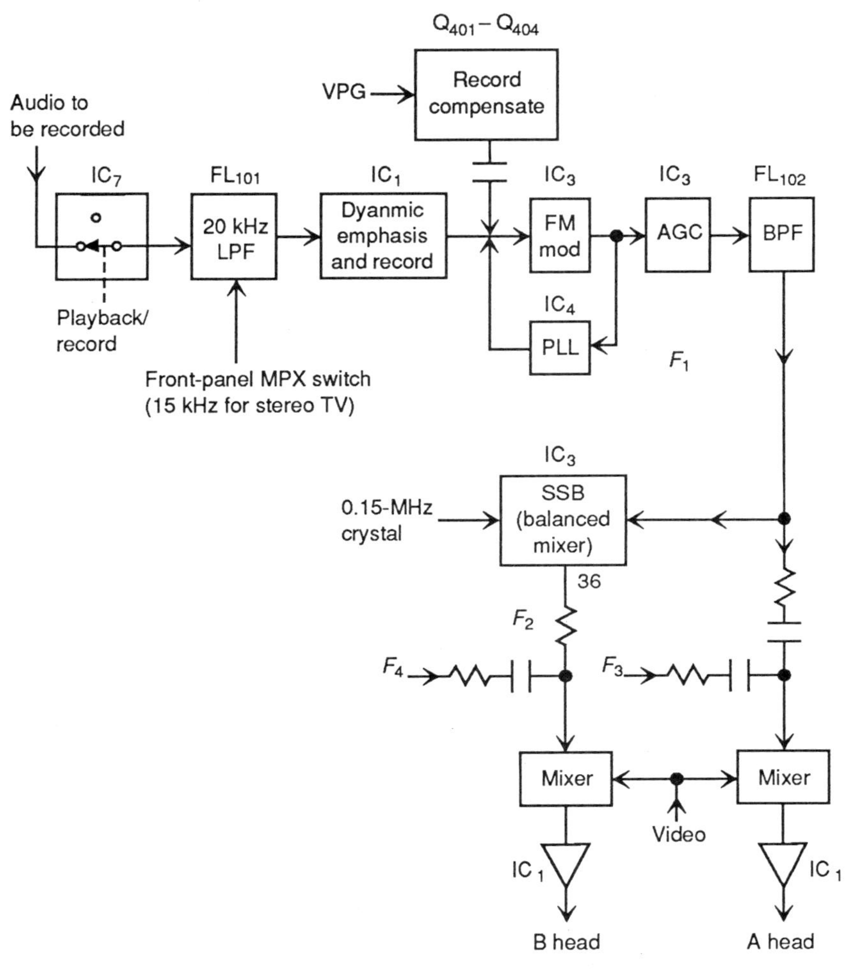

FIGURE 11.5 Audio record sequence in Beta hifi.

FL_{102} to produce the F_1 carrier (Fig. 11.4). Some of the F_1 signal is sampled by the PLL detector in IC_4 to phase-lock the FM modulator.

The F_1 signal is then added to a crystal-controlled 0.15-MHz signal in the single sideband (SSB) circuit of IC_3. (The SSB circuit is essentially a balanced mixer that is similar to those used in SSB transmitters.)

Mixing of the F_1 signal and the 0.15-MHz reference produces the F_2 signal at pin 36 of IC_3. The F_2 carrier is then mixed with a right-channel F_4 carrier and recorded on track B of the tape. For the A head, the left-channel F_1 is added to the right-channel F_3 signal and recorded on track A of the tape. (The four audio-signal carriers are mixed with video information and applied to the A and B heads through IC_1.)

11.3.4 Beta Audio Playback

Figure 11.6 shows the playback sequence in a typical Beta hifi VCR. The Super Beta audio-record sequence is similar.

The four FM audio carriers F_1 through F_4 are recovered separately from the tape using BPFs FL_1 through FL_4. The F_1 signal is filtered from the A track by FL_1 and added to the 0.15-MHz signal at the balanced mixer in IC_3. This produces one-half of the left channel. The other half of the audio signal is taken from the B head through BPF FL_2. The two halves are added together by the RF switching pulses (the VPG pulses at pin 39 of IC_2).

The demodulated audio signal for the left channel is processed by the hold circuit in IC_2 to remove some noise. A playback compensation circuit (Q_{401} through Q_{404}) adds a voltage shift to the audio signal (from pin 11 of IC_2) at the beginning of each TV field. (Again, if this circuit fails, there is a 30-Hz buzz in both channels.) The processed left-channel audio is filtered by FL_1 (controlled by the front-panel MPX switch, as discussed for record) to remove noise.

The left-channel signal is then amplified and deemphasized by the dynamic-emphasis circuit of IC_1. This restores the high-to-low frequency relationship of the original signal. Once the stereo audio signals are restored to the original condition, the signals are distributed to rear-panel audio jacks through a series of

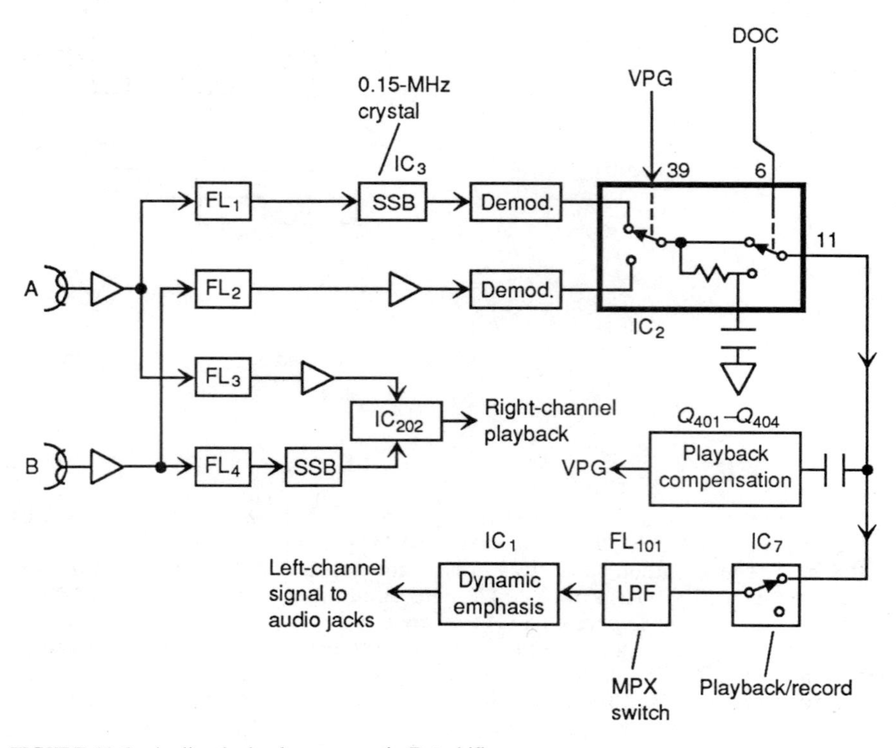

FIGURE 11.6 Audio playback sequence in Beta hifi.

switches, muting circuits, and selection circuits (similar to those found on most present-day VCRs).

11.4 8-MM AUDIO

Figure 11.7 shows the frequency spectrum for the format used in 8-mm camcorders. Figure 11.8 shows the 8-mm tape format and video-head configurations. At one time, 8-mm VCRs (that can play the 8-mm cassettes directly) were under design but have since been abandoned in favor of the popular VHS and Beta formats. However, 8-mm camcorders are in wide use.

We will not go into the full description of the 8-mm circuits here, since we are interested in audio. (*Lenk's Video Handbook* describes the 8-mm circuit functions and characteristics in boring detail.) However, there are three important points to consider in 8-mm audio.

First, as shown in Fig. 11.8, a single channel of mono FM is recorded on tape along with the video and tracking signals (in the same tape area as the video and tracking signals). This provides the audio 8-mm for camcorders in present use.

Second, there is an area on tape reserved for pulse code modulation (PCM) audio and tracking signals. This PCM area is 1.25-mm below the video, FM-audio, and tracking area. The PCM area is not required for present-day 8-mm operation but is reserved for possible stereo operation. A tape recorded with PCM audio (if such tapes become a reality) will have the FM audio (mixed with video and tracking) to make all 8-mm tapes compatible.

Third, there is another audio track at the bottom of the tape. Like the cue track at the top of the tape, the bottom audio track is 0.6-mm wide and is sepa-

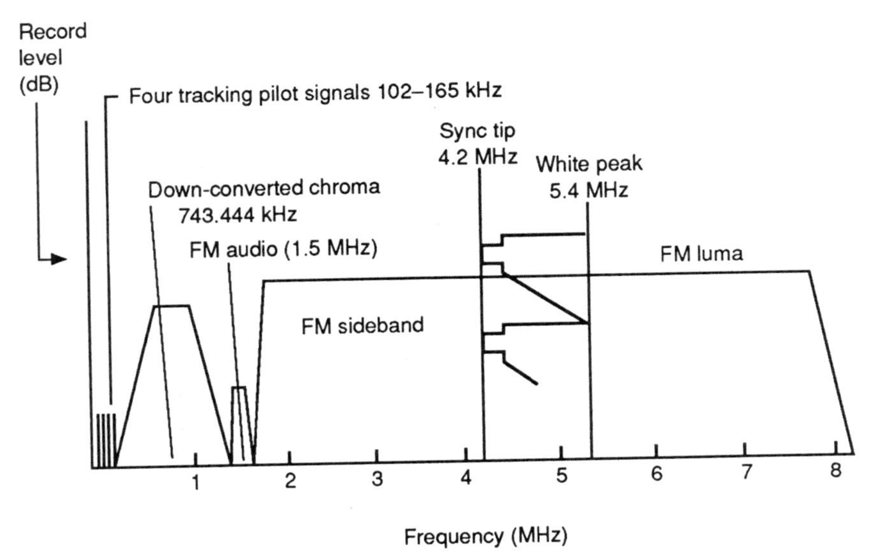

FIGURE 11.7 NTSC spectrum for the 8-mm format (pickup).

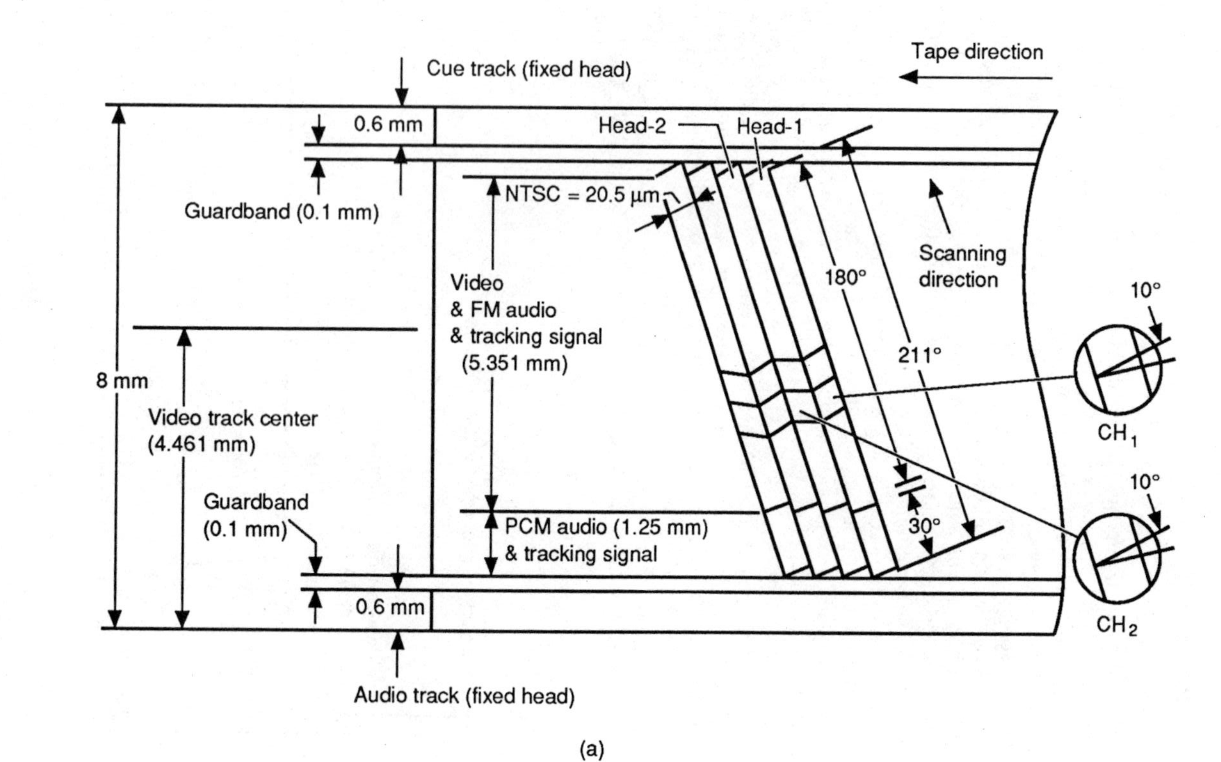

(a)

FIGURE 11.8 The 8-mm tape format and video-head configurations (pickup).

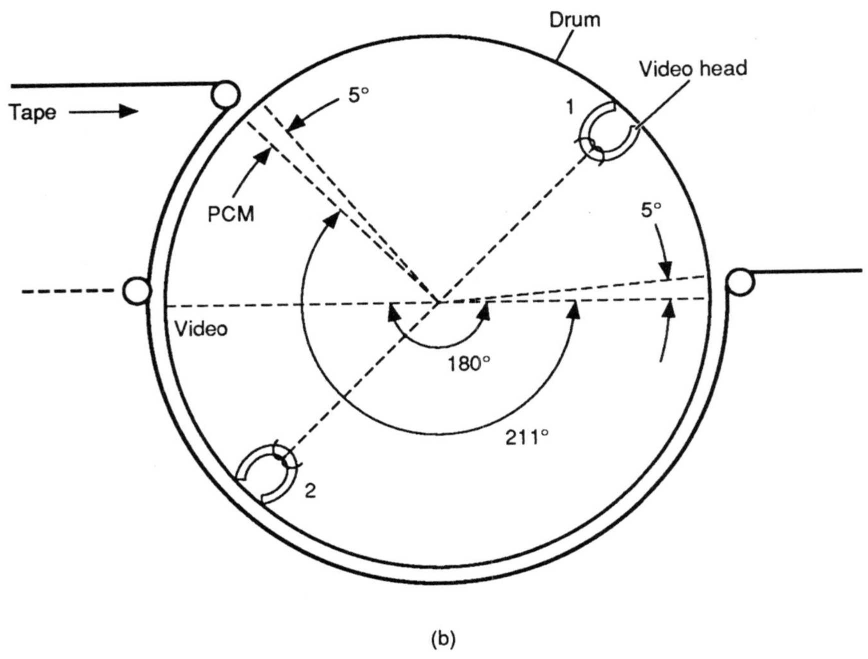

FIGURE 11.8 (*Continued*)

rated by a 0.1-mm guardband. Again, this bottom audio track cannot be recorded without also recording the full video and FM audio and tracking (to maintain compatibility on all 8-mm equipment (camcorders and possible future 8-mm VCRs).

11.4.1 Audio Record in 8 mm

The 8-mm format requires that the audio signal be recorded on an FM carrier along with the video signal on the video track. This gives the audio in the 8-mm format a much higher-quality sound. The carrier used resembles that in a Beta hifi (Sec. 11.3) except that four carriers are not needed since the audio is mono. Only one carrier is used for present-day 8 mm.

Figure 11.9 shows the 8-mm audio-record functions. Most of these functions are performed by a single IC. Microphone audio is applied through a microphone amp and an HPF to an AGC amp within the main IC. The gain of the AGC amp is controlled by an AGC detector. In turn, the AGC detector is controlled by an AGC time-constant switch. A front-panel cam/line switch changes the characteristics of the AGC amp for line or camera recording.

The output from the AGC amp is applied to an LPF, which removes high-frequency (ultrasonic) noise. Although the ultrasonic noise cannot be heard, the noise can affect the noise-reduction circuits. From the LPF, the signal passes through a hold-2 circuit, which is used for dropout compensation and correction

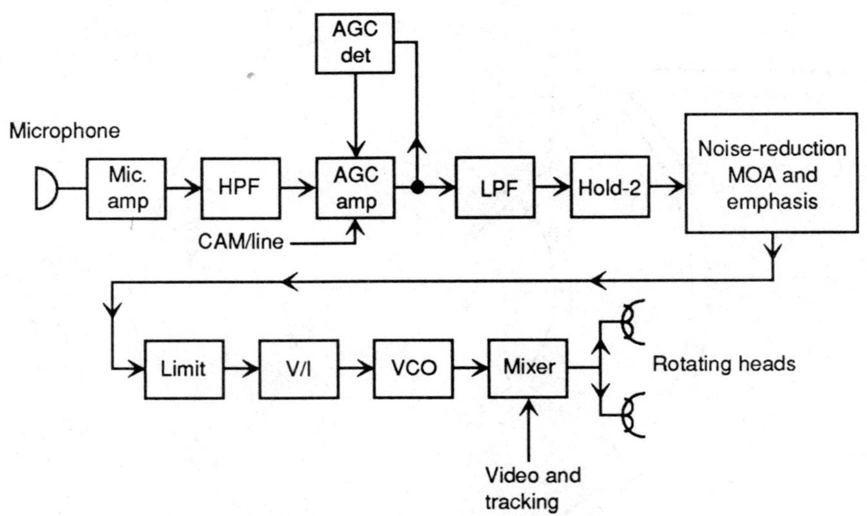

FIGURE 11.9 Audio record sequence in 8 mm.

in playback (but is not used in record). The signal from hold-2 is applied to a noise-reduction circuit.

The NR circuit consists of two sections: a main operational amplifier, or MOA, (which has different functions in record and playback modes), and an expansion network consisting of weighting filters, emphasis circuits, and a VCA.

In record, the output of the expansion circuit is fed back directly into the MOA, which acts as a compression amp to compensate for the expansion network. This compression of the input audio signal reduces the dynamic range required on tape.

The output from the NR circuit is limited to prevent overmodulation of the FM carrier and is applied to a voltage-to-current (V/I) converter. This converter controls the FM frequency of the VCO and produces a modulated carrier that is similar to that of Beta hifi. The deviation and center frequency of the carrier are set by adjustments external to the IC. The modulated audio carrier is mixed with video and tracking information and applied to the heads.

11.4.2 Audio Playback in 8 mm

As shown in Fig. 11.10, the audio FM (AFM) signal which has been recorded on tape is picked up by the video heads along with the video and tracking information. The AFM is separated from other signals by a 1.5-MHz BPF and applied to an FM phase detector through an amplifier and limiter. The phase detector also receives a fixed 1.5-MHz output from the VCO. The resulting error component represents the audio, which is applied through hold-1, an LPF, and hold-2 to a noise-reduction circuit.

Switching pulses (which have been doubled) are applied to hold-1 to remove the switching noise present when the RF switching pulse switches the carrier between the heads. If a dropout is detected, the dropout detector applies a pulse to

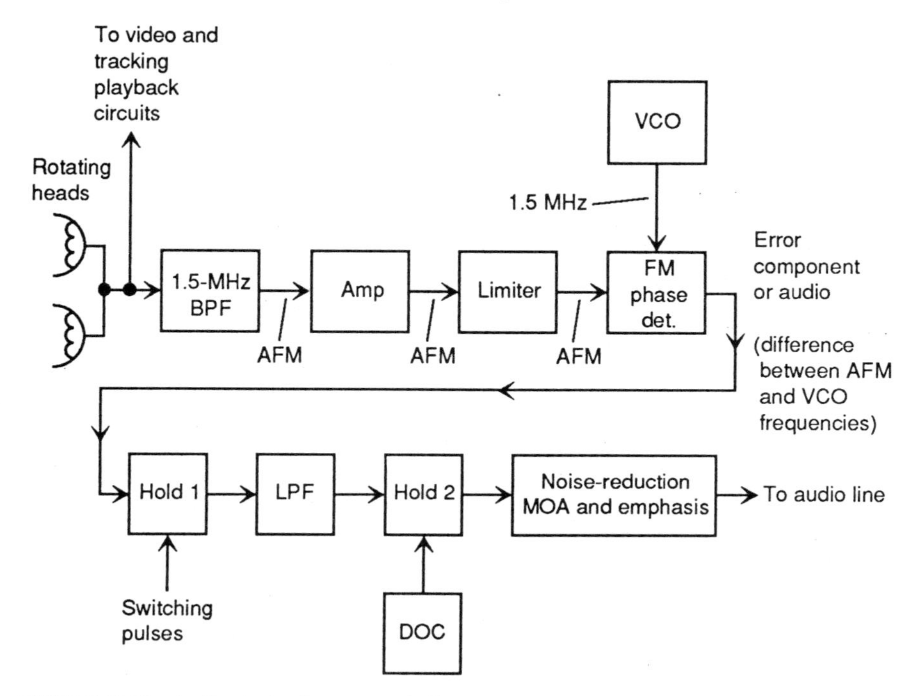

FIGURE 11.10 Audio playback sequence in 8 mm.

hold-2 and maintains the existing output level for the duration of the dropout. (On most 8-mm circuits, if a dropout becomes excessive, a mute signal is developed, and audio is muted further along in the audio path.)

The output from hold-2 is applied to the noise-reduction circuit (including the MOA, which now acts as a unity-gain amplifier). The audio signal is restored to the original form and applied to the audio line through a series of switches, muting circuits, and selection circuits.

11.5 DIGITAL AUDIO TAPE (DAT)

Figure 11.11 shows the block diagram of a typical DAT player. The DAT format is a combination of technologies found in VCRs and CD players. The tape drive (or transport) mechanism is quite similar to that of VCRs, using servos for both the capstan and cylinder servo system, as well as reel sensors, dew sensors, tape-end detectors, loading motors, etc.

The tape is recorded and played back by rotating heads that are similar to those in VHS hifi, Beta hifi, and 8-mm VCRs. If you understand the VCR technologies, you will recognize the tape-drive circuits and mechanism. Of course, there is no video, and the tape-drive mechanism is much smaller (even smaller than 8-mm and VHS-C). Figure 11.12 shows the DAT tape format and rotating audio-head configurations.

As in the case of a CD player, most of the audio processing circuits are con-

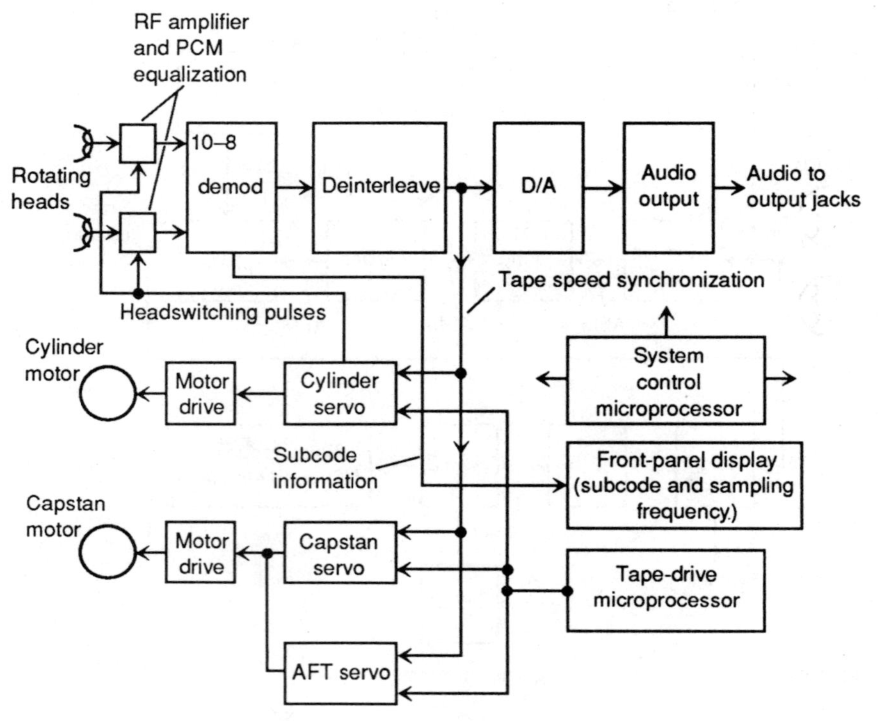

FIGURE 11.11 Block diagram of a typical DAT player.

tained in a few ICs. The following is a summary of the signal flow from tape heads to the player audio output.

11.5.1 Quantization and Interleaving

The information is recorded on tape in digital PCM form that is similar to that used in CDs. That is, the audio information is first converted (*quantized* or *quantitized*, whichever you prefer) into a digital format (digital bit stream) where each sampling point on the audio waveform is converted to a series of digital 1s and 0s. The digital bits are then *interleaved* (or scrambled) in a specific order using a specific code (which also includes error correction for the digital bit stream). The interleaved digital bits are then applied to tape through the usual amplifiers and heads (quite similar to the 8-mm format).

11.5.2 Sampling Frequencies

The sampling frequency for DAT analog-to-digital conversion is the same as for CD, or 44.1 kHz, when playing a commercially prerecorded tape. A 48-kHz sam-

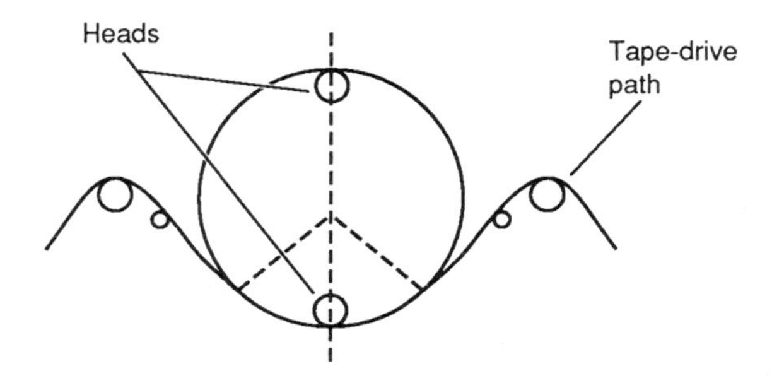

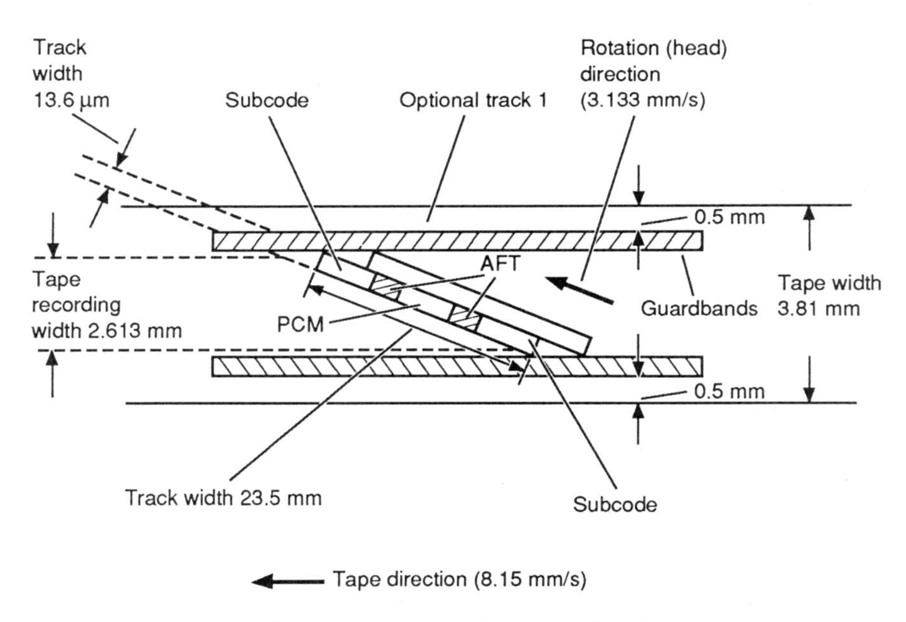

FIGURE 11.12 DAT tape format and rotating audio-head configurations.

pling rate is used when playing a DAT recorded in the digital broadcast system (DBS) and transmitted through satellites. (The sampling rate for 8 mm is 31.5 kHz.)

11.5.3 Eight-to-Ten Conversion

As discussed in Chap. 8, the conventional 8-bit digital system is converted to a 14-bit system in CDs (to increase resolution). In a CD player, this is known as eight-to-fourteen modulation (EFM). In DAT technology, the 8-bit system is converted to 10 bits (but, for some strange reason, is not known as ETM).

11.5.4 DAT Format

As shown in Fig. 11.12, the information recorded on each DAT track consists of three parts, or areas. First is the *audio information* in PCM. Second is ATF (*auto track finding*) information used to synchronize the tape-drive mechanism (similar to that used in 8 mm). The third area contains *subcode information* which shows such data as the beginning of a track (for music search), skip functions which locate the beginning and end of music, program number codes, index number codes, elapsed time (from the beginning of the DAT, program time code, and table of contents (TOC) which lists all selections on tape. The subcode area is used primarily for commercial DATs. Note that the subcode area for a DAT is over 4 times that of the corresponding area on a CD (the disk directory).

11.5.5 DAT Player Operation

In the DAT player of Fig. 11.11, the PCM information taken from the tape by the heads is amplified and equalized before application to the audio-processing circuits. Note that the cylinder servo provides head-switching pulses to the input amplifier circuits. In turn, part of the playback signal is returned to the servos to synchronize tape speed with the information being taken from the tape.

11.5.6 Audio Processing

The playback signal is first applied to a ten-to-eight demodulation circuit which returns the digital audio back to an 8-bit system from the 10-bit system used during record. The 8-bit playback signal is then de-interleaved (unscrambled) back to the original order of quantization, and any errors are corrected. The processed digital audio signal is then filtered and converted back to analog form (true audio) through D/A converters. The processed audio is amplified and applied to output connectors.

11.5.7 Microprocessor Control

Note that all functions in DAT recorder/players and DAT players are usually performed automatically in a few ICs, under control of a microprocessor (the system-control microprocessor). A separate microprocessor is used for the tape-drive mechanism.

11.5.8 Subcode and Sampling-Frequency Display

In a typical DAT player, the subcode information is extracted after ten-to-eight demodulation and applied to a front-panel display (to show what is being played, elapsed time, etc.). Front-panel indicators also show which of the three sampling frequencies is being used. The audio-processing ICs also detect which sampling frequency is recorded and apply the corresponding sampling-pulse information to the tape-drive servos. In effect, the DAT circuits recognize the type of tape being played (based on sampling rate). It is therefore unnecessary for the user to set switches to play a given type of tape.

11.6 HIFI AUDIO-TAPE TROUBLESHOOTING

Figure 11.13 is a troubleshooting diagram for hifi audio-tape circuits. Although a VHS VCR is shown, the approach also applies generally to Beta hifi and Super Beta. Note that there are six circled numbers on Fig. 11.13. These numbers represent the points to be checked when troubleshooting the hifi circuits. The following notes describe the checks to be made at each point.

11.6.1 Playback Troubleshooting

The first point to check is at the output of the bandpass filter. Check playback operation by playing a *hifi* audio test tape (not a regular audio test tape) while monitoring the BPF output (or FM demodulator/expander input). This can be done at TP_5 and TP_6. If the FM carrier is absent or abnormal, suspect IC_1, IC_3, the FM audio heads, or the BPF FL_1 and FL_2.

Also check adjustment of RT_1 and RT_2. Before you condemn any of the parts, recheck the *front-panel tracking control.* A misadjusted tracking control can affect hifi audio (as is the case with video) since both hifi audio heads and video

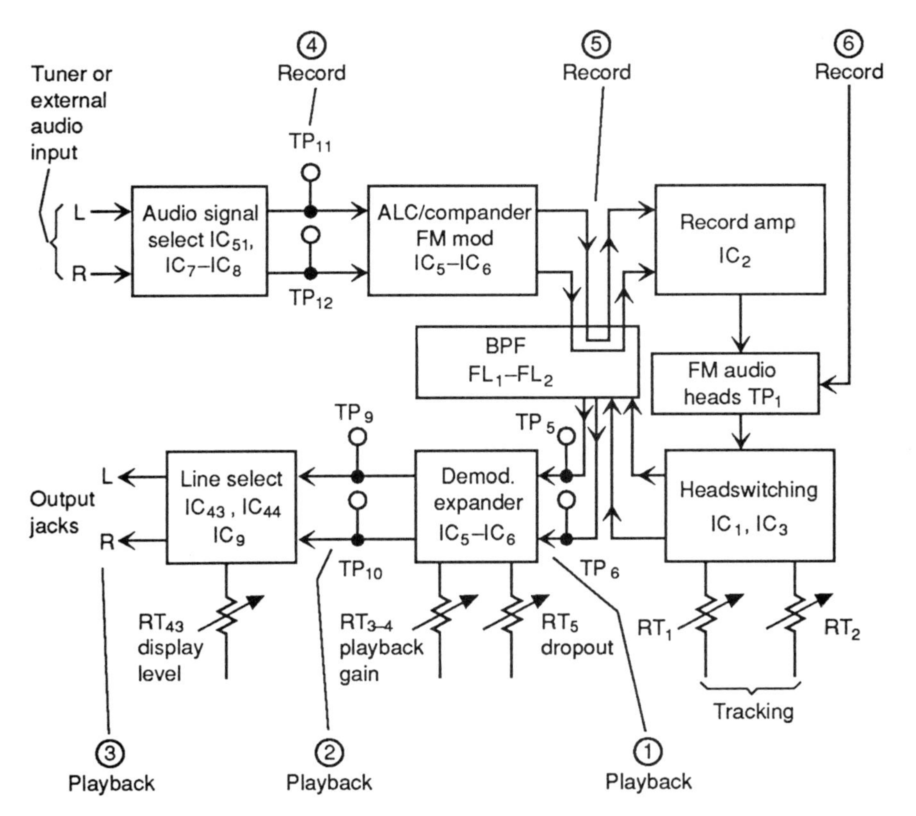

FIGURE 11.13 Universal troubleshooting diagram for hifi audio-tape circuits.

heads are on the same cylinder. The tracking controls (and internal tracking adjustments) have no effect on the linear (nonhifi) audio recorded on the fixed linear head.

The second point to check is at the output of the FM demodulator/expander. Again, check playback operation by playing a hifi audio test tape while monitoring the demodulated (and processed) audio. This can be done at TP_9 and TP_{10}. If the demodulated audio is absent or abnormal, suspect IC_5 and IC_6. Also check the adjustment of RT_3, RT_4, and RT_5.

The third point to check is at the rear-panel audio-output (line-output) jacks. Again, check playback operation by playing a hifi audio test tape while monitoring the audio. This can be done at the jacks. If the audio is absent or abnormal at the jacks, with good audio at TP_9 and TP_{10}, suspect IC_{43} and IC_{44}, IC_9, or the related circuits.

If the audio is good at the jacks but the front-panel audio-level indications are absent or abnormal, check the adjustment of RT_{43}. Likewise, if the audio is good at the jacks but the RF-modulator output is absent or abnormal, suspect circuits between the playback audio output and the RF-modulator input (usually a series of switches, relays, and amplifier-buffers).

11.6.2 Record Troubleshooting

The fourth point to check is at the output of the audio-select circuits. Either the tuner or rear-panel audio input can be selected to check record operation. Check for record audio at TP_{11} and TP_{12}. If good audio is available (from either the tuner or audio-input jacks) but the audio at TP_{11} and TP_{12} is absent or abnormal, suspect IC_{51}, IC_7, or IC_8. Also check settings of the front-panel record level and automatic limiter control (ALC) switches and controls.

The fifth point to check is at the output of the BPFs F_1 and F_2. These output should be FM carriers (1.3 and 1.7 MHz for VHS hifi). If the FM carriers are absent or abnormal, with good audio at TP_{11} and TP_{12}, suspect IC_5, IC_6, FL_1, or FL_2.

The sixth point to check is at the audio heads. Again, check for FM carriers at TP_1. If the FM carriers are absent or abnormal, with good carriers at the output of FL_1 and FL_2, suspect IC_2 and the related circuits.

11.6.3 Hifi (FM-Audio) Head Troubleshooting

The hifi or FM-audio heads are narrower than conventional linear (nonrotating) audio heads. Because of the very small tolerance for head position (a few millionths of a meter), it is possible that dust or other material compressed between the upper and lower cylinder assemblies may cause improper audio operation. So, if operation is not correct after head replacement, do not assume that the replacement heads are defective.

Remove the replacement heads and carefully clean both contact surfaces, then reinstall the replacements. It may be necessary to repeat this cleaning more than once, even with new replacement heads. Always follow the service literature for head replacement and/or cleaning procedures.

Index

ABOUT THE AUTHOR

For over 39 years, **John D. Lenk** has been a self-employed consulting technical writer spcializing in practical troubleshooting guides. A long-time writer of international best-sellers in the electronics field, he is the author of 71 books on electronics which together have sold more than 1 million copies and have been translated into eight languages. Mr. Lenk's guides regularly become classics in their fields and his most recent books include *Complete Guide to Modern VCR Troubleshooting and Repair, Digital Television, Lenk's Video Handbook*, and *Practical Solid-State Troubleshooting*, which sold over 100,000 copies.